THE ROLE OF DUST IN DENSE REGIONS OF INTERSTELLAR MATTER

THE ROLE OF DUST IN DENSE REGIONS OF INTERSTELLAR MATTER

Proceedings of the Jena Workshop, held in Georgenthal, G.D.R., March 10–14, 1986

Edited by

THOMAS HENNING and BRINGFRIED STECKLUM
Universitäts-Sternwarte Jena

Reprinted from Astrophysics and Space Science, Vol. 128, No. 1

D. Reidel Publishing Company

A MEMBER OF THE KLUWER ACADEMIC PUBLISHERS GROUP

Dordrecht / Boston / Lancaster / Tokyo

Library of Congress Cataloging-in-Publication Data

The role of dust in dense regions of interstellar matter.

This work is cataloged under LC No. 86–29845

CIP-data appear on separate card.
ISBN-13: 978-94-010-8184-9   e-ISBN-13: 978-94-009-3785-7
DOI: 10.1007/978-94-009-3785-7

Published by D. Reidel Publishing Company,
P.O. Box 17, 3300 AA Dordrecht, Holland.

Sold and distributed in the U.S.A. and Canada
by Kluwer Academic Publishers,
101 Philip Drive, Norwell, MA 02061, U.S.A.

In all other countries, sold and distributed
by Kluwer Academic Publishers Group,
P.O. Box 322, 3300 AH Dordrecht, Holland.

Die Astronomie äusserte er, ist mir deswegen so wert,
weil sie die einzige aller Wissenschaften ist, die auf
allgemein anerkannten, unbestreitbaren Basen ruht,
mithin mit voller Sicherheit immer weiter durch die
Unendlichkeit fortschreitet. Getrennt durch Länder
und Meere teilen die Astronomen, diese geselligsten
aller Einsiedler, sich ihre Elemente mit und können
darauf wie auf Felsen fortbauen.

Conversation between J. W. van
Goethe and chancellor F. van
Müller, 16th December, 1812

Participants of the Jena Workshop.

1. B. Stecklum	2. C. Friedemann	3. H.-E. Fröhlich
4. L. K. Haikala	5. G. Rüdiger	6. J. Dorschner
7. J. Gürtler	8. R. Schmidt	9. J. M. Greenberg
10. P. G. Mezger	11. H. J. Habing	12. M. Wolf
13. M. Šolc	14. R. Chini	15. J. Staude
16. W. Krätschmer	17. R. Tschaepe	18. S. Klose
19. H.-G. Reimann	20. W. Pfau	21. H. Zimmermann
22. Z. Pintèr	23. J. Palouš	24. R. Schielicke
25. Th. Henning	26. H. Meusinger	

LIST OF PARTICIPANTS

E. Chini, Max-Planck-Institute for Radioastronomy, Bonn.

J. Dorschner, University Observatory, Jena.

C. Friedemann, University Observatory, Jena.

H.-E. Fröhlich, Central Institute for Astrophysics, Academy of Sciences of the GDR, Potsdam-Babelsberg.

J. M. Greenberg, University of Leiden, Leiden.

J. Gürtler, University Observatory, Jena.

H. J. Habing, University of Leiden, Leiden.

L. K. Haikala, University of Helsinki, Helsinki.

Th. Henning, University Observatory, Jena.

S. Klose, University Observatory, Jena.

W. Krätschmer, Max-Planck-Institute for Nuclear Physics, Heidelberg.

E. Krügel, Max-Planck-Institute for Radioastronomy, Bonn.

H. Meusinger, Karl Schwarzschild Observatory, Tautenburg.

P. G. Mezger, Max-Planck-Institute for Radioastronomy, Bonn.

J. Palouš, Institute for Astronomy, Czechoslovak Academy of Sciences, Prague.

W. Pfau, University Observatory, Jena.

Z. Pintèr, Loránd Eötvös University, Budapest.

H.-G. Reimann, University Observatory, Jena.

G. Rüdiger, Central Institute for Astrophysics, Academy of Sciences of the GDR, Potsdam-Babelsberg.

R. Schielicke, University Observatory, Jena.

K.-H. Schmidt, Central Institute for Astrophysics, Academy of Sciences of the GDR, Potsdam-Babelsberg.

R. Schmidt, Central Institute for Astrophysics, Academy of Sciences of the GDR, Potsdam-Babelsberg.

M. Šolc, Charles University, Prague.

J. Staude, Max-Planck-Institute for Astronomy, Heidelberg.

B. Stecklum, University Observatory, Jena.

R. Tschaepe, Central Institute for Astrophysics, Academy of Sciences of the GDR, Potsdam-Babelsberg.

M. Wolf, Charles University, Prague.

H. Zimmermann, University Observatory, Jena.

TABLE OF CONTENTS

PREFACE

TH. HENNING and B. STECKLUM

University Observatory, Jena

This workshop was organized by the University Observatory Jena and devoted to the physics and chemistry of dense regions of interstellar matter. It was especially dealing with the properties of interstellar dust grains and star formation in those regions. This field of research was opened in Jena already in the 1950s and an early *IAU Colloquium* on the topic of interstellar matter was held here in 1969.

Since that time, the subject of interstellar matter has grown into a much more important part of astrophysics than it was in 1969. Now we are beginning to understand the process of star formation in a greater detail. The discovery of many interstellar molecules by radioastronomers opened the new field of interstellar chemistry. In addition, the application of the new techniques of infrared astronomy led to the discovery of several absorption bands, e.g., the 3.1 µm ice band, which pointed to the existence of grain mantles. More recently, the detection of infrared emission lines was the first hint to the existence of a new component of interstellar matter. All these things were discussed extensively during the workshop. The very successful IRAS mission, which was also a subject of this workshop, gave us many new insights and unexpected findings, e.g., the detection of infrared cirrus clouds.

The study of interstellar dust is important not only by itself but also in connection with the star formation process. The dust mainly determines the thermal and dynamical structure of interstellar clouds. Consequently, the workshop dealt also with problems of star formation. The discovery of the Becklin–Neugebauer object in the Orion molecular cloud in 1967 and later on similar objects embedded within dense regions of molecular clouds enabled the study of very early stages of stellar evolution. While the BN objects appear to be most likely fully formed OB stars, the IRAS data and the use of submillimetre telescopes should render the detection of real protostars possible. These questions were a topic of the workshop just as bipolar nebulae which are related to both early and late stages of stellar evolution. Besides those local aspects of star formation some lectures dealt with global aspects such as the initial mass function, the star formation rate and the formation of OB associations.

We wish to thank all lecturers for their efforts in preparing the manuscripts of their lectures and sending them in time. We feel that the discussions in their edited form cannot reflect the exciting atmosphere prevailing during the days of the workshop. Nevertheless, it is hoped that they will help to make the scientific contents of the papers more easily accessible.

We would be glad if the proceedings come out with some new problems to work on.

Astrophysics and Space Science **128** (1986) 1.

INTRODUCTION

J. M. GREENBERG

Laboratory Astrophysics, University of Leiden, The Netherlands

I am honored to be asked to introduce this second 'Jena' meeting on interstellar dust at Georgenthal and I thank the organizers for making this possible. For some of us, interstellar dust had not only provided the justification to come to this delightful place and to meet with old friends – it has been our scientific bread.

The research on dust has increased exponentially in the last years, after a long period during which, if it was not swept under the rug, its general relevance to astronomy was not universally recognized. It has been interesting to watch the comings and goings of ideas about dust since our first Jena gathering in 1969. Of course, by then we had certainly come a long way from the time when a colleague of Eddington could say that dust is like a ghost – something you do not believe in but which you are nevertheless afraid of. Controversies remain about what dust really is but there now exists a substantial body of observational and laboratory studies to provide a firm basis for our theoretical concepts and scientific discussions.

It was just about the time of your last meeting that many of the important new observational techniques in the ultraviolet and infrared were beginning to have an impact on our direct knowledge of dust. Equally important for the subject was the application of radioastronomical methods to the discovery of interstellar molecules and the burgeoning of theories of interstellar chemistry. Major discoveries were on the verge of being made.

The infrared spectrum of the BN object showed indeed that there was some solid H_2O in space as well as silicates. The study of these silicates has evolved as an important aspect of the dust in relation to circumstellar as well as interstellar regions not only through the observations but largely, as well, as a result of the theoretical and laboratory work of the Jena group. Those of us who have followed your work welcome this opportunity to get a first hand report on your progress in this as well as in other aspects of dust.

Laboratory methods are clearly providing a range of new possibilities in the interpretations, based largely on ground based infrared observations and space ultraviolet observations, of the complex story of dust evolution. However, what we shall also hear at this meeting is the major thrust in our knowledge of dust which already now and well into the future are provided by the remarkable achievements of the IRAS.

From the program it appears that in the next couple of days our reports and discussions on the latest theories and laboratory investigations of dust and such related topics as star formation, chemical evolution, and circumstellar shells will provide us with the scientific nourishment we came for.

At this time I predict an exciting and stimulating meeting and I hope that the time between now and our next meeting will be shorter than the last.

Astrophysics and Space Science **128** (1986) 3.

DUST MODELS CONFRONTED WITH OBSERVATIONS*

E. KRÜGEL

Max-Planck-Institut für Radioastronomie, Bonn, F.R.G.

(Received 23 June, 1986)

Abstract. We critically discuss the two major models for interstellar dust grains developed by Greenberg and Mathis and their co-workers. So far, knowledge about the dust has been obtained mainly from extinction and polarization measurements in the UV, visual and near IR. Analysis of dust emission at longer wavelengths generally requires an evaluation of the radiative transfer. Such a procedure is necessary to assess optical grain properties at long wavelengths and also to interpret the emission from the different kinds of IR objects. The 10 μm silicate resonance and the submm dust absorptivity are presented as two examples. We then briefly summarize the implications which follow from the existence of extremely small dust grains.

1. Introduction

The two observational pillars, on which our conception of the dust rests, are extinction and polarization. Figure 1 shows the mean interstellar extinction curve plotted in a form that can be readily interpreted as the ratio of the extinction at wavelength λ to that in visual at 0.55 μm. Variations of this curve in the far UV, at 0.22 μm and in the IR are still a matter of some controversy, but by and large the curve is adequately explained by the present dust models.

The basic empirical relation for linear polarization was found by Serkowski (1973). Let $\lambda_{\max}$ be the wavelength at which the polarization attains its maximum $p_{\max}$, then all curves have the same functional form

$$p = p_{\max} \exp\left\{ - K \ln^2(\lambda_{\max}/\lambda)\right\}, \tag{1}$$

where K was initially proposed to equal 1.15. Later, an increase of K with $\lambda_{\max}$ was noted (Codina-Landaberry and Magalhaes, 1976) and

$$K = 1.7\lambda_{\max} \quad (\lambda_{\max} \text{ in μm}), \tag{2}$$

it the now accepted form (Wilking *et al.*, 1980), where K may have values between 0.5 and 1.4. Equation (2) is excellently confirmed in high precision observations by Schulz and Lenzen (1983). In those few cases where circular polarization could be measured, it changes sign at a wavelength

$$\lambda_{\text{cir}} = \lambda_{\max} \tag{3}$$

(Martin and Angel, 1976), from which one can draw the important conclusion that the polarizing particles are dielectrics. Equation (1) is reproduced by the present dust

---

* Paper presented at a Workshop on 'The Role of Dust in Dense Regions of Interstellar Matter', held at Georgenthal, G.D.R., in March 1986.

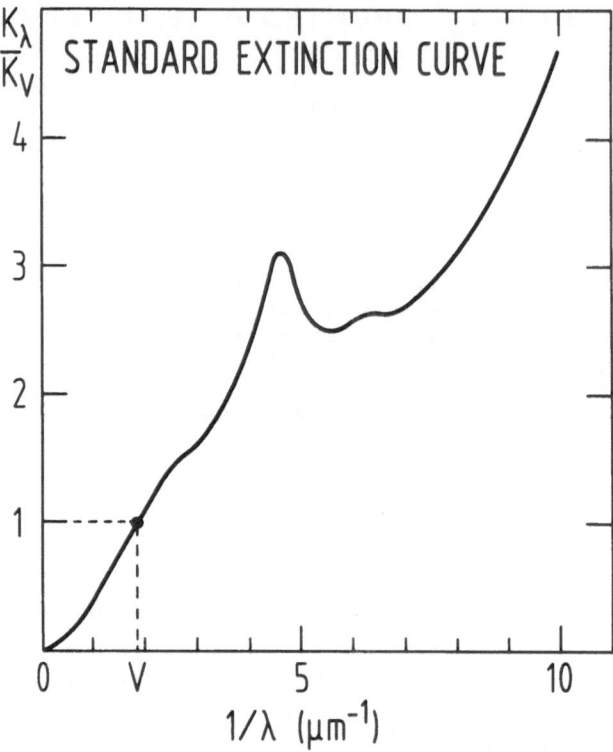

Fig. 1.   The average interstellar extinction curve after Savage and Mathis (1979) plotted as the ratio of the
extinction coefficient at a wavelength λ to that in the visual at 0.55 μm.

models; Equation (2) could be explained by Aannestad and Greenberg (1983) with
suprathermally aligned coated grains and Equation (3) has also been theoretically
understood (Martin, 1974).

It was claimed by Whittet and Breda (1978) that extinction and polarization are linked
by the relation

$$R = 5.6\lambda_{max}, \quad \lambda_{max} \text{ in } \mu m, \quad R = \frac{A_V}{E_{B-V}}. \tag{4}$$

However, there are several arguments against it: (a) R is remarkably uniform in the
diffuse medium, whereas λ_{max} shows considerable scatter. (b) In those few sources
where R has been reliably determined to be different from the standard value of 3.1,
Equation (4) does not hold. The best studied example is M17. Here the mean λ_{max} of
0.57 μm (Schulz et al., 1981) would according to Equation (4) result in $R = 3.2$, whereas
$R = 4.9$ is observed (Chini and Krügel, 1983). (c) From theory λ_{max} is tightly related to
the size of the polarizing grains, but R is not (Chini and Krügel, 1983).

2. Dust Models

We discuss the two major dust models developed by Mathis and co-workers (Mathis *et al.*, 1977; Mathis, 1979; Mathis and Wallenhorst, 1981) on one side and by Greenberg and associates (Hong and Greenberg, 1980; Greenberg and Chlewicki, 1983; Aannestad and Greenberg, 1983) on the other. Both have silicate (olivine) and graphite particles and both are, of course, subject to the conditions of cosmic abundances.

Greenberg proposes tiny (~ 0.01 µm) silicates to explain the rise in the extinction curve in the far UV ($\lambda^{-1} \geq 6$ µm^{-1}), small (~ 0.01 µm) graphites to produce the 0.22 µm hump and coated silicates for the visual region and for polarization. The bare silicates and graphites do not display a size distribution, but the coated particles are distributed in size according to

$$n(a) \propto \exp\left\{ -5\left(\frac{a - a_c}{a_i}\right)^3 \right\}, \tag{5}$$

where $a_i \approx 0.3$ µm is a cut-off and $a_c \approx 0.05$ µm the core radius. Equation (5) follows from a very physical balance equation, in which grain growth is due to accretion, and destruction is proportional to the geometrical grain surface. The composition of the coatings is the result of a whole history: silicate seedlings blown off from a late-type giant enter a dense molecular cloud and the most abundant and condensible elements stick to their surface to form a layer of 'dirty ice'. The grains are then illuminated with UV photons ($h\nu \geq 5$ eV) (supposedly even in molecular clouds!) that break up the chemical bonds in the dirty ice to form radicals. The ice mantles are occasionally heated by grain-grain collisions of relative speed $\lesssim 100$ m s^{-1} (which is plausible for a turbulent medium), whereupon the radicals combine to build new and complex refractory molecules that can survive in the diffuse interstellar medium. The recombination of the radicals releases heat which sometimes leads to a runaway effect and the whole mantle evaporates. (This would be a solution to the problem why we see molecules at all in dense cloud cores.)

We mention two weak points of the preceding scenario. The first was put forward by Mathis. In Greenberg's model the coated particles are responsible for polarization and scattering in the visible. As the polarizing grains are very likely dielectrics, their albedo should be much larger (~ 0.9) than what is observed in H II regions and reflection nebulae. However, in the very strong and hard radiation field around O and B stars the coated grains may be modified. Furthermore, a particle of 0.1 µm radius and $m = 1.6 - i0.05$ has an albedo of 0.73 (as low as 0.57, if $m = 1.6 - i0.1$) and would still be considered a dielectric. The second weak point of this dust model concerns its application and comes ironically from its merit htat it is embedded in a evolutionary scheme. This leads often to more qualitative than quantitative descriptions. We, therefore, use the Mathis model for applications (see also Chini, this issue).

In the Mathis model for the diffuse medium the silicate and graphite particles have a power law size distribution with an exponent of ~ 3.5. Upper and lower boundaries are not well defined, but lie around 0.01 and 0.3 µm, respectively. This model was derived in the effort to fit the extinction curve without any *a priori* assumptions about grain size or their evolution. Which mechanism could lead to the exponent of 3.5? One is repeated collisional fragmentation (Dohnanyi, 1969; Hellyer, 1970) which has been successfully applied for explaining the mass spectrum of meteorites. In the case of interstellar grains, however, the situation is more complicated. The outcome of their mutual collisions in the atmospheres of cool giants is generally not clear (Salpeter, 1974). In addition, with the typical values for the stellar wind of a giant the number of collisions a grain will experience is only of order unity. It is also unclear how the grain size is modified once the particles are in the interstellar medium. It was pointed out by Greenberg that the Mathis model also does have problems in explaining the variations in the observed far UV extinction without changing the position of the 0.22 µm hump (see also Fitzpatrick and Massa, 1986).

The dielectric functions used in both models for silicate and graphite are those compiled by Draine and Lee (1984). These authors re-evaluated the data in the literature and checked their consistency with the Kramers–Krönig relation. Draine (1985) also tabulated the functions. For the optical constants of grain mantles the data for dirty ice by Bertie *et al.* (1969) are generally recommended.

The most unsatisfactory aspect of the models by Mathis and Greenberg is that they exclude each other, while a definite test to decide among them has not yet been devised. Further research is needed.

3. Radiative Transfer

The extinction curve cannot be used for wavelengths $\lambda \geq 5$ µm, because there are no good standard stars and because at these wavelengths the grains also start to emit. In the IR one, therefore, has to calculate the radiative transfer to obtain information about the dust. We solve it in the following way: within the framework of the Mathis model we select five silicate and four graphite grain sizes in such a way that, with their cross sections calculated from Mie theory, they adequately reproduce the observed interstellar extinction curve (Figure 2). We then solve the radiative transport in spherical symmetry

$$\mu \frac{\partial I_\nu}{\partial r} + \frac{1 - \mu^2}{r} \frac{\partial I_\nu}{\partial \mu} = \sum_i \left\{ - \sigma_{i\nu}^a (I_\nu - B_\nu(T_{id})) - \sigma_{i\nu}^s (1 - g_{i\nu}) (I_\nu - J_\nu) \right\} , \quad (6)$$

where the index i refers to the grain component, $g_{i\nu}$ is the asymmetry factor, J_ν the mean intensity, B_ν the Planck function, T_{id} the dust temperature, and all other symbols have their usual meaning. The boundary conditions describe the flux at the stellar surface and at the cloud edge. The grain temperatures are derived from

$$\int \sigma_{i\nu}^a (B_\nu(T_{id}) - J_\nu) \, d\nu = 0 . \quad (7)$$

We show in Figure 3 the total mass-extinction coefficient and the mass-absorption

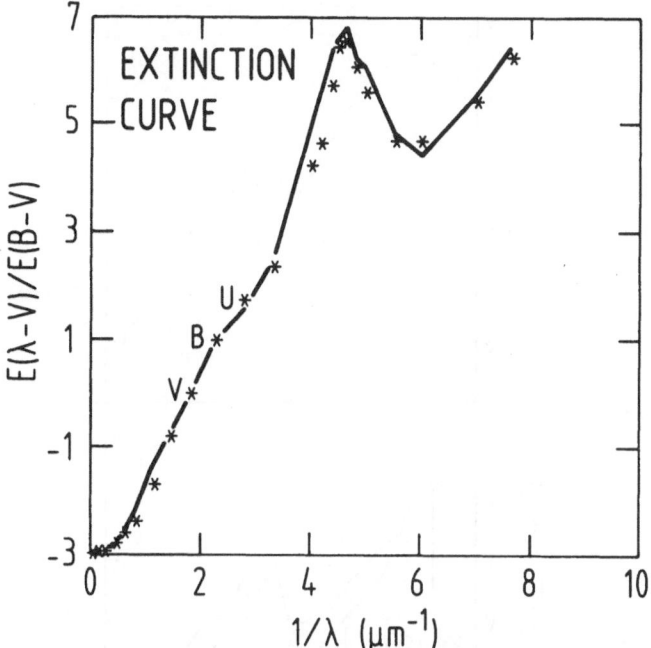

Fig. 2. A simple dust model after Mathis *et al.* (1977) consisting of graphite spheres with radii from 0.015 to 0.12 μm and silicate spheres with radii from 0.015 to 0.24 μm approximates the observed mean interstellar extinction curve (asterisks) well between 0.1 and 5 μm. The solid line is the model dust and is calculated from Mie theory. Altogether only nine kinds of grains differing in size and composition are used. For the applications discussed here and in the paper by Chini (this issue) this dust model is used.

coefficient for the graphite and silicate mixture over the whole wavelength range. The results with the dielectric functions by Draine (1985) are also plotted. We use very similar optical constants, except that in the submm regions ours give about three times larger values. Such values one would expect for ice-coated particles.

4. Dust Properties from Radiative Transfer

As an example, where knowledge of the density and temperature structure of the dust is needed, we discuss the analysis of the 10 μm silicate resonance (see Chini *et al.*, 1986). Where the dust is hot, this feature is in emission, at the cooler edge of the cloud it turns into absorption. The grain temperature in the cloud cannot be determined from the resonance alone. Instead one has to model the full spectrum, which may extend from the visual to 1 mm. Such a model, carried out according to the description of Section 3, can yield a distribution of the grain densities and their temperatures, which is consistent with all observations. Keeping this distribution fixed, one can then determine the optical constants of silicate in the resonance. This is shown in Figures 4 and 5 for the source NGC 7538(E). The dots in Figure 4(a) refer to IR measurements and are connected by

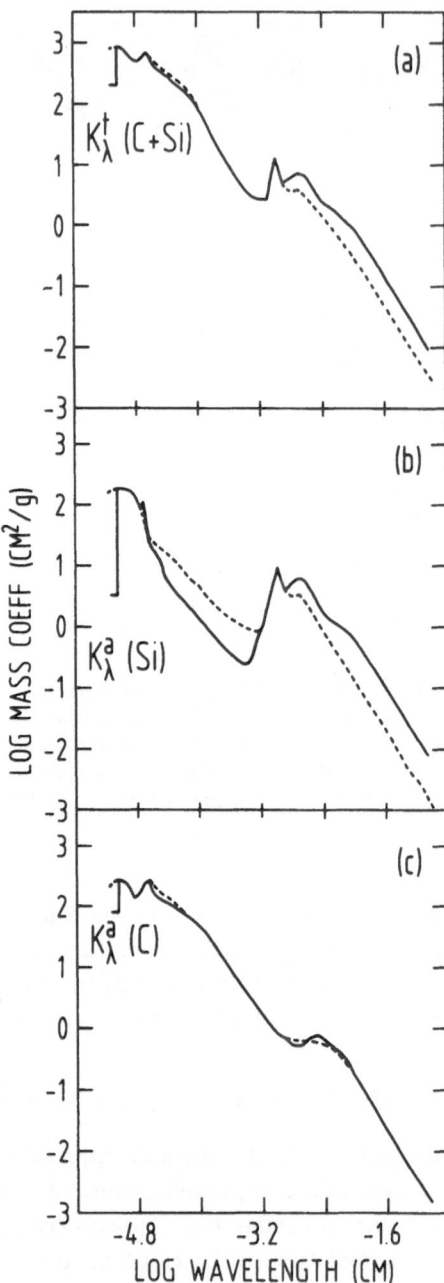

Fig. 3.   (a) The total mass extinction (absorption plus scattering) coefficient K_λ^t of silicate and graphite in cm^2 per g of interstellar matter of the dust described in Figures 1 and 2. Values with the dielectric functions of Draine (1985) are shown as dashed lines. If the 'Draine dust' has ice-coatings it would give values closer to the solid line. (b) and (c) The mass absorption coefficients of graphite and silicate alone.

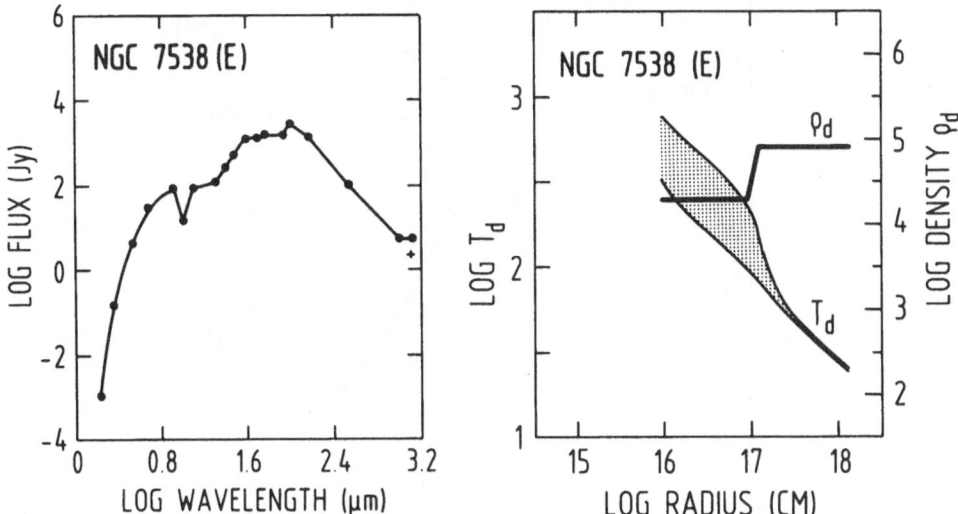

Fig. 4.   (a) The observed IR spectrum (dots) of NGC 7538(E) is fit at all wavelengths by a model to within a factor of two or better except at 1.3 mm (cross). The spectrum looks bumpy because of different apertures used in the observations. (b) The density and temperature structure of the dust in the model. The densities can under normal conditions be interpreted as number densities of atomic hydrogen.

a line. The values of the fit are not shown when they agree with the observation within a factor of two; otherwise they are marked by crosses. Figure 4(b) depicts the density and the temperature distribution of the dust as a function of distance from the star (because there are different kinds of grians there is a range of temperatures). Figure 5 shows a choice of the optical constant $m = n - ik$ that fits the spectrophotometry in the 10 μm resonance by Willner et al. (1982). The values are slight variations of the dielectric functions of Krätschmer (1980) obtained from laboratory measurements of sputtered olivine.

The far IR grain properties have also to be tested for consistency with observations. This can be done by checking in calculations of the kind presented in Figure 4 whether reasonable model dust distributions for the source can reproduce the measurements. At least, one can say that the far IR optical dust constants are not in contradiction with observations. With larger wavelengths the values become increasingly uncertain. The submm absorptivity Q_v is usually only approximated in the form $Q_v \propto v^m$, where suggested values for m range from 1 to 3. In the ideal case of the Rayleigh–Jeans limit the ratio of the flux densities at two submm wavelengths λ_1 and λ_2 is

$$\frac{S_1}{S_2} = \left(\frac{\lambda_2}{\lambda_1}\right)^{m+2}, \tag{8}$$

provided that both observations have been carried out with the same beam. In practice, $hc/kT\lambda \approx 1$ and one has to use in Equation (8) the Planck function and thus one needs an estimate of the dust temperature in the region that emits the submm radiation. Such

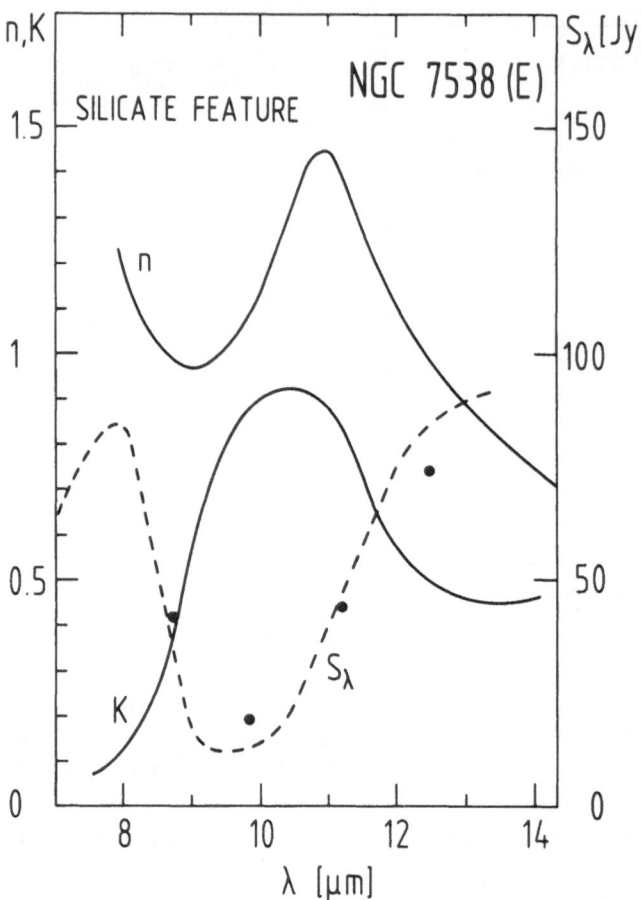

Fig. 5. The optical constant n and k for silicate in the 10 µm resonance which fit the spectrophotometry (dashed curve) by Willner *et al.* (1982) for NGC 7538(E). Exactly the same density and temperature structure is used as in Figure 4. Our fit agrees with the dashed curve to within 10% and is, therefore, not explicitly shown. The full dots are narrow band photometry by Werner *et al.* (1979).

an estimate can be obtained from a model that fits the total spectrum. Using 350 and 1300 µm observations of twelve star forming regions, we obtained in such a manner a mean $m = 2.04 \pm 0.4$ (Chini *et al.*, 1986).

Of course, there are other classes of objects where their study should include detailed radiative transfer to increase our knowledge about the dust. I only mention here reflection nebulae, and especially the very interesting near IR reflection nebulae as in Orion BN/KL (Werner *et al.*, 1983) or Cep A (Lenzen *et al.*, 1984). I expect that near IR scattering contains the best information about the sizes of the large grains.

5. Very Small Particles

In the past three years evidence has accumulated that there exists an additional component of very small particles with sizes (~ 10 Å) intermediate between those of the largest observed interstellar molecules and the smallest classical grains (~ 100 Å). This evidence comes mainly from the study of reflection nebulae where (a) the flux ratio $S_{2.2 \, \mu m}/S_{0.55 \, \mu m}$ is independent of distance from the central star, (b) the dust colour temperature is almost 1000 K and also independent of distance, (c) there are strong emission features between 4 and 13 µm (Sellgren et al., 1983, 1985; Sellgren, 1984). These data, as well as the flux ratios $S_{12 \, \mu m}/S_{100 \, \mu m}$ greater than 0.2 observed by IRAS in Cirrus clouds and others, could not be explained by classical grains. Leger and Puget (1984) pointed out that the IR emission from coronene ($C_{24}H_{12}$) is almost identical to the observed IR features. Therefore, coronene, which consists of seven concentrically-positioned aromatic cycles and is very stable, was advocated to be responsible for the emission in the reflection nebulae. Draine and Anderson (1985), on the other hand, considered very small (~ 10 Å) grains which experience stochastic temperature fluctuations after the absorption of a UV photon. They assumed that these grains could still be treated with classical Mie theory. However, this is rather doubtful and does not explain the IR emission features. I consider polycyclic aromatic hydrocarbons (PAHs) the more likely candidates.

Omont (1986) stressed the importance of the PAHs for the interstellar chemistry because of their large surface area ($\sim 50\%$ of all dust) and chemical activity due to electric charge and radical sites. For the IR emission from dust embedded objects the existence of the PAHs will probably not lead to dramatic changes because their optical surface is small ($\sim 1\%$ of all dust). Furthermore, in regions shielded from UV photons, the PAHs are probably not excited at all.

6. Conclusions

The observations of normal extinction and polarization are explained by the two incompatible dust models of Greenberg and Mathis. Both have to be modified or extended to incorporate the particles that produce the near IR emission features and the radiation from IRAS cirrus clouds. The study of the dust, which is based on the extinction curve, focuses on the short wavelength region (< 5 µm). Analyzing the optical dust properties at larger wavelengths requires evaluation of the radiative transfer. The underlying astronomical objects range from protostars to H II regions, from near-IR reflection nebulae to mass loss giants. They cannot be understood without knowledge of the dust in the whole IR spectrum. The field is promising because of fast progress in observations regarding spectral coverage, spectral and spatial resolution and sensitivity.

References

Aannestad, P. and Greenberg, M.: 1983, Astrophys. J. 272, 551.
Bertie, J., Labbe, H., and Whalley, E.: 1969, J. Chem. Phys. 50, 4501.

Chini, R. and Krügel, E.: 1983, *Astron. Astrophys.* **117**, 289.
Chini, R., Krügel, E., and Kreysa, E.: 1986, *Astron. Astrophys.* (in press).
Codina-Landaberry, S. and Magalhaes, A.: 1976, *Astron. Astrophys.* **49**, 407.
Dohnanyi, J.: 1969, *J. Geophys. Res.* **74**, 2531.
Draine, B.: 1985, *Astrophys. J. Suppl.* **57**, 587.
Draine, B. and Anderson, N.: 1985, *Astrophys. J.* **292**, 494.
Draine, B. and Lee, H.: 1984, *Astrophys. J.* **285**, 89.
Fitzpatrick, E. and Massa, D.: 1986, *Astrophys. J.* (in press).
Greenberg, M. and Chlewicki, G.: 1983, *Astrophys. J.* **272**, 563.
Hellyer, B.: 1970, *Monthly Notices Roy. Astron. Soc.* **148**, 383.
Hong, S. and Greenberg, M.: 1980, *Astron. Astrophys.* **88**, 194.
Krätschmer, W.: 1980, in I. Halliday and B. A. McIntosh (eds.), 'Solid Particles in the Solar System', *IAU Symp.* **90**, 351.
Leger, A. and Puget, J.: 1984, *Astron. Astrophys.* **137**, L5.
Lenzen, R., Hodapp, K., and Solf, J.: 1984, *Astron. Astrophys.* **137**, 202.
Martin, P.: 1974, *Astrophys. J.* **187**, 461.
Martin, P. and Angel, J.: 1976, *Astrophys. J.* **207**, 126.
Mathis, J.: 1979, *Astrophys. J.* **232**, 747.
Mathis, J. and Wallenhorst, S.: 1981, *Astrophys. J.* **244**, 483.
Mathis, J., Rumpl, W., and Nordsieck, K.: 1977, *Astrophys. J.* **217**, 425.
Omont, A.: 1986, *Astron. Astrophys.* (submitted).
Salpeter, E.: 1974, *Astrophys. J.* **193**, 579.
Savage, B. and Mathis, J.: 1979, *Ann. Rev. Astron. Astrophys.* **17**, 73.
Sellgren, K.: 1984, *Astrophys. J.* **277**, 623.
Sellgren, K., Allamandola, L., Bregman, J., Werner, M., and Wooden, D.: 1985, *Astrophys. J.* **299**, 416.
Sellgren, K., Werner, M., and Dinerstein, H.: 1983, *Astrophys. J.* **271**, L13.
Serkowski, K.: 1973, in J. M. Greenberg and H. C. Van de Hulst (eds.), 'Interstellar Dust and Related Topics', *IAU Symp.* **52**, 145.
Werner, M., Becklin, E., Gatley, I., Matthews, K., Neugebauer, G., and Wynn-Williams, C.: 1979, *Monthly Notices Roy. Astron. Soc.* **188**, 463.
Werner, M., Dinerstein, H., and Capps, R.: 1983, *Astrophys. J.* **265**, L13.
Wilking, B., Lebofsky, M., Martin, P., Rieke, G., and Kemp, J.: 1980, *Astrophys. J.* **235**, 905.
Whittet, D. and Breda, I.: 1978, *Astron. Astrophys.* **66**, 57.
Willner, S. *et al.*: 1982, *Astrophys. J.* **253**, 174.

Discussion

J. M. Greenberg: I have a comment with regard to the albedo of the core-mantle particles. I believe that Mathis *et al.* calculated this exceedingly high albedo for core-mantle particles using a mantle with a pure real or exceedingly small imaginary part of the index of refraction. It turned out by our measurements in the laboratory that the mantles in the diffuse clouds have a moderate but not large imaginary part of the index. It does not make the metals but the dielectrics which have an imaginary part of the index of about 0.1 or 0.2 and, e.g., an imaginary part of 0.05 in the visual. The albedo is quite consistent for the overall particle distribution one has observed in diffuse clouds. We found no particular difference between our albedo in the visual and what Mathis *et al.* obtain. So, both are in the error bars which are observed.

E. Krügel: These numbers are very easily checked. I would be somewhat surprised that Mathis *et al.* took the wrong values.

J. M. Greenberg: Do not be surprised. He took an assumed number for which he had no particular reason to assume it.

E. Krügel: If the albedo is of the order of 60%, this is what always is expected in this wavelength range, no one would have a problem. If the numbers should turn out to be, consistent with circular polarization, 90% or more, then unless the grains are modified in H II regions or reflection nebulae, we would have troubles.

J. M. Greenberg: One talks about the 3.1 µm feature in relationship to the amount of extinction. To say that the most important thing is the thickness of the mantle is not quite correct. As a matter of fact the dilution by the water ice is so important so that the two come together. So, when you talk about relating the strength of the ice band to the thickness of the mantle then this is really not correct.

E. Krügel: I can comment for the source which was shown here. R. Chini and I obtained the 3.1 µm fit. There the result is the following: you can fit the absorption dip at 3.1 µm without difficulties, but someone has to tell you how thick the mantle is.

J. M. Greenberg: You misunderstand the point. But when I gave my talk you saw what I was talking about.

H. J. Habing: I have one comment that is about the 12 to 60 µm excess or the ratio. De Vries in Leiden, he is doing his thesis, measured very faint clouds which you can see on modern Schmidt plates, southern sky Schmidt plates, where he measured them in extinction by counting stars and from scattered light at 6000 Å and you can compare that with the infrared. You find that there is a one-to-one correspondence, wherever the 100 µm emission there is a cloud. Now the remarkable thing is that he has two clouds which look optically very identical, one has 0.7 mag of extinction, the other has 1.1 or so, but otherwise there are no major differences between them, the reflected light is the same. But one has strong 12 µm emission and the other one does not have any. That is a riddle.

E. Krügel: I assume the scattering clouds have nothing to do with the source of light.
H. J. Habing: The scattered light is diffuse galactic light and there is no ultraviolet involved.

P. G. Mezger: We are talking about the PAHs. These are very small particles which, however, are proposed to emit primarily at 100 µm. The basic thing is if you integrate over the total infrared emission of external galaxies or our own galaxy something between 10 and 15% of the total emission is coming out in a second hump which peaks around 10 µm where certainly not all this is coming from the very small grains.

J. Dorschner: My question concerns such PAHs. Of course I agree with you that they are very attractive for explaining a lot of observational facts, but my problem is that such compounds need carbon double bonds. What we observe in interstellar molecules mostly are no double bounds. We observe threefold bonds or single bonds. How would you solve this problem?

E. Krügel: I am not the expert. But one way of producing and circumventing interstellar chemistry would be the shattering of graphite particles into smaller and smaller units and those which survive may be structures like coronene.

J. Dorschner: But then you must have strong evidence that graphite, the mineralogical variety of graphite, exactly exists. This is not clear, I think.

E. Krügel: I think from the discussion about the 220 nm hump that something very close to graphite produces it. My conception always is that true graphite is in interstellar space and if it is then shattered in shocks, one could possibly produce components like coronene.

J. Dorschner: I hope you are right.

J. M. Greenberg: I will get back to the point of matching the extinction curve, the calculations look very much like the extinction curve. Since I am not going to talk about this in my lecture I would like to make some strong remarks about this. There is no model today which can produce a fully satisfactory extinction curve which is characterized by several very important features. One very important thing is that the hump is always in the same position. The only way which one can accomplish that is if the particles which produce the hump are invariant, i.e., invariant in optical properties. Within the framework of a particle size distribution where something produces the hump, something produces the far UV, and something else produces the visual one can accomplish that without changing any of the optical properties. Within the framework of the Mathis model the position of the hump in the average ISM is matched by taking a specific size distribution of particles, small and large particles. In this model the large particles contribute substantially to the far UV extinction when you change the size distribution and if you conserve the far UV, which one can surely do, you move the hulp beyond the observed position. I think this is a very important criterion which you should not easily ignore.

E. Krügel: I see your point, but you only shift the problem. If you say, to fix the wavelength of the hump the grains have to be invariant in size, then I would like to know what fixes the size of the carbon grains.

P. G. Mezger: I should like to mention there was a workshop on these small grains in Liège where chemists and astronomers were together and apparently there was no agreement. The chemists were saying of actually PAHs should be the small particles then one would expect much more lines in the NIR which are not seen. So, it has not yet completely been settled.

THE ROLE OF GRAINS IN MOLECULAR CHEMICAL EVOLUTION*

J. M. GREENBERG

University of Leiden, Laboratory Astrophysics, Leiden, The Netherlands

(Received 27 June, 1986)

Abstract. The observations of dust gas in diffuse and molecular clouds are shown to reflect not only their current state but their past history. The interpretation of infrared spectra of dust in molecular clouds using appropriate core-mantle grains shows that: (1) the kinds and amounts of ices, (2) the relative proportion of such important interstellar molecules as H_2O and CO, (3) the evidence for the less abundant solid species $X—C \equiv N$, COS, H_2S, and (4) the thermal history of the dust may all be demonstrated quantitatively from laboratory analog studies of ultraviolet photoprocessing of relevant ices and from theoretical studies of gas-dust interactions. In diffuse clouds the dust is shown to consist predominantly of refractory organic compounds which originate as residues of the photoprocessing of volatile ices in molecular clouds and which undergo further physical and chemical evolution in the diffuse clouds.

1. Introduction

Chemical evolution in the space between the stars follows a complex series of steps involving ions, electrons, atoms, and molecules in the gas and on and in the small solid interstellar dust grains. The role of grains is a very important one in determining the relative amounts of some of the most important molecules. For example, although the H_2O on grains is observed to be more abundant than CO (van de Bult *et al.*, 1985; Whittet *et al.*, 1985; Lacy *et al.*, 1984) just the opposite is true in the gas (Mann and Williams, 1980).

The principal processes in the overall chemical evolution scheme involving the grains are accretion and ultraviolet photoprocessing of grain mantles. The study of these processes has been made possible by the creation of laboratory conditions which simulate those relevant to the evolution of grains in interstellar space. The key experimental components are the low temperature, low pressure, and the ultraviolet. These have been amply discussed in Hagen *et al.* (1979) and Van IJzendoorn (1985). We have studied infrared spectra of various molecular mixture both qualitatively and quantitatively, finding for some solid mixtures the values of the complex index of refraction $m = m' - im''$ as a function of temperature. We have produced stored radicals and studed their diffusion and, in some cases, the explosive reactions caused by sudden heating (d'Hendecourt *et al.*, 1982). All of these results have been applied, as seen in the following selected brief summaries, to simulating grains conditions which may be either directly or indirectly observed by astronomical means.

---

* Review paper presented at a Workshop on 'The Role of Dust in Dense Regions of Interstellar Matter', held at Georgenthal, G.D.R., in March 1986.

A basic fact which emerges is that the evolution of grains is a cyclic one in which the grains pass sequentially from shielded molecular cloud conditions to exposed diffuse cloud conditions many times during their full lifetime (Greenberg, 1982a, b, 1986).

2. The Ice Band

The 3.07 μm OH stretch gets special attention because it is by far the dominant infrared feature of grain mantles. It has proven to be not only a probe of the chemical but also the physical evolution of dust.

Two examples of ice absorption observations (Whittet *et al.*, 1983) are shown in Figure 1 and compared with a reference absorption of crystalline H_2O. We note that the width and peak absorption of the grains in the line-of-sight to HL Tau is closer to that of crystalline ice than that of Elias 16. The values of m' and m'' as applied to the calculation of the absorption cross sections of various shapes of particles show that the absorption structures shown in Figure 1 are best matched by long particles whether cylindrical or spheroidal (van de Bult *et al.*, 1985; Greenberg *et al.*, 1983) and that the shape of the particle is almost as important as the choice of the ice mixture in determining the degree of matching. One may deduce from the absorption strength, width and peak positions of the ice band in Elias 16 and HL Tau that the fractional amount of H_2O in the grain mantles is greater than 50% and, in fact is probably as much

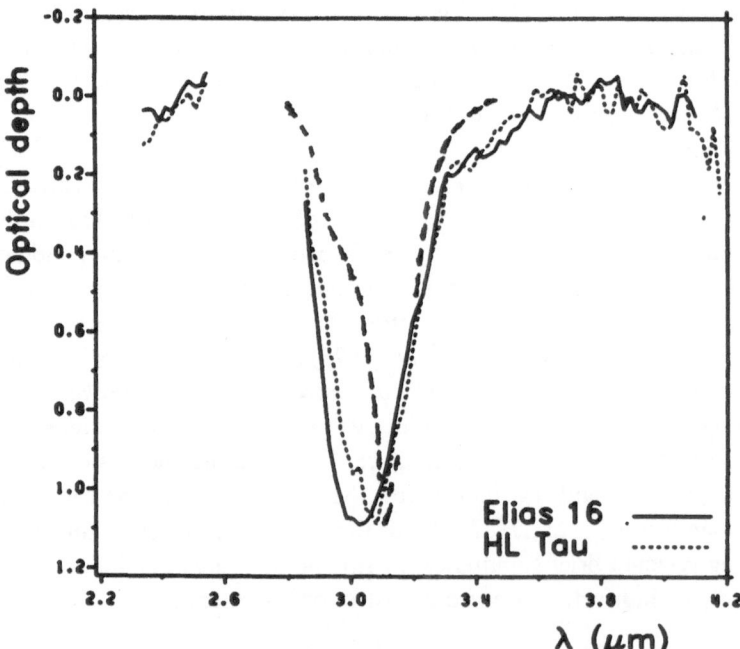

Fig. 1.   Comparison of the observed spectra of HLK Tau and Elias 16 normalized to the same optical depth and with straight lines connecting the points. Dashed curve is absorption by crystalline ice (Bertie *et al.*, 1986).

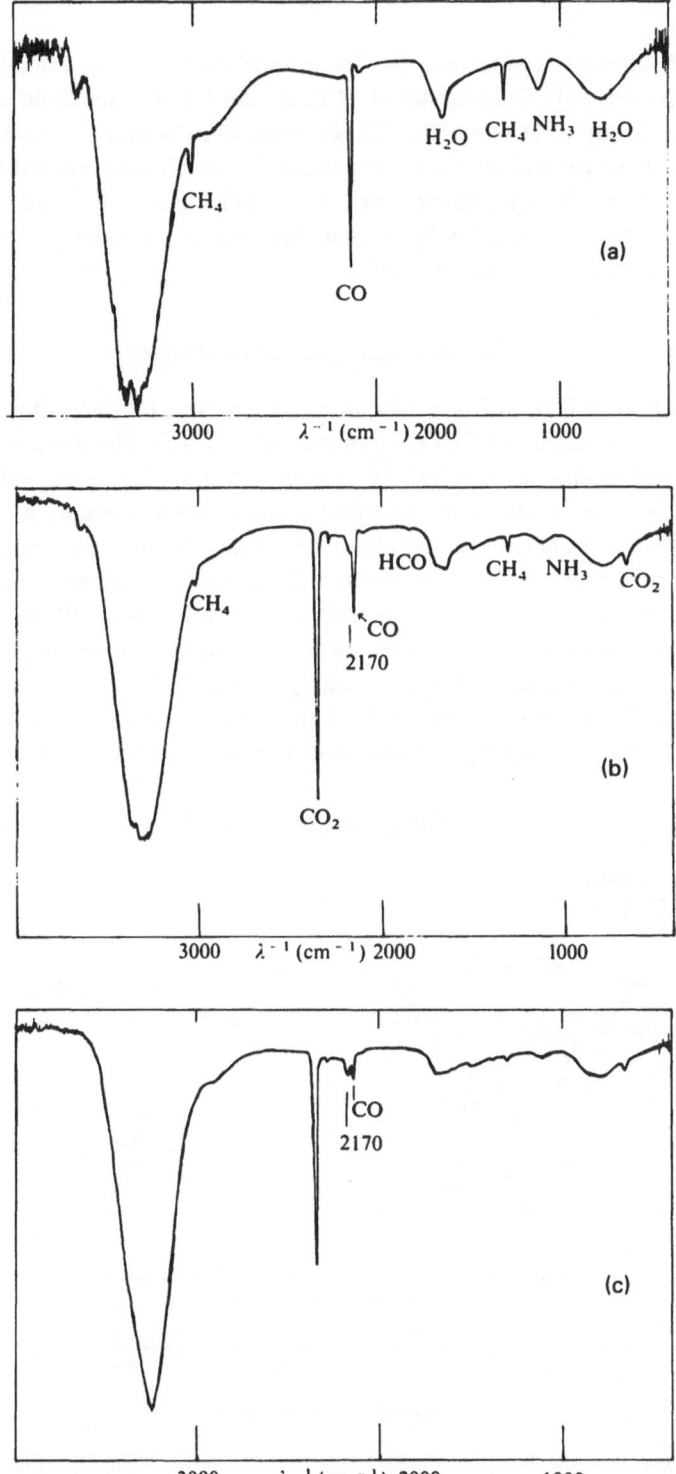

Fig. 2. Spectra of a water rich mixture ($H_2O : CO : NH_3 : CH_4 = 6 : 2 : 1 : 1$). (a) Non-irradiated 10 K; (b) photolyzed; (c) photolyzed and warmed up to 95 K.

as 60–70%. Furthermore, while the Elias 16 ice band is characteristic of mantles formed and kept at the very low temperature of $\lesssim 15$ K, the HL Tau ice band clearly shows annealing by warmup to at least 50 K. Since Elias 16 is behind the Taurus molecular cloud it samples the general average of dust mantles which have formed there. On the other hand HL Tau is known to have a high degree of local activity probably in the form of an accretion disk (Cohen, 1983) so that the dust in its vicinity is likely to have undergone a significant amount of heating.

3. Photoprocessing of Grain Mantles

The ultraviolet irradiation of a water rich ice mixture ($H_2O : CO : NH_3 : CH_4 = 6 : 2 : 1 : 1$) is shown in Figure 2 (d'Hendecourt et al., 1986). The initial stage is that of a deposit without irradiation which shows infrared absorption features belonging to the initial molecular constituents. Next shown is the result of 5 hr irradiation corresponding to about a million years in a molecular cloud. Many new features are seen among which the obviously strong one of CO_2. The weak feature of HCO indicates about 0.1 to 1% radical concentration. The short wavelength wing at 4.62 μm (2170 cm^{-1}) appearing on the CO absorption at 4.67 μm has particular significance because it persists at relatively high temperatures while CO is readily evaporated as shown in the spectrum after the mixture has been warmed to 95 K. This effect has also been observed in W33A (see Figure 3) where the 4.62 μm absorption has been identified as belonging to a

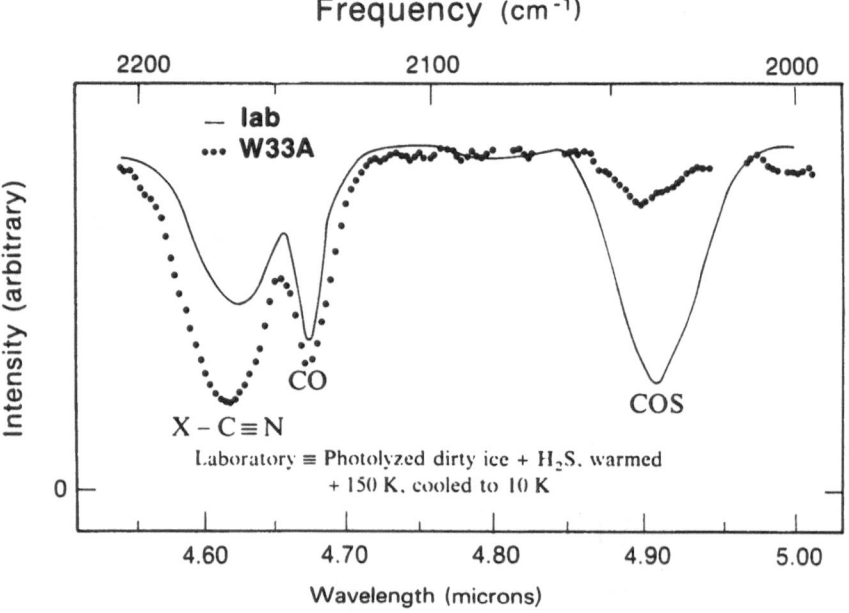

Fig. 3.   Laboratory spectrum of a frozen mixture of simple molecules, including H_2S, CO, and NH_3, which has been subjected to UV-irradiation at 10 K and then warmed to 85 K. The spectrum of W33A is shown for comparison with its baseline flattened.

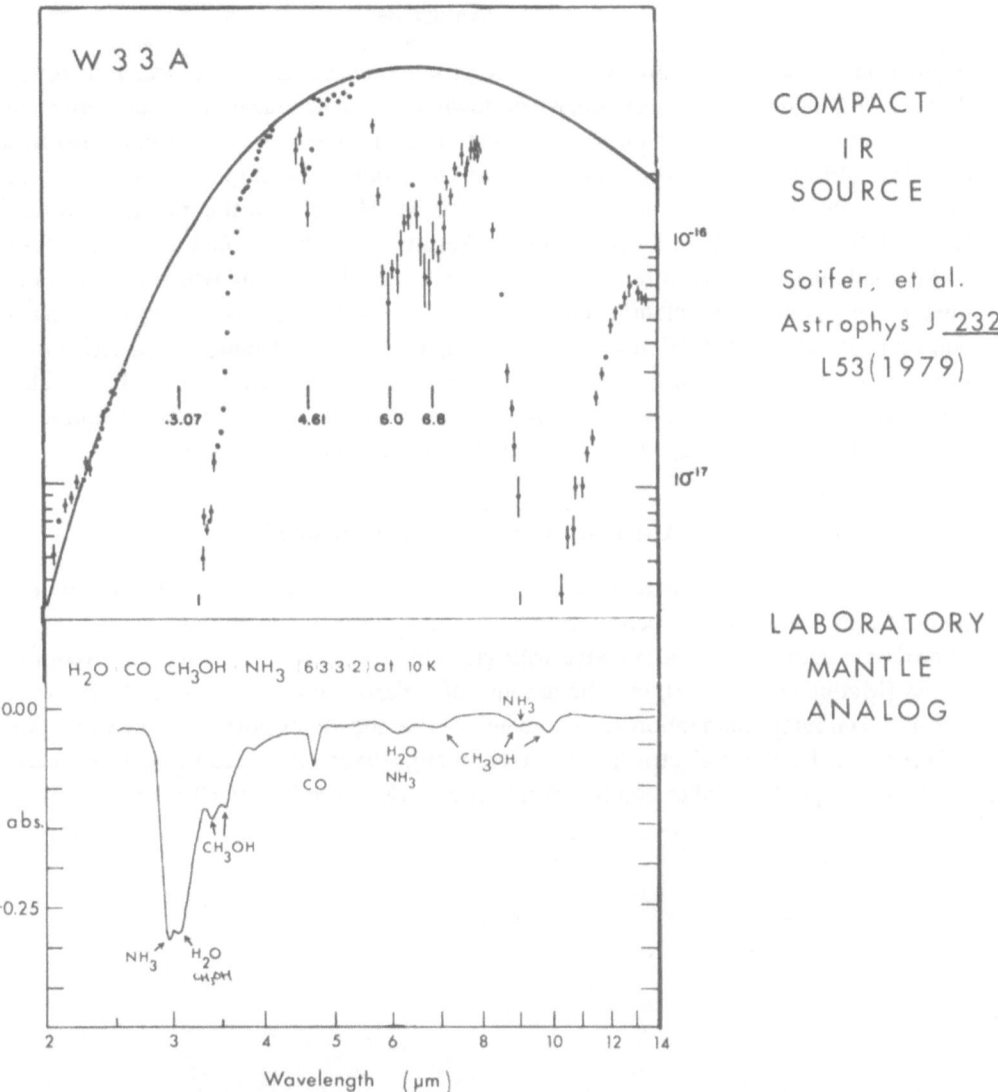

Fig. 4. Spectrum of a compact source W33A showing several features in addition to the strong H_2O and silicate absorptions at 3.07 and 9.5 μm. For comparison is shown a spectrum of a laboratory mantle analog mixture.

molecule (as yet unidentified) containing the cyanogen group —C≡N (Lacy *et al.*, 1984). The deeply dust imbedded object W33A has also provided us with evidence of photoprocessing of grain mantles containing H_2S (seen at 3.9 μm) and producing COS (carbonyl sulfide). We note how the amount of CO relative to H_2O in W33A is indeed small when we see the full spectrum shown in Figure 4 (Soifer *et al.*, 1979).

4. Grain Explosions

Irradiation at low temperatures leads to the production *and* storage of free radicals in the ices. This highly reactive mixture, when slowly warmed produces chemiluminescence and thermoluminescence which have been studied in some detail in various mixtures (Van IJzendoorn, 1985). When the grains are maintained at temperatures $T_d \lesssim 15$ K essentially all the radicals remain for times $\sim 10^8$ yr. However, if the dust temperature is suddenly raised to about 27 K, laboratory results show that enough radicals diffuse and recombine (releasing stored energy) to lead to a chain reaction which explodes the mixture. In an interstellar cloud such temperatures may be created either by grain-grain collisions at velocities > 40 m s^{-1} (Greenberg, 1979; d'Hendecourt *et al.*, 1983) in a turbulent medium or by bombardment by cosmic-ray iron ions (Léger *et al.*, 1985). This explosive desorption of molecules from the dust mantles appears to be adequate to maintain the gas phase species against total depletion onto the grains.

5. Organic Refractory Grain Mantles

Cosmic abundance arguments show that silicates alone are inadequate to provide the amount and wavelength dependence of extinction (Greenberg and Hong, 1975). An appropriate combination of silicates with graphite may satisfy the extinction requirements (Mathis *et al.*, 1979), but the amount of polarization and the relation between linear and circular polarization seem to demand that the major portion of the extinction be produced by the dialectric grians (Greenberg, 1986; Aannestad and Greenberg, 1983). Although this implies that the silicate grains which produce the 9.5 μm absorption

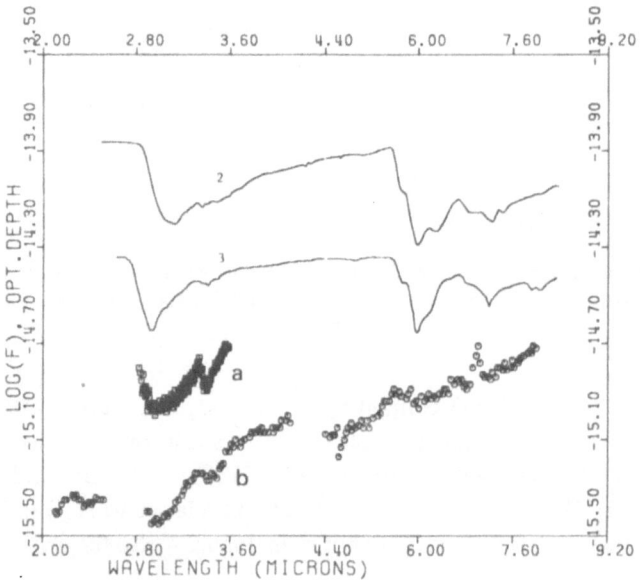

Fig. 5a.

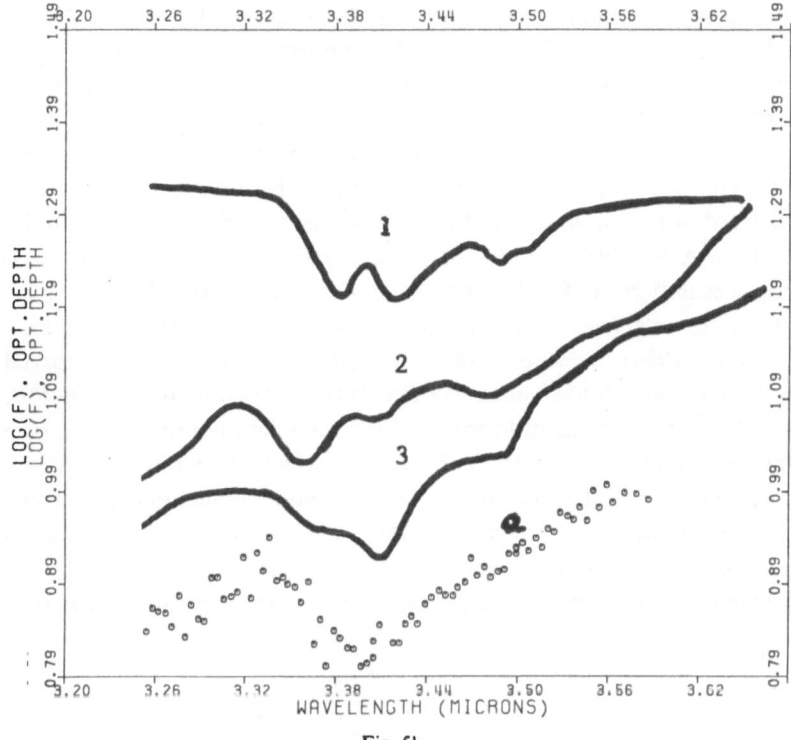

Fig. 5b.

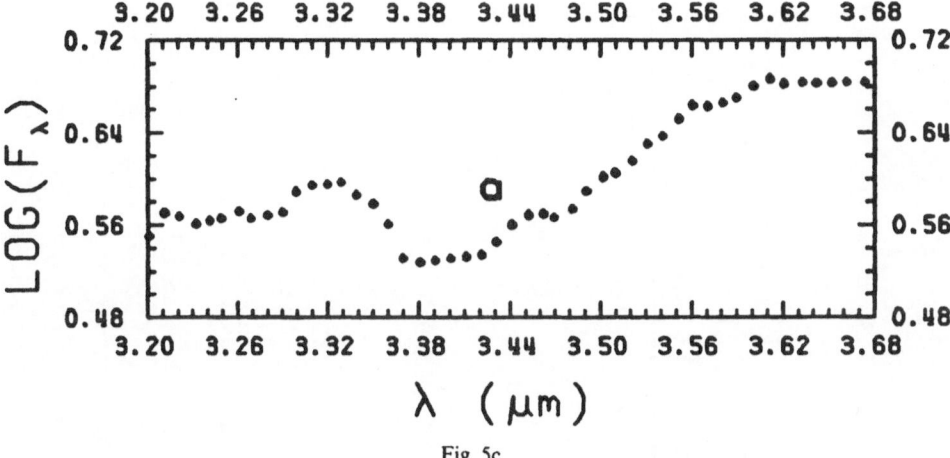

Fig. 5c.

Fig. 5a–c. Infrared absorption spectra of several organic refractory residues and the galactic center sources IRS7 and Sgr AW. (1) Residue of CH_4, (2) residue of $CO : H_2O : CH_4 : NH_3 = 2 : 2 : 1 : 2$, (3) residue of $CO : H_2O : NH_3 = 5 : 5 : 1$, (a) IRS7, (b) Sgr AW. Separate curve for 3.4 μm feature IRS7 is fully smoothed and shows a peak absorption somewhere between that of residues 2 and 3.

must be coated with mantles which contain the organics O, C, and N, there is no evidence that the normally abundant H_2O of molecular cloud grain mantles exist in adequate amount in the diffuse cloud medium. Thus, another form of grain mantle which contains O, C, and N must exist. This form is produced as the result of ultraviolet photoprocessing of ices leading to organic refractory molecular mixtures (Greenberg et al., 1972; Hagen et al., 1979). The direct evidence for organic residues in space was finally exhibited in the observation of the 3.4 μm feature toward objects in the galactic center as in Sgr AW (Willner et al., 1979). This feature has been further observed with much better resolution in IRS 7 (Butchart et al., 1986). Laboratory residues whether produced by ultraviolet (see Figure 5) or proton bombardment (Moore and Donn, 1982; Strazzulla et al., 1984) have not produced a precise match to the observed 3.4 μm structure although the UV irradiated residues appear closest to the observations. On the other hand the measured rate of formation by ultraviolet photoprocessing in molecular clouds has been shown to provide an absorption strength which is adequate to account for the mantles on the dust grains in the diffuse cloud medium along the distance to the galactic center (Schutte and Greenberg, 1986). The variety of absorptions in the 3 μm region observed towards the galactic center (Figure 6) shows that some, even though not much, H_2O must exist along that line-of-sight implying the presence or a bit of molecular

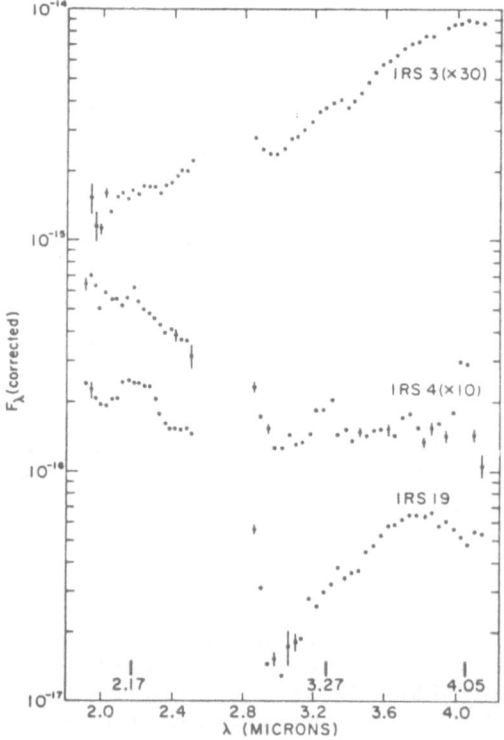

Fig. 6. Spectra of galactic center sources corrected for interstellar extinction. Note the variety of absorptions peaked at about 3 μm evidently not due to normal H_2O.

cloud dust extinction (Schutte and Greenberg, 1986). In any case, the presence of organic refractory mantles is clealry proven so that, in general, the grains in molecular clouds must consist of silicate cores with layered mantles, the inner portions being the residues which have been produced in molecular clouds and which have undergone long term irradiation in the diffuse cloud medium. The outer portion in molecular clouds is an irradiated icy mixture and in diffuse clouds is the latest formed residue.

6. Grain Mantle and Gas Evolution in Molecular Clouds

The interplay between the atoms and molecules in the gas and in the grains consists of essentially five basic steps: (1) ion-molecule reactions in the gas, (2) atom and molecule sticking on grains, (3) grain surface reactions, (4) ultraviolet irradiation and production of stored radicals, (5) grain explosions replenishing the gas (d'Hendecourt et $al.$, 1985). The state of the grains and gas is a time-dependent one $even$ if one maintains constant cloud conditions. The time-dependence is, of course, further modified if we allow the cloud itself to evolve or change. This remains to be considered in the future. For the moment I present a small sample of the results applicable to a rather dense cloud of density $n_H = 2 \times 10^4$ cm^{-3} and a mean ultraviolet attenuation relative to the diffuse cloud medium of $\sim 3 \times 10^{-4}$. This leads to an ultraviolet irradiation time for the production of 1% radicals of less than 10^5 yr which is adequate to provide explosive mixtures between grain-grain collision times.

In Figures 7(a), (b), and (c) the time-dependence of the mantle components show several very important consequences.

(1) At early times the grain mantles are dominated by H_2O, CH_4, and NH_3 in relative amounts more or less as given by Van de Hulst (1949).

(2) At intermediate times (10^5-10^7 yr) the mantle remains dominated by H_2O and would ultimately contain a very large CO concentration if we neglect the photo-processing which leads to a large concentration of CO_2 (an effect neglected in this calculation).

(3) At longer times, H_2O + CO_2 remain the dominant constituents with no significant amount of CH_4 or NH_3.

The mantle composition is at all $times$ very different from the gas composition and notably this is exhibited by the fact that the H_2O in the dust is never less than 10 times that in the gas so that water ice is not $accreted$ on grains it is created on grains. This is why the H_2O in grain mantles is always observed to be larger than the CO even though, in the gas, the CO is by far the dominant molecular constituent. The theoretically predicted drop in the H_2O mantle fraction in dense clouds for $t > 10^7$ yr is probably prevented from generally occurring because such clouds are not likely to be stable over times as long as this either as a consequence of cloud-cloud collisions or cloud contraction. Thus a canonically stable value of the fraction of H_2O in molecular cloud dust outer mantles is probably about 60–70% as already noted for Elias 16 and even for the B.N. ice mantles where they exist (Greenberg, 1982b).

The solid CO which has been observed (Whittet et $al.$, 1985) in Elias 16 representing

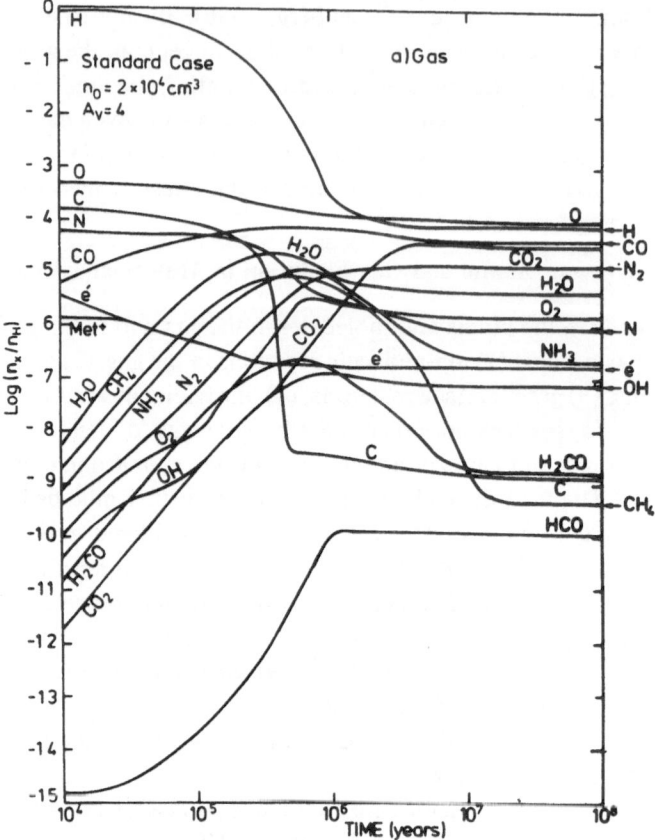

Fig. 7a.   Time-evolution of the abundances of the molecular components of the gas in a cloud with density
$n_{\rm H} = 2 \times 10^4$ cm^{-3} and extinction $A_V = 4$ mag.

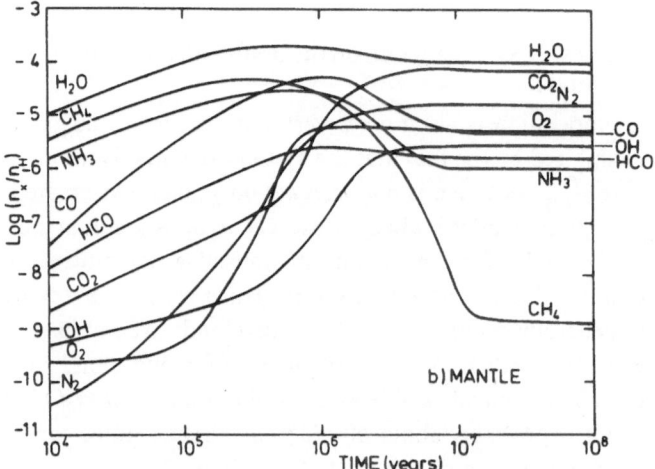

Fig. 7b.   Same as Figure 7(a) for the grain mantle.

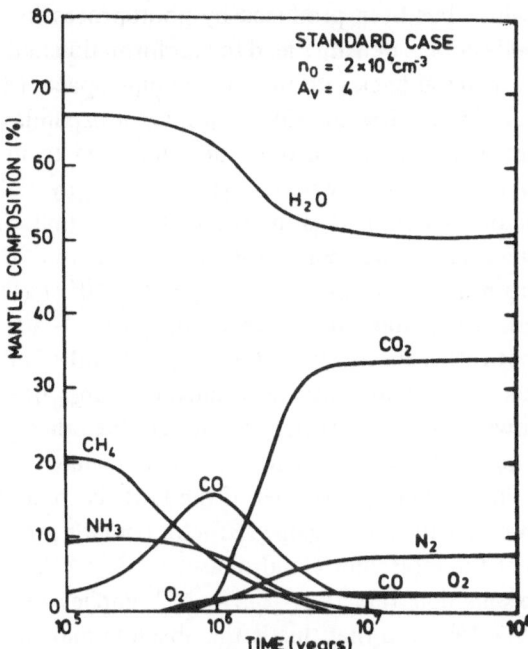

Fig. 7c. Relative fractions of the major volatile components of the grain mantles in a cloud with $n_H = 2 \times 10^4$ cm^{-3} and extinction $A_V = 4$ mag.

normal molecular cloud dust is less abundant than the solid H_2O even though the CO/H_2O in the gas is $\gg 1$. It was shown by van de Bult *et al.* (1985) that the fraction of H_2O in the outer icy grain mantles is about 60% and is unheated. The relative amounts of H_2O and CO are given by the ratio $n_{H_2O}/n_{CO} = (\sigma_{CO}/\sigma_{H_2O})(\tau_{H_2O}/\tau_{CO})$ where the cross section per CO molecule is about $\sigma_{CO} \simeq 1.3 \times 10^{-18}$ cm^2 (Lacy *et al.*, 1984) and the cross section per H_2O molecule is $\sigma'_{H_2O} \simeq ((f - 0.15)/0.85))\sigma^{pure}_{H_2O}$ where $f = H_2O$ fraction in the grains and 'pure' stands for pure amorphous ice. The reduction factor $(f - 0.15)/0.85$ results from the statistical proportion of polymeric ice (Greenberg *et al.*, 1983). We derive $\sigma^{pure}_{H_2O} = 0.6 \times 10^{-18}$ cm^2 by inserting the measured absorptive part of the index of refraction, $m''(3.07) = 0.477$, in the relation, $4\pi m''/\lambda = s(Mm_H)^{-1}\sigma^{pure}_{H_2O}$, where $s = 1$ g cm^{-3}, $M = 18$, $m_H =$ hydrogen mass. This gives $\sigma_{H_2O} = 0.32 \times 10^{-18}$ cm^2 and $n_{H_2O}/n_{CO} \simeq 5$. This is consistent with the values obtained for a wide variety of cloud conditions (d'Hendecourt *et al.*, 1985). In HL Tau, as expected from the evidence for heating of the icy mantle (van de Bult *et al.*, 1985), the solid CO is substantially reduced, being $\tau_{CO} < 0.05$, leading to $n_{H_2O}/n_{CO} > 20$.

7. Cyclic Evolution of Grains

The evolutionary picture of dust which is emerging is a cyclic one in which the particles find themselves alternately in diffuse clouds and in molecular clouds. A small silicate core captured within a molecular cloud gradually builds up an inner mantle of organic refractory material which has been produced by photoprocessing of the volatile ices. Within the dense clouds critical densities lead to star formation and subsequent ejection of some of the cloud material back into the surrounding space. Much of this material finding itself in a very tenuous low-density environment expands to the diffuse cloud phase. Dust particles in the diffuse medium are subjected to numerous destructive processes which rapidly erode the outer volatile mantles away and, in fact, erode part of the organic refractory material. It is important to note that without their organic refractory mantles the silicate cores could not survive. The rate of destruction of pure silicate grain leads to a maximum lifetime of $\tau_{Sil} \simeq 4 \times 10^8$ yr (Draine and Salpeter, 1979) which converts to a mass loss rate of $d\rho_{Sil}/dt = -5 \times 10^{-43}$ g cm^{-3} s^{-1}. Assuming a mass loss rate from M-stars of $1\, M_\odot$ yr^{-1} and a full cosmic abundance silicate production leads to a production rate for silicates of $d\rho_{Sil}/dt < 10^{-45}$ g cm^{-3} s^{-1} which is 100 times lower than the destruction rate. On the other hand, the production rate of the O.R. of $d\rho_{O.R.}/dt \simeq 10^{-41}$ g cm^{-3} s^{-1} is adequate to replenish the mantle material lost in the diffuse cloud phase even if the O.R. is somewhat less tough than the silicates. Therefore, silicate core-organic refractory mantle grains survive the diffuse cloud phase to reenter the molecular cloud phase.

The mean star production rate of 1–$2\, M_\odot$ yr^{-1} implies an interstellar medium turnover time of $\sim 5 \times 10^9$ yr so that this is the absolute maximum lifetime of a dust particle no matter how resistant to destruction. If we use a mean molecular cloud-diffuse cloud period of 2×10^8 yr (10^8 yr in each) then a typical grain anywhere in space will have undergone at least 20 cycles so that, for example, the typical diffuse cloud dust particle age is $\gtrsim 10^9$ yr and consists of a mix of particles which have undergone a wide variety of photoprocessing. Note that the organic refractory mantles are subjected to the highest photoprocessing rates in the diffuse cloud phase. This would imply that the organic refractory mantle on a grain is not a homogeneous substance. A result of sequential formation (in the molecular cloud phase) and of intense photoprocessing (in the diffuse cloud phase) would lead to a layering in which the innermost layers have been the most irradiated and the outermost layer is first generation organic refractory.

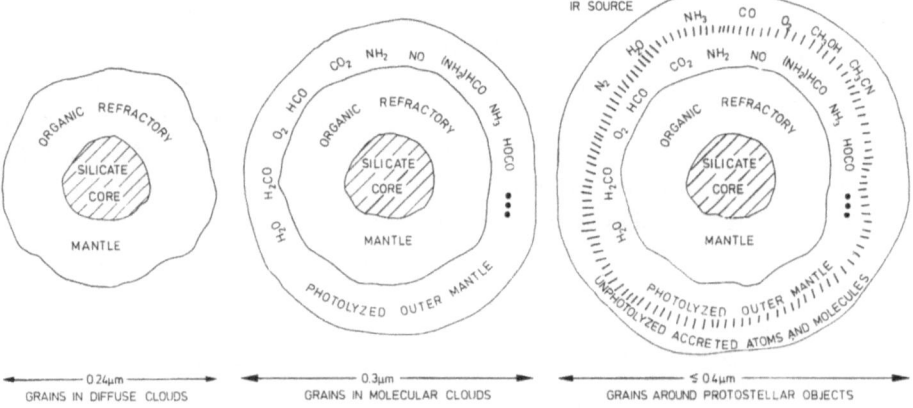

Fig. 8.   Cross-section of grains at various stages of evolution.

Because of this kind of layering, and the fact that the grains are of vairous ages leads one to expect average homogeneity of diffuse cloud grains both in size and structure which is observed as a uniformity in the visual extinction curve and a rather structureless 3.4 μm feature. In other words, diffuse cloud grains represent a steady-state average of grains of a multiplicity of chemical and physical histories.

A schematic representation of grains in the various regions of space is shown in Figure 8. In the final stage of cloud condensation we may expect that all remaining (condensable) molecules will have accreted onto the dust. In addition, the very small ($\lesssim 0.01$ μm) particles will be collected and trapped within the outer volatile mantle.

References

Aanestad, P. A. and Greenberg, J. M.: 1983, *Astrophys. J.* **272**, 551.

Butchart, I., McFadzean, A. D., Whittet, D. C. B., Geballe, T. R., and Greenberg, J. M.: 1986, *Astron. Astrophys.* **154**, L5.

Cohen, M.: 1983, *Astrophys. J.* **270**, L69.

d'Hendecourt, L. B., Allamandola, L. J., Baas, F., and Greenberg, J. M.: 1982, *Astron. Astrophys.* **109**, L12.

d'Hendecourt, L. B., Allamandola, L. J., and Greenberg, J. M.: 1985, *Astron. Astrophys.* **152**, 130.

d'Hendecourt, L. B., Allamandola, L. J., Grim, R. A., and Greenberg, J. M.: 1986, *Astron. Astrophys.* **158**, 119.

Draine, B. T. and Salpeter, E. E.: 1979, *Astrophys. J.* **231**, 438.

Greenberg, J. M.: 1973, in M. A. Gordon and L. E. Snyder (eds.), *Molecules in the Galactic Environment*, Wiley, New York, p. 94.

Greenberg, J. M.: 1979, in B. E. Westerlund (ed.), *Stars and Star Systems*, D. Reidel Publ. Co., Dordrecht, Holland, p. 173.

Greenberg, J. M.: 1982a, in J. E. Beckman and J. P. Phillips (eds.), *Submillimetre Wave Astronomy*, Cambridge Univ. Press, Cambridge, p. 261.

Greenberg, J. M.: 1982b, in L. L. Wilkening (ed.), *Comets*, Univ. of Arizona Press, Tucson, p. 131.

Greenberg, J. M.: 1986, in F. Israel (ed.), *Proceedings of IRAS Symposium 'Light on Dark Matter'*, D. Reidel Publ. Co., Dordrecht, Holland.

Greenberg, J. M. and Hong, S. S.: 1975, in G. B. Field and A. G. W. Cameron (eds.), *The Dusty Universe*, Neal Watson Academic Publ., p. 31.

Greenberg, J. M., van de Bult, C. E. P. M., and Allamandola, L. J.: 1983, *J. Phys. Chem.* **87**, 4243.

Greenberg, J. M. Yencha, A. J., Corbett, J. W., and Frisch, H. L.: 1972, *Mém. Soc. Roy. Soc. Liège* **3**, 425.

Lacy, J. H., Baas, F., Allamandola, L. J., Persson, S. E., McGregor, P. J., Lonsdale, C. J., Geballe, T. R., and van de Bult, C. E. P. M.: 1984, *Astrophys. J.* **276**, 533.

Léger, A., Jura, M., and Omont, A.: 1985, *Astron. Astrophys.* **144**, 147.

Mann, A. P. C. and Williams, D. A.: 1980, *Nature* **283**, 721.

Mathis, J. S., Rumpl, W., and Nordsieck, K. H.: 1979, *Astrophys. J.* **217**, 425.

Moore, M. H. and Donn, B.: 1982, *Astrophys. J.* **257**, L47.

Schutte, W. and Greenberg, J. M.: 1986, in F. Israel (ed.), *IRAS Symp. 'Light on Dark Matter'*, D. Reidel Publ. Co., Dordrecht, Holland.

Soifer, R. W., Puetter, R. C., Russell, R. W., Willner, S. P., Harvey, R. M., and Gillett, F. C.: 1979, *Astrophys. J.* **232**, L53.

Strazzulla, G., Cataliotti, R. S., Calcagno, L., and Foti, G.: 1984, *Astron. Astrophys.* **133**, 77.

Van de Bult, C. E. P. M., Greenberg, J. M., and Whittet, D. C. B.: 1985, *Monthly Notices Roy. Astron. Soc.* **214**, 289.

Van de Hulst, H. C.: 1949, *Rech. Astron. Utrecht* **11**, part 2.

Van IJzendoorn, L. J.: 1985, Ph.D. Thesis, Univ. of Leiden, Holland.

Whittet, D. C. B., Bode, M. F., Longmore, A. J., Baines, D. W. T., and Evans, A.: 1983, *Nature* **303**, 298.

Whittet, D. C. B., Longmore, A. J., and McFadzean, A. D.: 1985, *Monthly Notices Roy. Astron. Soc.* **216**, 45P.

Willner, S. P., Russell, R. W., Puetter, R. C., Soifer, B. T., and Harvey, P. N.: 1979, *Astrophys. J.* **229**, L65.

Discussion

P. G. Mezger: Would these small grains without the silicate core be the same material which in earlier talks we referred to as yellow stuff?

J. M. Greenberg: That is right in part. At least those which give the 220 nm hump, although they require further photoprocessing.

P. G. Mezger: For me the thing would be more convincing if you could produce an extinction curve as far as 1000 μm.

J. M. Greenberg: We met some difficult problems. I am not sure that the PAHs are important enough, I agree with you they probably are not. What I do believe is that small particles may actually be responsible for the near-infrared extinction. That is we also have carbon particles in our distribution and in our grain models we have used graphite because we do not have anything better yet. No matter which carbon you use problems of getting the near-infrared extinction remain.

But the question is: Do these 220 nm particles produce that near-infrared extinction? The answer to that is that I do not know yet. I know about the specific absorptive properties (3.4, 3.0 μm), but I do not know what the overall absorption properties are, let us say at 4 or 10 μm.

I believe that there is carbon in space, however I found very great difficulties with graphite for a number of reasons. We see sufficient evidence for smoky stars to believe that carbon is coming out from stars. The question is whether it is graphite. At a meeting in Les Houches two weeks ago nobody really believed in graphite. They were talking about hydrogenated amorphous carbon.

To summarize the situation: We do not know what is producing the variability in the NIR. I think the observed differences are real.

E. Krügel: There is a very interesting point about the accretion on the grains which you mentioned. In a dense cloud the accretion is so effective that all the molecules should have gone and you solved the riddle in that you say that we have some mechanism by which they evaporate from the grains. You say that you obtain the kinetic energies for evaporation from grain-grain collisions with velocities of the order of 0.1 km s^{-1}. I assume this is for pure inelastic collisions. But there are some cloud cores where molecules are observed with very narrow lines and of course the grains cannot have thermal velocities, they can only have turbulent velocities. But in these very dense cold cores there is very little turbulence. I have the suspicion that you run into trouble not only from the energetic point of view but if the velocities are too low you do not get enough collisions to get the balance which is observed.

J. M. Greenberg: First of all I do not know hold old the cores are. If you were talking about 10^6 yr there is no problem, you would still see them. In terms of the turbulence there is one way to answer. If we do not observe turbulence the mechanism of grain-grain collisions cannot work. The question then remains: Is there another possible triggering mechanism? Jura and Léger proposed the possibility of explosions following the passage of cosmic iron particles. They calculated the flux on the grains in a galactic cloud and apparently the cross sections are comparable to those of grain-grain collisions.

Th. Henning: Is there any observation of the absorption from mantles at the long-wavelength wing of the silicate band?

J. M. Greenberg: Yes, I believe so. There is some evidence that there is in some regions an existence of an extra ice absorption.

H. J. Habing: From IRAS data we have nothing conclusive yet. We have some problems with the determination of a 'standard' continuum.

INTERSTELLAR GAS DEPLETION AND DUST PARAMETERS*

H. ZIMMERMANN

Jena University Observatory, G.D.R.

(Received 27 June, 1986)

Abstract. It is shown that on theoretical grounds the relative abundances of the elements in the interstellar gas phase should be correlated with the wavelength of maximum interstellar polarization λ_{max}. If these correlations can be determined by observations, then there is the possibility to determine the relative abundances of the heavier elements within the mantles as well as within the cores of interstellar dust grains, at least in principle.

The observational data available up to now confirm the existence of such correlations between λ_{max} and the interstellar gas phase abundances of titanium, iron, magnesium, and carbon. Statements about the chemical composition of the dust particles are not yet possible. For this there are observations of the interstellar gas depletion needed, especially in such lines of sight where λ_{max} has extreme values.

1. Introduction

After the first suggestions regarding the formation of interstellar dust particles by Lindblad (1935) and the developing of these ideas in the 1940s by ter Haar (1943), Kramers and ter Haar (1946), and van de Hulst (1949), different theories about grain formation and evolution were developed. Nevertheless, up to now there is no generally accepted dust model regarding the chemical composition of the grains.

Most theories of grain formation and evolution start with the assumption that interstellar dust particles are formed somewhere in the atmospheres of late-type giant stars and ejected from there into interstellar space. The then interstellar dust grains can grow by condensation of interstellar gas particles on to the grains, forming a mantle around the initial bare grains. The rate of growth depends on the physical conditions the grains meet in the interstellar space as well as on the physical properties of the particles themselves. A dust grain may consist, therefore, of a refractory core, the bare grain, surrounded by a mantle of volatile material, likely in some ice-type form. All the grain material is taken from the gas phase. If it is assumed that the solar ('cosmic') abundances of the elements apply to a medium free of dust in the Galaxy at the distance of the Sun from the galactic centre, then any short-fall in the gas phase abundances measured spectroscopically may be interpreted as implying presence in grains. The assumption of solar abundances in a dust free medium is supported by comparisons with, for instance, early-type stars.

In interstellar space a dust grain may not only grow. Under unfavorable conditions the mantles may also be reduced or destroyed by thermal sputtering, essentially in connection with high-velocity shocks of the interstellar clouds in which the grains are

* Paper presented at a Workshop on 'The Role of Dust in Dense Regions of Interstellar Matter', held at Georgenthal, G.D.R., in March 1986.

situated (e.g., Duley, 1982, and references therein). It is generally believed that the refractory cores are scarcely affected by this process. If a_0 is the initial radius of the bare grains and Δa the mantle thickness of the particle, then Δa is time dependent. Because we do not know the course of life of an individual grain nor of a multitude of grains, and because we do not know the exact formation and evolution processes of dust particles, we can not say either what radius $a = a_0 + \Delta a(t)$ an individual grain will have at the time t after leaving its place of birth or what mean radius $\bar{a} = \bar{a}_0 + \overline{\Delta a(t)}$ the dust grains within a cloud will have, where $\bar{a}_0$ is the mean radius of the bare grains and $\overline{\Delta a(t)}$ the mean mantle thickness at the time t.

2. Theoretical Considerations

If the abundance of an element X_i relative to hydrogen in the atmosphere of a star before grain formation is given by the solar value

$$n(X_i)/n(\mathrm{H}) = n(X_i)/n(\mathrm{H})|_{\odot},$$

and the relative abundance in an interstellar gas cloud containing only bare dust particles by

$$n(X_i)/n(\mathrm{H})|_c,$$

then

$$\varkappa_c(X_i) = \left.\frac{n(X_i)}{n(\mathrm{H})}\right|_c \left/ \left.\frac{n(X_i)}{n(\mathrm{H})}\right|_{\odot}\right.$$

describes the depletion of the element X_i in the interstellar gas phase (with the above mentioned condition). The missing elements are incorporated into the bare grains. If one could determine the values $\varkappa_c(X_i)$ by observations one would know the overall element composition of the cores of the interstellar dust grains.

With the incorporation of interstellar gas particles into the mantles we get

$$\overline{\Delta a(t)} = f_1\left(\sum_i [\varkappa_c(X_i) - \varkappa(X_i; t)]\right) \tag{1}$$

where $\varkappa(X_i; t)$ describes the gas depletion within an interstellar cloud of the element X_i, which depends on the time t the grains spent in interstellar space. To sum is over all elements, $i = 1, 2, \ldots$. The expression

$$\sum_i [\varkappa_c(X_i) - \varkappa(X_i; t)]$$

gives the mantle composition regarding elements more massive than hydrogen. Under undisturbed conditions the $\varkappa(X_i; t)$ decrease with time t, whereas $\overline{\Delta a(t)}$ increases. In other cases, when sputtering occurs, the mantle thickness declines and the relative abundance of the elements in the gas phase climb up. Therefore, the function f_1 is not necessarily monotonic in t, but it gives an unique relation between mantle thickness and gas depletion. This is true not only in considering the overall depletion

$$\sum_i [\alpha_c(X_i) - \alpha(X_i)],$$

but also if we consider a special element X_i,

$$\overline{\Delta a(t)} = f_1(\alpha_c(X_i) - \alpha(X_i; t)).$$

It is not to be expected that an element vaporizes from the mantle while all other elements condensate.

From theoretical considerations we expect that with changing mantle thickness the overall chemical composition of the coated dust grains changes and, therefore, the optical parameters of the grains will be modified. This should have observable effects. The interpretation of such observations is generally very difficult and mostly rather indirect. But there is one observable parameter, the wavelength of maximum interstellar polarization λ_{max} which depends in an unique way on the mean grain radius. With increasing a, λ_{max} is shifted to longer wavelengths. λ_{max} also depends on the chemical composition of the grains. Given a constant mean grain radius, λ_{max} will vary continuously and monotonically with a continuous changing of the composition. Therefore, we expect that

$$\lambda_{max} = f_2\left(\bar{a}_0 + \overline{\Delta a}; \sum_i [\alpha_c(X_i) - \alpha(X_i)]\right).$$

In this expression the time is no longer relevant. Using equation (1) we get

$$\lambda_{max} = f_3\left(\sum_i [\alpha_c(X_i) - \alpha(X_i)]\right). \tag{2}$$

That means, we should expect a strict correlation between the wavelength of maximum interstellar polarization and interstellar gas depletion, where we can consider either all elements together or an individual element as representative for all.

The importance of Equation (2) lies in the rather strict independence of λ_{max} of the unknown history of the particles. After a grain has endured a sequence of growing and sputtering phases or chemical explosions in the mantle material, the mantle composition will be somewhat different from that shortly after the first beginning of mantle growth (Greenberg, 1986). But if the mean radius of the particles is the same, the effect on λ_{max} should be very small.

The two quantities λ_{max} and $\alpha(X_i) = N(X_i)/N(H)$ are directly observable. Instead of volume densities we have to use column densities $N(X_i)$ and $N(H)$ of the element X_i and hydrogen, respectively. The mantles will contain some hydrogen (e.g., bound in different ices) but the ratio of these condensed hydrogen atoms to the atoms remaining in the gas is always small. The mean depletion of the elements forming ices, e.g., oxygen, is of the order of 0.1 (Harris et al., 1984). With a solar abundance of these elements relative to hydrogen smaller than 7×10^{-4} (oxygen, the most abundant massive element), hydrogen is reduced by less than 10^{-3}, if each of these heavier condensed atoms bound the appropriate number of hydrogen atoms. Therefore, the column density of hydrogen (atomic and molecular) is practically uneffected by accretion and $\alpha(X_i)$ gives indeed in a very good approximation the relative abundance of the element X_i in the gas phase.

With decreasing mantle thickness (i.e., smaller gas depletion) λ_{max} changes to smaller wavelengths. In the extreme case of bare particles λ_{max} reaches a minimum, if one supposes that the bare core particles also give rise to interstellar polarization. At this minimum the difference $\alpha_c(X_i) - \alpha(X_i)$ vanishes, and the observable quantity $N(X_i)/N(H)$ equals $\alpha_c(X_i)$. That means, it should be possible, at least in principle, to determine the overall composition of bare grains by observations, if one can determine the minimum of λ_{max} by observations, and for such lines-of-sight the gas depletion for as many elements as possible. When $\alpha_c(X_i)$ is known then observations of $N(X_i)/N(H)$ for lines-of-sight with large λ_{max} give the possibility to calculate $\alpha_c(X_i) - \alpha(X_i)$, i.e., to determine the relative abundance of the heavier elements in the grain mantles. Unfortunately, the amount of hydrogen incorporated in the mantles cannot be determined in this way.

TABLE I

Wavelength of maximum interstellar polarization (λ_{max} in μm) and relative interstellar gas abundances

HD	Name	λ_{max}	$\alpha(Ti)$	$\alpha(Fe)$	$\alpha(Mg)$	$\alpha(C)$
2905	κ Cas	0.53 ± 0.01	0.0126	0.0224	0.39 ± 0.08	4.8 ± 3.7
24398	ζ Per	0.54 ± 0.03	0.0022	0.0045	0.092	0.82
24912	ξ Per	0.58 ± 0.01	0.0036	–	0.15 ± 0.06	–
30614	α Cam	0.46 ± 0.01	0.0257	–	0.58 ± 0.12	0.36
37903		0.71 ± 0.02	–	0.0065	–	–
38771	κ Ori	0.51 ± 0.02	0.0071	–	0.49 ± 0.05	0.73 ± 0.30
40111	139 Tau	0.68 ± 0.05	0.0234	–	0.62 ± 0.07	–
64760		0.58 ± 0.02	–	–	0.63 ± 0.07	–
74375		0.57 ± 0.01	–	0.0245	–	0.42 ± 0.26
74575	α Pyx	0.52 ± 0.01	–	–	0.87 ± 0.43	–
112244		0.69 ± 0.02	–	0.0209	–	–
113904	θ Mus	0.55 ± 0.01:	–	0.0372	0.41 ± 0.13	–
135591		0.59 ± 0.01	–	0.0214	0.43 ± 0.20	–
141637	1 Sco	0.54 ± 0.03	–	0.0126	0.35 ± 0.04	0.15 ± 0.04
144217	β^1 Sco	0.61 ± 0.02	0.0023	0.0151	0.24 ± 0.03	–
144470	ω^1 Sco	0.59 ± 0.01	0.0028	0.0102	0.30 ± 0.03	0.18 ± 0.11
145502	ν Sco	0.70 ± 0.03	0.0034	0.0123	0.41 ± 0.16	0.36 ± 0.27
147165	σ Sco	0.56 ± 0.02	0.0026	0.0145	0.27 ± 0.06	–
147889		0.80 ± 0.01	–	0.0079	–	–
147933	ρ Oph	0.68 ± 0.01	–	0.0032	0.06 ± 0.01	0.37 ± 0.31
149038	μ Nor	0.59 ± 0.02:	–	–	1.00 ± 0.40	–
149757	ζ Oph	0.59 ± 0.01	0.0039	0.0102	0.32	0.64 ± 0.21
150898		0.57 ± 0.01	–	0.0355	–	–
155806		0.56 ± 0.01	–	0.0195	0.50 ± 0.10	–
164353	67 Oph	0.53 ± 0.02	0.0087	–	–	–
165024	θ Ara	0.57 ± 0.01:	–	0.0257	0.43 ± 0.05	1.88 ± 0.90
184915	κ Aql	0.56 ± 0.02	–	0.0141	0.24 ± 0.07	0.52 ± 0.33

3. Observational Data

To prove whether the relation described by Equation (2) is confirmed by observations, a compilation was done for such lines-of-sight for which determinations of interstellar gas depletion are given in the literature as well as determinations of λ_{max}. Unfortunately, from the multitude of determinations of interstellar gas depletion only a small number can be used for a proof, essentially because of missing polarization observations. Further, for a given element X_i there must exist enough observations (more than about 10) of $x(X_i)$ and λ_{max}, distributed over a larger range of λ_{max}, to get a reliable correlation. In a sample too small it may not be clear whether observed variations from star to star in the interstellar abundance of an element are real or simply a result of limited accuracy of the measurements. Lines-of-sight possibly influenced by circumstellar polarization are of small use and were omitted in the compilation. In Table I the columns 1 and 2 give the observed stars. The wavelengths of maximum interstellar polarization (column 3) are taken from Serkowski *et al.* (1975) or Coyne *et al.* (1974) as well as the mean errors. Mean errors followed by a colon may be underestimated. In such cases where more than one determination of λ_{max} are given by Serkowski *et al.* (1975), the mean erros were calculated with the quantities given by these authors. Columns 4 to 7 present the relative interstellar gas abundance of titanium, iron, magnesium, and carbon. The values for titanium are taken from Stokes (1978), in the case of 67 Oph from Hobbs (1984), for iron from Savage and Bohlin (1979), in the cases of HD 37903 and HD 147889 from Zeippen *et al.* (1977). The magnesium abundances were calculated with the column densities of magnesium and hydrogen given by Murray *et al.* (1984). A solar abundance of magnesium relative to hydrogen of 3.5×10^{-5} was assumed, the value used by these authors too. For ζ Oph the magnesium abundance was taken from Morton (1974) and for ζ Per from Snow (1977). In the case of carbon upper and lower limits of the abundance were taken from Figure 3 of the paper by Jenkins *et al.* (1983) and straight averages calculated. This is a very crude method. But the uncertainties of the values given by these authors are also very high. Only such lines-of-sight were chosen, where lower as well as upper limits are given. For ζ Per the carbon abundance was taken from Snow (1977), for x Cam from Tarafdar *et al.* (1983).

4. Correlations and Discussion

Figures 1 to 4 show the correlations between the relative gas phase abundances of the different elements and λ_{max}. The error bars show the mean errors, if known, of the different observations. The uncertainties of the relative interstellar magnesium abundances were determined from the mean errors of the magnesium column densities given by Murray *et al.* (1984). For carbon the same uncertainties of the abundances were used as given by Jenkins *et al.* (1983).

The dashed lines in Figures 1 to 4 denote the least-squares fits of the data. In the case of carbon the determinations of the relative abundances in the lines-of-sight to the stars κ Cas and ϑ Ara give considerably higher values than the solar abundance. This is

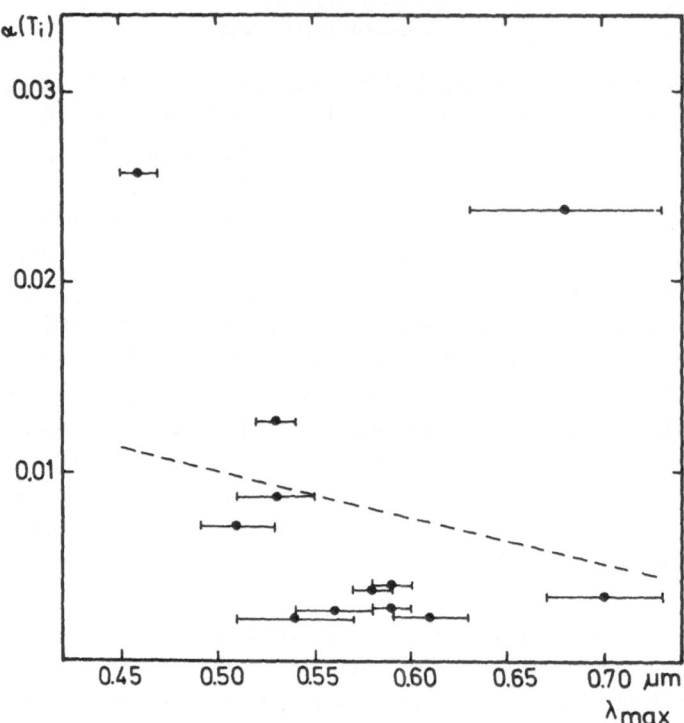

Fig. 1.   Plot of the interstellar gas-phase abundance of titanium against the wavelength of maximum interstellar polarization λ_{max}. The dashed line gives the least-squares fit of the data.

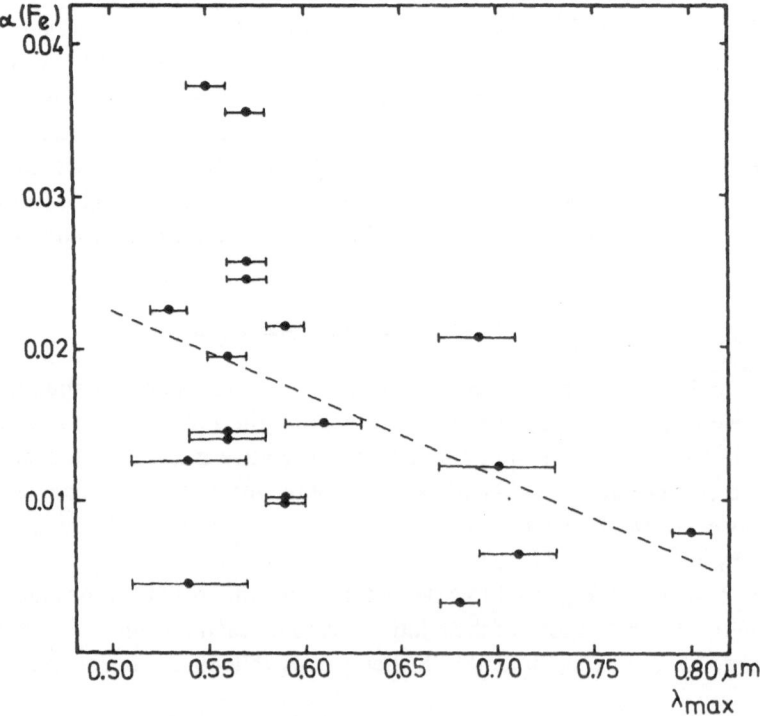

Fig. 2.   The same as Figure 1 but for iron.

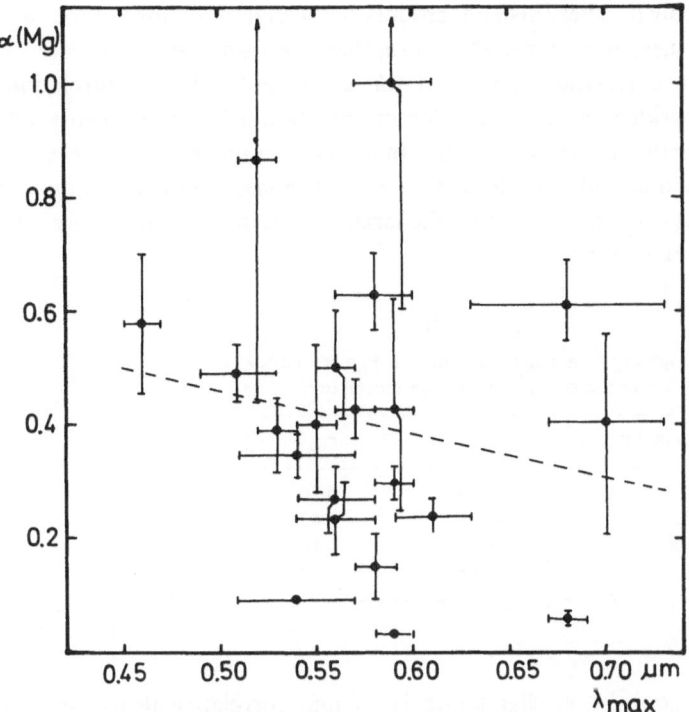

Fig. 3.   The same as Figure 1 but for magnesium.

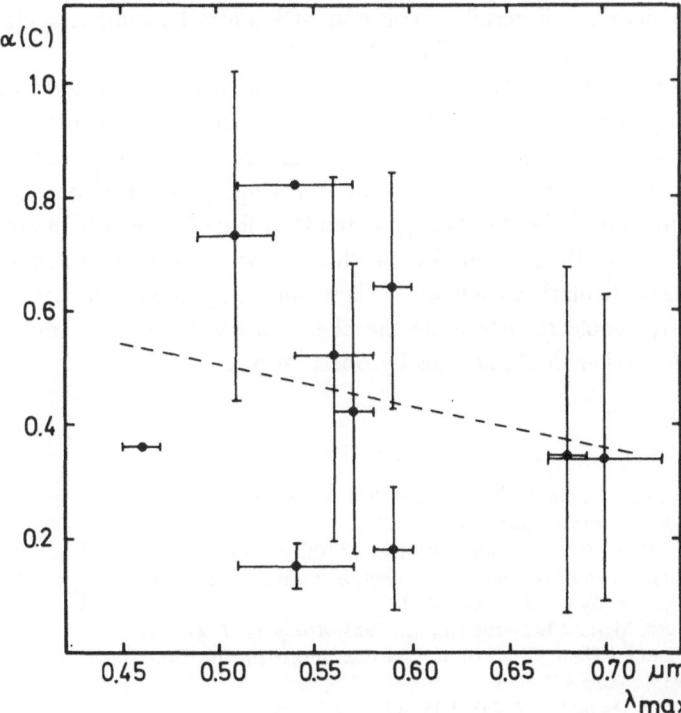

Fig. 4.   The same as Figure 1 but for carbon. The data for the lines-of-sight towards κ Cas and ϑ Ara are omitted, see text.

caused probably by a very high uncertainty of the determinations of the column densities · of carbon in these directions. Therefore, these two stars were omitted. In Table II the slopes of the correlation lines and the coefficients of the correlations are given. Obviously, besides the large scatter of the data all four elements obey the same correlation: with increasing λ_{max} the abundances in the interstellar gas phase decrease. This corresponds with the theoretical expectations. The correlation coefficients are small, but this is mainly caused by the large uncertainties of the column densities of the different atoms or ions.

TABLE II

Slope m of the regression lines in Figures 1 to 4
and the coefficient r of the correlations

Element	m	r
Ti	− 0.024	0.20
Fe	− 0.055	0.42
Mg	− 0.78	0.18
C	− 0.68	0.22

The idea that the interstellar gas depletion may correlate with λ_{max} is not new. Already Savage and Bohlin (1979) considered this problem for iron. They stated, that there is no systematic tendency discernible. This conclusion may be somewhat effected by their use of $\log \alpha(Fe)$ (in their notation $\log \delta$) instead of $\alpha(Fe)$.

Statements about the chemical composition of the mantles and the cores of the dust grains are not possible with the data available up to now. This is because the number of lines-of-sight for which $\alpha(X_i)$ as well as λ_{max} are determined are too small. But above all, with the data at hand it is not possible to get $\alpha_c(X_i)$ because there are no data for $\lambda_{max} < 0.46$ μm. The full range of λ_{max} extends at least between 0.33 μm and 0.89 μm (Wilking et al., 1980, 1982). Therefore, further observations of interstellar gas depletion should include particularly such lines-of-sight where λ_{max} has extreme values. With such data it seems possible to determine the chemical composition of the cores and the mantles of interstellar dust particles by observations.

References

Coyne, G. V., Gehrels, T., and Serkowski, K.: 1974, *Astron. J.* **79**, 581.
Duley, W. W.: 1982, *Astrophys. Space Sci.* **85**, 221.
Greenberg, J. M.: 1986, *Astrophys. Space Sci* **128**, 17 (this issue).
Harris, A. W., Gry, C., and Bromage, G. A.: 1984, *Astrophys. J.* **284**, 157.
Hobbs, L. M.: 1984, *Astrophys. J. Suppl.* **56**, 315.
Jenkins, E. B., Jura, M., and Loewenstein, M.: 1983, *Astrophys. J.* **270**, 88.
Kramers, H. A. and ter Haar, D.: 1946, *Bull. Astron. Inst. Neth.* **10**, 137.
Lindblad, B.: 1935, *Nature* **135**, 133.
Morton, D. C.: 1974, *Astrophys. J.* **193**, L35.

Murray, J. M., Dufton, P. L., Hibbert, A., and York, D. G.: 1984, *Astrophys. J.* **282**, 481.
Savage, B. D. and Bohlin, R. C.: 1979, *Astrophys. J.* **229**, 136.
Serkowski, K., Mathewson, D. S., and Ford, V. L.: 1975, *Astrophys. J.* **196**, 261.
Snow, T. P.: 1977, *Astrophys. J.* **216**, 724.
Stokes, G. M.: 1978, *Astrophys. J. Suppl.* **36**, 115.
Tarafdar, S., Prasad, S., and Huntress, W. T.: 1983, *Astrophys. J.* **267**, 156.
ter Haar, D.: 1943, *Bull. Astron. Inst. Neth.* **10**, 1.
van de Hulst, H. C.: 1949, *Rech. Astron. Obs. Utrecht* **11**, Part 2.
Wilking, B. A., Lebowsky, M. J., and Rieke, G. H.: 1982, *Astron. J.* **87**, 695.
Wilking, B. A., Lebowsky, M. J., Martin, P. G. Rieke, G. H., and Kemp, J. C.: 1980, *Astrophys. J.* **225**, 905.
Zeippen, C. J., Seaton, M. J., and Morton, D. C.: 1977, *Monthly Notices Roy. Astron. Soc.* **181**, 527.

Discussion

P. G. Mezger: In the moment you just establish the correlation between the grain radius and the depletion, you have to say something about the mantle structure.

H. Zimmermann: Yes, if you take only the observations here and if you have such a tendency, then you can read what elements are more incorporated in the mantles.

H. J. Habing: One of the elements that has been notoriously depleted for a long time is calcium. You did not show any diagram.

H. Zimmermann: You can say nothing about this because you have mostly only the column densities of Ca II. Determinations of the total calcium depletion exist only for few stars.

E. Krügel: Did I understand correctly that you see also a tendency for carbon and λ_{max}?

H. Zimmermann: Yes.

E. Krügel: Would you interpret this that carbon has something to do with polarization?

H. Zimmermann: No. What I say is that we have a connection between polarization, the wavelenth of maximum polarization, and the mean grain size, dependent on mantle growth.

E. Krügel: It does not mean that the carbons are polarizing?

J. M. Greenberg: I think there is a misunderstanding. The correlation between depletion and λ_{max}, if there is a correlation, shows us if the carbons or whatever we are going to talk is depleted to make the particles bigger. Therefore, the carbon is producing the polarization because it is making the particles bigger. That is the assumption, I think.

H. Zimmermann: Would you agree?

J. M. Greenberg: I would say if the particles are larger then the λ_{max} should shift and if the particles are growing by accretion then the question is what atoms are there depleted. I really agree in principle. But it is my opinion that the carbon depletion is exceedingly difficult to measure. It is an observational problem to correlate the carbon.

H. Zimmermann: This is an observational problem, indeed. In future we must look very carefully to decide this. But I think we should make such observations. We should select the stars firstly from the polarization and then the abundance. Here, in this list, you have first the abundance and then the polarization and then you have a very bad correlation.

ON THE GEOMETRICAL CROSS-SECTION OF AN N-MER*

S. KLOSE

Universitäts-Sternwarte Jena, G.D.R.

(Received 23 June, 1986)

Abstract. We have modeled the growth of an N-mer by means of Monte-Carlo calculations. The geometrical cross-section of the N-mer is measured and compared with the one obtained if one assumes that the N-mer is of spherical form. A difference up to a factor of 2 was found for $N \leq 30$.

We consider the process that a Monomer is growing to an N-mer. We assume this process to occur in the following way. At the beginning there is only one Monomer. This Monomer then collides with other Monomers of equal size, mass, etc. The collision probability is a function of the geometrical cross-section of the growing N-mer. To calculate this cross-section the procedure generally used is the following. Let the volume of one Monomer be V_0, then the volume of the N-mer is $V = NV_0$ (supposed that there are no voids within the N-mer). Now one introduces an effective radius by setting $V = 4\pi r^3/3$. Thus, $Nr_0^3 = r^3$, where r_0 is the effective radius of a Monomer. The geometrical cross-section of the N-mer is $A(N) = N^{2/3}\pi r_0^2$. For small values of N r_0 cannot be neglected against r. Then $A(N) = \pi r_0^2(N^{2/3} + 4N^{1/3} + 4)$, because $r = N^{1/3}r_0 + 2r_0$, as it is shown in Figure 1.

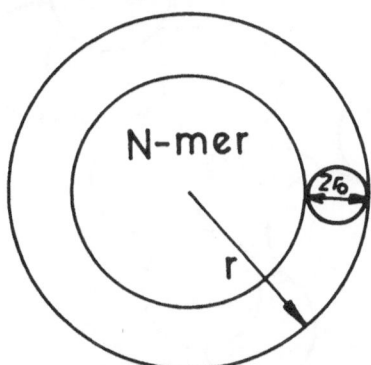

Fig. 1. For small N r_0 cannot be neglected against r.

We will now apply this method to the case of a non-symmetric N-mer and compare the geometrical cross-section obtained with the one really measured. We, therefore, modeled a grain growth: i.e., the evolution of such an N-mer in a simple way. The following assumptions were made:

* Paper presented at a Workshop on 'The Role of Dust in Dense Regions of Interstellar Matter', held at Georgenthal, G.D.R., in March 1986.

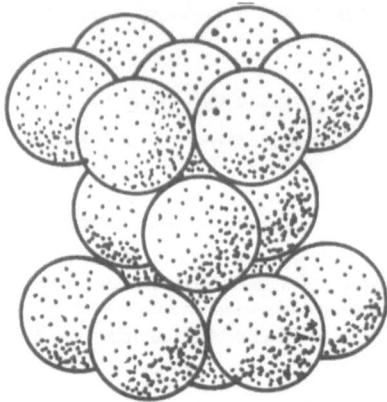

Fig. 2.   The places which can be taken by a Monomer represent a hexagonale lattice.

(1) All Monomers are spheres and they will not be destroyed during the sticking process to the N-mer.

(2) Every Monomer is surrounded by 12 potential sticking points. A potential sticking point *can* be taken, but must not be taken. These potential sticking points represent a crystalline structure (hexagonal lattice).

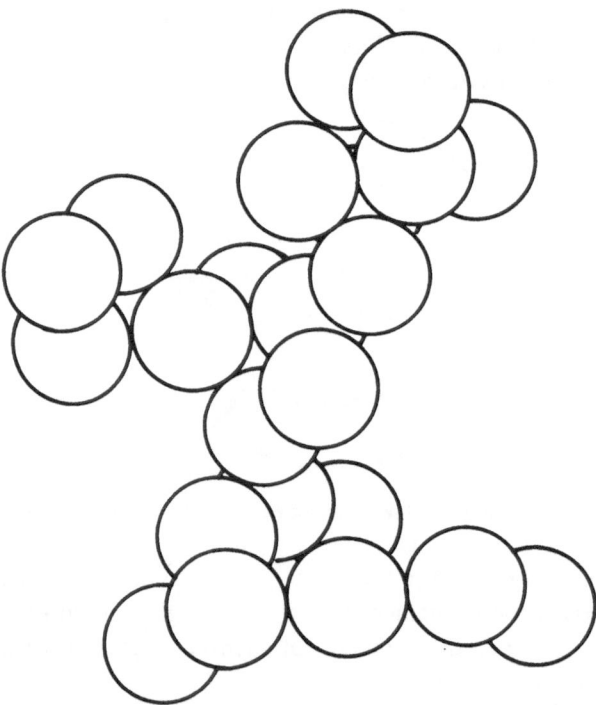

Fig. 3.   An N-mer with $N = 30$ seen from one direction of incidence of a Monomer. The N-mer is consisting of 30 equal spherical Monomers of radius r_0. Consider that the geometrical cross-section of the N-mer is larger than its projected area because the inciding Monomers have a radius r_0, too.

(3) All potential sticking points at the surface of the N-mer can be taken with the same probability. There is no process which destroys the N-mer or changes the places of the sticked Monomers.

(4) The directions of incidence of the Monomers are isotropically distributed.

The last assumption was realized by random values. The N-mer is growing by building up a hexagonale configuration. A typical configuration is shown in Figure 3. We measured the really projected geometrical cross-section of the N-mer for six different directions of incidence of a Monomer and determined its mean geometrical cross-section: $A(N) = \frac{1}{6} \Sigma_{i=1}^{6} A_i(N)$. These six directions are given by the upper part of a dodecahedron (Figure 4).

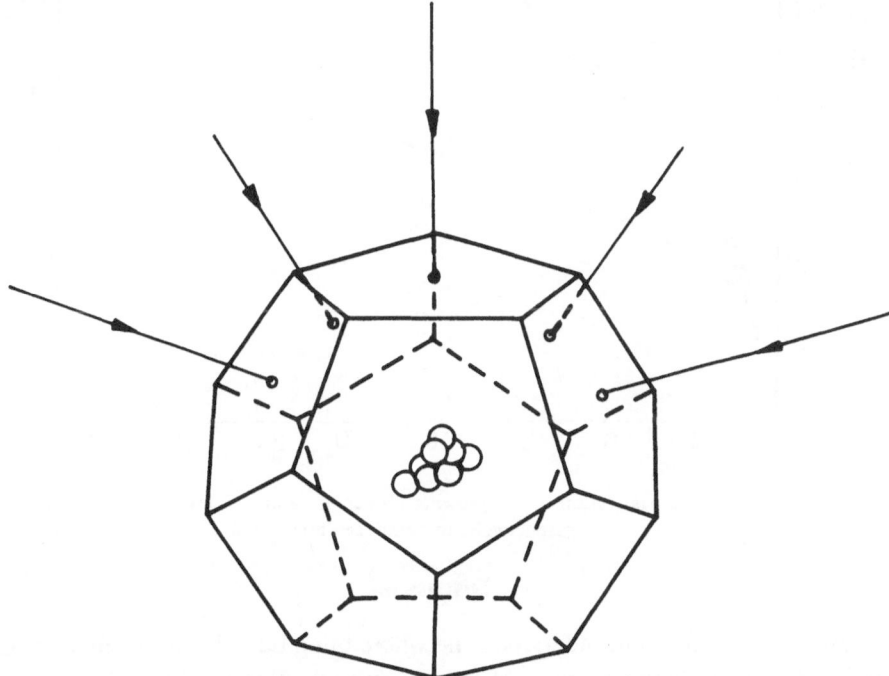

Fig. 4. We calculated the mean geometrical cross-section of the growing N-mer basing on its different cross-sections for 6 directions of incidence of a Monomer.

The results obtained are presented in Figure 5. Curve 1 represents the geometrical cross-section of the N-mer predicted from the method generally used. Curve 2 is the result obtained from our measurements. The really measured geometrical cross-section of the N-mer is larger than the approximated one up to a factor of 2 ($N \leq 30$). It is clear in a more detailed study more physical foundations must be taken into account.

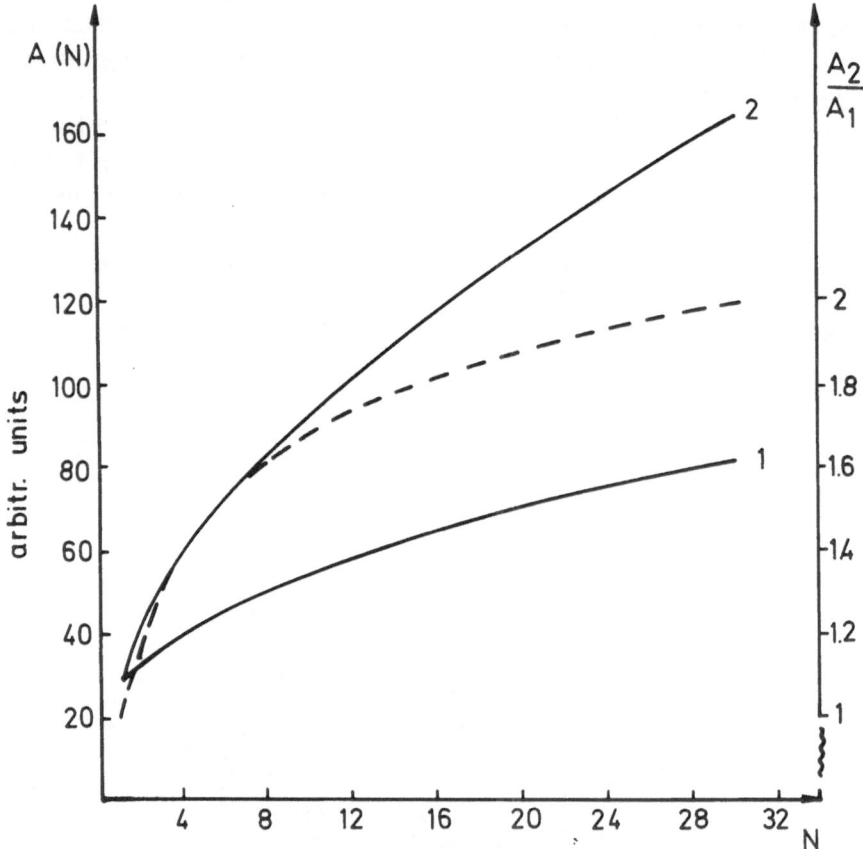

Fig. 5. The geometrical cross-section of the growing N-mer. A_1 and A_2 correspond to curve 1 and 2, respectively. In detail, see text.

Discussion

H. J. Habing: Maybe I misunderstood the whole thing but I have the impression you were building crystalline silicates and not amorphous materials?

S. Klose: This is correct. But the aim of this paper was not to model a crystal growth. I only wanted to show how the geometrical cross-section of a *non*-symmetric (see Figure 3) N-mer depends on the value of N. The N-mer's I constructed are non-symmetric because (1) N is small (≤ 30) and (2) I have not applied crystal physics (assumption No. 3).

B. Stecklum: I have the impression that the two curves are deviating increasingly for larger N but of course they have to come together for very large N. How large has N to be?

S. Klose: My computating time was to short to perform calculations for larger N. On the other hand for larger N more physical processes must be taken into account.

EXPERIMENTAL INVESTIGATIONS OF ASTRONOMICALLY IMPORTANT INTERSTELLAR SILICATES*

J. DORSCHNER and TH. HENNING

Universitäts-Sternwarte Jena, G.D.R.

(Received 27 June, 1986)

Abstract. In this paper methods and results of laboratory experiments for the investigation of the silicate component of interstellar dust are reviewed. In Section 2 basic properties expected for astronomically important interstellar silicates (AIIS) are discussed. Chemical constraints coming from the abundance of elements, from the depletion in the interstellar gas and from theoretical calculations of the condensation processes point to magnesium silicates. Some basic structural properties of interstellar silicates, the expected high degree of lattice disorder and spectral features expected for interstellar silicate grains are discussed. In Section 3 a review on laboratory investigations of AIIS is given. Physical and chemical methods for producing amorphous silicates are summarized. Important measurements of optical data for AIIS are listed. Spectral characteristics of amorphous silicates produced in order to simulate the interstellar dust silicates are discussed. From the comparison of the observed MIR silicate bands with those of the experimentally produced silicates it is concluded that at least two types of dust silicates exist in interstellar space: molecular-cloud silicate (suggested to be of pyroxene-type) and late-type star silicate (suggested to be of olivine-type). The mass absorption coefficient at the 10 μm peak of both types of silicate grains amounts to 3000 cm² g⁻¹ and the ratio of 20 to 10 μm peaks amounts to about 0.5. Finally, open questions in connection with laboratory experiments are mentioned and recommendations for future experiments are given.

1. Introduction

The advent of infrared observational techniques in astronomy in the nineteen sixties rendered it possible to prove the occurrence of silicates even beyond the solar system. For theoretical reasons, interstellar grains consisting of silicates similar to those of the most common meteorites have been postulated nearly 20 years ago by Dorschner (1968). As a matter of fact, the first observational evidence for the presence of silicates in extra-solar particulates was found only a short time later. Infrared excess radiation observed at 10 μm wavelength in the spectra of oxygen-rich late-type stars (Gillett *et al.*, 1968) and in the spectrum of the Ney–Allen nebula in the Orion Trapezium region (Stein and Gillett, 1969) turned out to be an emission band of circumstellar and interstellar silicate particles, respectively. Independently, Woolf and Ney (1969) and Dorschner (1971, submitted in 1969 on the *2nd IAU Colloquium on Interstellar Dust*) identified these observed excesses with stretching vibrations of silicon-oxygen tetrahedra and pointed to the probable existence of bending modes at about 20 μm as an *experimentum crucis* of the silicate hypothesis. First evidence for the 20 μm band due to silicate particles was presented by Hackwell *et al.* (1970), who discovered it as absorption band together with the 10 μm absorption band in the spectrum of the galactic centre, and by Low and

---

* Paper presented at a Workshop on 'The Role of Dust in Dense Regions of Interstellar Matter', held at Georgenthal, G.D.R., in March 1986.

Swamy (1970), who found indications for 20 μm emission in the spectrum of the circumstellar shell around α Orionis, where the 10 μm emission had been already detected in 1969. Nearly at the same time, the 10 μm emission feature was found in the spectrum of the comet Bennett (Maas *et al.*, 1970).

Because of the small absorption efficiency of silicate particles even in the peak of the 10 μm band silicate absorption features are only found in the spectra of O-rich stars that are surrounded by extremely thick envelopes, in hot spots (e.g., BN objects) in molecular clouds and in some sources behind optically thick interstellar dust clouds (e.g., IR sources in the galactic centre). On the other hand, lots of silicate emission sources, mostly optically thin to moderately thick shells around O-rich late-type stars and pre-Main-Sequence objects, e.g., T Tauri stars, have been found. Among the objects with the 10 μm emission feature the Orion Trapezium nebula plays a distinguished role because its 10 μm profile represents the unique case, where, up to now, molecular cloud dust can be observed under the favourable conditions of small optical thickness.

The low-resolution spectra gained by IRAS (Beichmann *et al.*, 1985) have strongly enlarged the number of sources with silicate features. More than 1800 stars with O-rich shells showing 10 μm emission bands and more than 300 sources with the 10 μm feature in absorption have been detected. Many of these spectra also show the corresponding 20 μm feature.

These middle infrared (MIR) silicate bands have extremely broad profiles which do not show any indications that they are composed of several blended components. The only exception where a sort of internal band structure becomes visible are sources with silicate self-absorption, e.g., IRC + 40448 and OMC2–IRS3. As radiative transfer calculations (Henning, 1983) suggest silicate emission features show weaker profile variations with changing optical depth of the 10 μm band than absorption features do. This result may explain why the observed emission features are very similar from one source to the other one, whereas the absorption features show some more variations of the profiles from source to source.

In contrast to the structureless appearance of the observed silicate bands terrestrial, lunar and meteoritic silicates show rich structure in their bands caused by blending of several components. In principle, this striking difference has prevented any mineralogical identification of the interstellar silicate. Revealing the nature of the latter and procuring realistic parameters of the interstellar dust grains has become the task of interdisciplinary experimental work of astrophysicists, mineralogists, and chemists on astronomically important interstellar silicates (abbreviated by AIIS throughout this paper). AIIS form a subdivision of astronomically important silicates (AIS). In addition to AIIS, AIS include also silicates suggested to be present on (and in) celestial bodies (e.g., small bodies of the solar system).

2. Basic Properties Expected for AIIS

2.1. CHEMICAL CONSTRAINTS

In order to manage the identification problem of interstellar and circumstellar silicate particulates it does not suffice to find silicates that allow a satisfactory representation of such important observational data as the extinction and polarization curves in terms of a multi-component dust model. Unfortunately, these observational data have proved not to be very diagnostic with respect to silicates. AIIS which can be considered true candidates for explaining the nature of the silicate dust component and which, therefore, are of importance for the simulation of observed dust properties by laboratory experiments must fulfil some more conditions.

A fundamental chemical constraint is set by the cosmic abundance of elements (see, e.g., Palme *et al.*, 1981). Generally, the total amount of silicates which can be formed by any processes in an environment containing gas of a composition according to the mean cosmic abundance of elements is strictly limited by silicon abundance. Really, this upper limit is, however, not reached because Si partially remains in the gas phase as element as well as in the form of gaseous silicon compounds, e.g., SiO. By the way, the observed SiO is an important whitness testifying for the possibility of chemical reactions leading to silicates. In gas of mean cosmic composition the abundances of Mg and Fe are nearly as large as that of Si, so that, on principle, the chemical premises for the occurrence of Mg/Fe silicates to the extent that is determined by Si are given. Next metals following Si, Mg, and Fe in the table of mean cosmic abundance of elements, i.e., Al, Ca, Ni, and Na, should play only a minor role for the chemical composition of the silicates because they are mere 'impurities'. Nevertheless, these elements may affect the optical properties of the grains to an extent that causes observable effects.

It is a very interesting fact that just these elements which are suggested to form silicate particles are strongly depleted in interstellar gas. Correlations found between the observed depletions and condensation temperature – first pointed out by Field (1974) – and between depletion and cloud density – first pointed out by Savage and Bohlin (1979) – can be interpreted in favour of the above suggestions. Whittet (1984) proposed a dust model based on these depletions and concluded that the depleted Si is locked up in silicates of the composition X_2SiO_4 and $XSiO_3$ (X = Mg or Fe) and any remaining metals are oxidized. If all available silicon is consumed this way, then the total silicate/oxide grain density in the interstellar matter should reach the order of magnitude 10^{-26} g cm^{-3}, which is consistent with other observational data.

Circumstellar shells around M-type stars and related objects like OH/IR stars, which are oxygen-rich (content of O larger than that of C), are the most favourable environment for the formation of silicate particles. The dust formation in the outflow of these stars (stellar wind) is subject to theoretical calculations. Most of this theoretical modelling of condensation processes in the past was based on strongly simplifying assumptions: equilibrium condensation, classical nucleation theory, neglecting chemical reaction network and hydrodynamics. Even more sophisticated modern calculations are sensitively dependent on uncertain material data so that the results are not yet very definitive. They, however, offer important constraints for the composition of the resulting silicate particles. New approaches to the primary condensation processes in the stellar winds of late M-type stars by Donn and Nuth (1985) and Gail and Sedlmayr (1986) point to the important role of SiO condensation initiating silicate dust formation. The latter authors expect that a more or less pure magnesium silicate of the composition Mg_2SiO_4 should form. This is in agreement with new calculations by Fadeyev and Henning (1986) and with conclusions drawn from the comparison of 10 μm bands of experimentally produced amorphous silicates of this composition (cf. Section 3.3) and observationally-based IR efficiencies of circumstellar silicate particles (Henning *et al.*, 1983). In the surroundings of very young objects particles of the composition $MgSiO_3$ rather than Mg_2SiO_4 should, however, be present as the 10 μm absorption bands of these sources indicate (Gürtler and Henning, 1986).

Chemical reactions of the primarily formed silicate grains with gaseous compounds, e.g., water vapour (resulting in layer-lattice silicates) or carbon monoxide (resulting in carbonates), can considerably modify the composition of the finally resulting dust particles. We, therefore, also expect the occurrence of hydrous silicates (layer-lattice silicates, cf. Section 2.3) and point in this connection to one possible chemical reaction discussed by Larimer and Anders (1967) in the framework of meteorite formation

$$5\ MgSiO_3 + H_2O \rightleftarrows Mg_3Si_4O_{10}(OH)_2 + Mg_2SiO_4 \,,$$

which is taking place at temperatures around 350 K. This reaction shows that there are even transmutations of $MgSiO_3$ to Mg_2SiO_4 and *vice versa* possible if water vapour and hydrous silicates (in this case, talc) are present. We, furthermore, concede that modifications of the composition of the silicate particles could be possible even in interstellar space and in molecular clouds. Thus, in spite of a unique origin of the silicate particles, several well-distinct sorts of silicate dust may appear (cf. Section 3.3).

The most primitive solar system solids, the carbonaceous chondrites, which have been studied thoroughly in the laboratory, contain matrices of hydrous silicates representing a low-temperature condensate as well as chondrules and aggregates of olivine $((Mg, Fe)_2SiO_4)$ and pyroxene $((Mg, Fe)SiO_3)$ that are high-temperature condensates. Even if this material seems to be much more primitive than that of planetary rocks, it is, however, hardly primitive enough to be compared with interstellar dust. It has been suggested, but by no means conclusively proved, that carbonaceous chondrites contain 'fossil' interstellar dust. As a matter of fact, with the exception of the chondrules that consist of glassy silicates the matrices as well as the other inclusions contain crystalline minerals, whereas interstellar silicates should be amorphous. Most of the interplanetary dust particles (IDP) that were collected by airplanes (Sandford and Walker, 1985) also contain crystalline minerals. In their present state IDP and grains in carbonaceous meteorites cannot serve as models for interstellar dust grains. Of course, it cannot be excluded that the IDP and the grains in the C-meteorites, if they were indeed interstellar silicates, recrystallized to their present state. What concerns IDP, this recrystallization could have taken place as late as when they were heated in the Earth's atmosphere. It is, however, interesting that new comets – e.g., Bennett, Kohoutek, and West – show very broad and structureless silicate features at 10 μm. This cometary silicate may possibly be compared with interstellar silicate dust.

2.2. DEGREE OF LATTICE DISORDER

The appearance of the 10 and 20 μm silicate bands unambiguously demonstrates that the lattices of silicate particles occurring in interstellar space as well as in stellar winds of M-type stars and also in the vicinity of very young and pre-Main-Sequence stars are highly disordered. In the literature the term 'amorphous' silicates is generally used in this connection. We should like to point out that this term is not very sharp. Here, we define the amorphous state, which should be a constitutive property of AIIS, only by the statement that long-range order of the lattice is lost (not the short-range order, so that there is a certain resemblance to fluids), but that, contrary to normal fluids, no short-time fluidity is reached (Henning and Svatoš, 1985). In any case, if we could subject the interstellar silicate to X-ray diffraction analysis, then the resulting diagrams should show no sharp peaks.

As the experimental simulation shows the vitreous state of silicate glasses can be considered a first approximation to what is expected to be the structural state of this dust silicate (cf. Section 3.3). Glasses are a special class of X-amorphous solids which underwent a glass transition (at the transition temperature some parameters – e.g., heat capacity and expansion coefficient – jump from the value typical of a solid to those typical of a liquid). The glassy state is a non-equilibrium state, which can persist for cosmological times as the lunar glasses show. The long-range disorder manifests itself by the loss of the regular arrangements of the basic building blocks of silicate lattices – i.e., the silicon-oxygen tetrahedra (cf. Section 2.3). The tetrahedra themselves are only slightly distorted. Bond lengths and angles are statistically distributed around their normal values, what causes strongly broadened bands (Wong and Angel, 1976).

Additionally to structural disorder typical of laboratory-produced glasses primitive cosmic silicates could also reveal a certain degree of chemical disorder, for instance, in the form of nonstochiometric ratios of elements contained. At least, this should be true for the primary condensates (cf., e.g., Donn and Nuth, 1985). Secondary modifications may change the things considerably.

The amorphous state, which must be reached if laboratory experiments with AIIS shall give a realistic simulation of interstellar silicates, represents a wide transitional field between the case of absolute chemical disorder (suggested to consist of unarranged elements and/or simple compounds, e.g., oxides, that could be ingredients for the synthesis of silicates) and the case of strict order of crystal lattices. This field is accessible only by laboratory experiments (Figure 2).

2.3. ARCHITECTURE OF AIIS LATTICES

The basic building stone common to all types of silicate lattices is the silicon-oxygen tetrahedron. In it the small Si^{4+}-ion (radius 0.42 Å) is fourfold coordinated by O^{2-}-ions (radius 1.40 Å) forming the corners of the tetrahedron. In the ideal, undisturbed case

nesosilicates

inosilicates

phyllosilicates

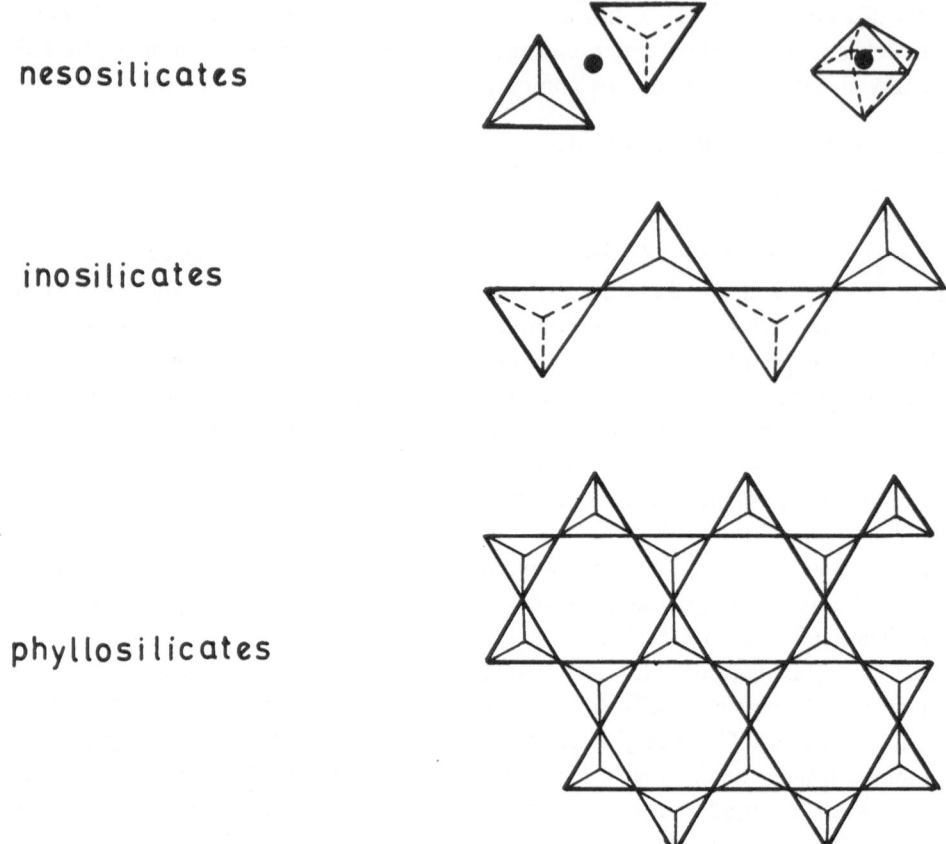

Fig. 1. Schematic representation of the connection of SiO$_4$ tetrahedra of three basic silicate types. With nesosilicates the octahedral coordination of metal ions (filled circle) is illustrated.

the edge from one corner to a neighbouring one is 2.62 Å long, the Si–O bond lengths and the O–Si–O bond angles amount to 1.60 Å and 109°28′16″, respectively. The coordination number 4 leading to tetrahedral symmetry in the arrangements of the oxygen ions is determined only by the ratio of the Si^{4+}- to the O^{2-}-radius, which amounts to 0.3.

If we consider metal ions in silicate lattices, which are important in connection with AIIS, for example, Mg^{2+} (radius 0.66 Å) and Fe^{2+} (radius 0.74 Å), then the radii ratios between these ions and O^{2-} get values around 0.5. In this case the coordination number becomes 6, that means, the oxygen ions form an octahedral cage around the metal ions. In our context it suffices to consider AIIS lattices composed of silicon-oxygen tetrahedra and metal-oxygen octahedra where the former are much tighter bonded than the latter.

Depending of the formation conditions SiO$_4$-tetrahedra can polymerize: one, two, three, or all of the four oxygen ions of a tetrahedron may be in common with neighbouring

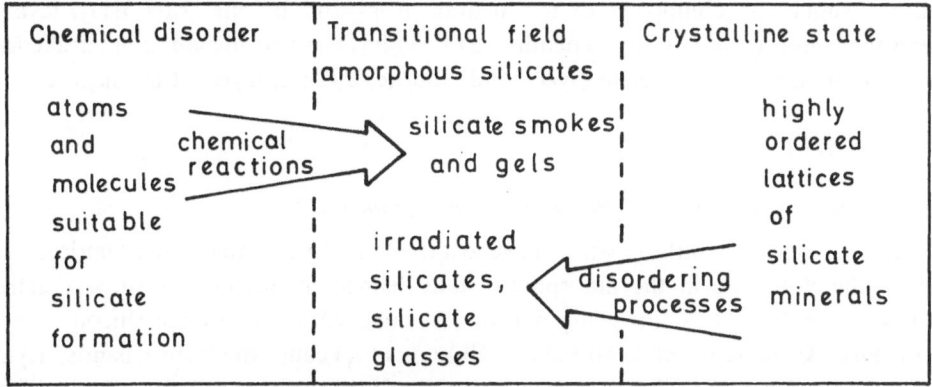

Fig. 2. Between highly-ordered lattices of silicate minerals and mere mixtures of silicate-constituting molecules the transitional field of amorphous silicates exists, which is accessible from both sides by laboratory experiments.

tetrahedra. In this way, various types of silicate lattices well-known from their occurrence in terrestrial rocks come about. What concerns AIIS it suffices to consider the following three types of silicate lattices:

(1) nesosilicates ('island' silicates),
(2) inosilicates ('chain' silicates),
(3) phyllosilicates ('leaf' silicates).

In Figure 1 the structure of these basic types is schematically illustrated. Nesosilicates consist of distinct SiO_4-tetrahedra (without polymerization) connected by metal ions in such a way that three of the oxygens of one tetrahedron together with three of the neighbouring one form an octahedron around the metal ion so that the oxygen octahedra come to pass, which were explained above. The most important representative of the nesosilicates is the mineral olivine $(Mg, Fe)_2SiO_4$ being a solid solution of the silicates Mg_2SiO_4 (forsterite) and Fe_2SiO_4 (fayalite).

Inosilicates consist of chains of polymerized tetrahedra bonded over two corners by bridging oxygens. Important representatives are the minerals of the pyroxene group, e.g., bronzite $(Mg, Fe)SiO_3$, a solid solution of the corresponding Mg and Fe end members enstatite and ferrosilite, respectively. Mg and Fe are again placed on octahedrally coordinated lattice sites. Olivine and bronzite are main minerals of the ordinary chondrites.

Phyllosilicates are also called layer-lattice silicates. They contain sheets of silicon-oxygen tetrahedra consisting of two-dimensional arrangements of polymerized SiO_4-groups that have three oxygens each in common with the neighbours. The arrangement is such that the bridging oxygens are situated in the same plane and the 'free' corners of the tetrahedra point to the same side of this plane where octahedrally coordinated Al and Mg ions are placed. The octahedra are formed not only by oxygen ions, but largely by hydroxyl groups, so that they establish sheets of hydroxides superimposed on tetrahedral sheets. Many phyllosilicates contain water of hydration. Because the cosmic abundance of Mg is much larger than that of Al, cosmically important layer-lattice silicates should be composed of tetrahedral layers and brucite, $Mg(OH)_2$, layers. Important representatives are serpentine, a (1 : 1)-layer-lattice silicate (one tetrahedral layer alternating by one brucite layer), and chlorite, a (2 : 2)-layer-lattice silicate.

2.4. SPECTRAL FEATURES

2.4.1. *IR-Active Vibrations of Silicon-Oxygen Tetrahedra*

All silicates inclusive of those which have lost the long-range order of their lattices and which, therefore, are in the amorphous state defined in Section 2.2 show spectral features near 10 and 20 µm being due to IR-active vibrations of the silicon-oxygen tetrahedra. Asymmetric Si–O stretching vibrations (v_3) cause the former bands, asymmetric O–Si–O bending vibrations (v_4) perpendicular to the bonds the latter ones. They are these features that have been observed by IR astronomy for almost 20 years as it was mentioned in the introduction of this review.

2.4.2. *Molecular Vibrations of OH Groups and Water of Hydration*

Hydroxyl groups and/or H_2O molecules in the lattices of phyllosilicates give rise to several interesting IR active vibrations. If such silicates play some role in interstellar dust grains, then the observations of such bands can be expected. We, therefore, include them in this overview. Table I contains the IR molecular vibrations of the H_2O molecule. As a constituent of hydrous silicates OH and H_2O cause the appearance of broad bands near 3 and 6 µm. Knacke and Krätschmer (1980) investigated such bands of grains of carbonaceous chondrites in the laboratory. Laboratory measurements of the reflection spectra of carbonaceous chondrites revealed the presence of the overtone bands at 1.4 and 1.9 µm (cf. Gaffey and McCord, 1979). A band at 3 µm suggested to be due to water in silicates has been found in the reflection spectra of several C- and U-type asteroids (Lebofsky, 1980). Although phyllosilicates have often been proposed as an important constituent of the interstellar dust (Zaikowski *et al.*, 1975; Dorschner *et al.*, 1978; Friedemann *et al.*, 1979), up to now no unambiguous observational evidence for their occurrence has been found (Knacke *et al.*, 1985).

TABLE I

NIR vibrations of free H_2O molecules

Type of vibration	Spectral designation	Wavelength (µm)
Symmetric O–H stretches	v_1	2.73
Asymmetric O–H stretches	v_3	2.66
H–O–H bending vibrations	v_2	6.27
Overtone vibrations	$2v_2$	3.13
	$2v_3$	1.4
	$v_2 + v_3$	1.9

2.4.3. Crystal Field Transitions

It is a well-known fact from terrestrial mineralogy that 3d ions in sixfold coordinated sites of silicate lattices (octahedral symmetry of the crystal field) cause the appearance of absorption lines in the visual and the NIR. From the cosmic abundance of elements iron ions could possibly play a major role in cosmic silicates. As a matter of fact, a strong band near 0.9 μm diagnostic to Fe^{2+} in pyroxene lattices and a still stronger band near 1.0 μm due to Fe^{2+} in olivine have been found in the reflection spectra of many asteroids and also of carbonaceous chondrites (Gaffey and McCord, 1979). These bands have, however, not been found in the interstellar extinction curve. In the disordered lattices of interstellar silicate grains these bands, at best, should have the appearance of very broad and shallow structures (comparable to VBS) which are difficult to detect. The suggestions by Huffman (1970), Manning (1970), and Dorschner (1970) that crystal field transitions of iron in silicate particles could account for prominent diffuse interstellar bands have not been confirmed.

2.4.4. Electronic Transitions of Ions in Silicate Lattices

On principle, electronic transitions in the atoms of interstellar silicate dust particles, which cause strong absorption features in the UV, are to be expected. Such bands could be very important in connection with the interpretation of the 220 nm hump of the interstellar extinction curve. Greenberg and Chłewicki (1983) thoroughly discussed which role silicate particles could play for the explanation of the hump and the FUV rise of the extinction curve. They concluded that, although no known graphite or silicate particles can reproduce the observed hump and FUV extinction satisfactorily, particles of related chemical composition cannot be excluded being suitable candidates.

It is well known that the early optimism by Huffman and Stapp (1971), who claimed to reproduce the 220 nm hump on the base of their measurements of the optical constants of enstatite, was not supported by subsequent measurements of Egan and Hilgeman (1975) and Lamy (1978). Gerastchenko (1975), who measured the extinction of small grains manufactured from the Saratov meteorite (a chondrite composed mainly of olivine and bronzite), could, however, reproduce the hump, at least in a qualitative manner. Laser-vaporization of olivine and enstatite (Stephens, 1980) did not give any indications for the presence of the hump in the spectra of the condensed particles.

The electronic transition in oxygen ions on low-coordinated sites, which was proposed by MacLean et al. (1982) on the base of measurements on magnesium oxide and irradiated silica deserves further attention, even if the oxide dust model by Duley et al. (1979) is not very likely and the discussion about the Mg depletion remains controversial. An important argument in favour of Duley's interpretation of the 220 nm hump, the feature observed at 6.5 μm^{-1} in the interstellar extinction curve, has collapsed because it turned out to be an instrumental effect in the TD-1 spectra. From the lacking MgO feature at 6.5 μm^{-1} an upper limit for the amount of MgO dust can be estimated. In their paper on this workshop, Friedemann and Gürtler conclude that more than 97% of MgO should be locked in other compounds (e.g., in silicates). Pure MgO particles should, therefore, play no important role.

2.4.5. *FIR Spectrum*

From the observations, there is no conclusive evidence for the occurrence of silicate bands beyond 30 µm. Such FIR bands may, however, be expected. We point in this connection to bands being due to vibrations of the metal ions in their octahedral oxygen cages (cf. Section 2.3). Measurements of serpentine and chlorite grains (Koike and Hasegawa, 1982) show that there are bands between 30 and 100 µm. They did, however, not interpret the cause of these bands.

Generally, the FIR dust absorption is a matter of controversy among observers as well as theoreticians. From theoretical standpoint, Seki and Yamamoto (1980) argue that the absorption efficiency of amorphous silicate grains with radii smaller than 0.01 µm should show a λ^{-1} dependence for $\lambda > 100$ µm, whereas those of crystalline and of larger amorphous grains should be proportional to λ^{-2}. We, thus, conclude that the efficiency factors depend not only on the grain chemistry, but also on internal structure and particle size. From the observations, at present no final decision on the power of λ is possible. In FIR only few laboratory experiments have been carried out. Koike and Hasegawa (1982) found extinction coefficients for particles of the phyllosilicate montmorillonite proportional to $\lambda^{-0.8}$.

3. Results of Laboratory Experiments with AIIS

3.1. SIMULATION OF THE AMORPHOUS STATE

The first problem to be met if one tries to simulate interstellar silicate dust in the laboratory in order to measure optical data is the production of such a degree of lattice disorder that the resulting material becomes amorphous enough to be compared with cosmic dust silicates. Producing the amorphous state both physical and chemical ways can be gone (cf. Figure 2). For disordering silicate lattices the following physical processes have been applied by the experimentalists:

(i) irradiation of a polished silicate surface by fast noble gas ions (Krätschmer and Huffman, 1979);

(ii) laser-vaporization of silicates and subsequent condensation to vitreous microdoplets (Stephens and Russell, 1979; Stephens, 1980);

(iii) arc-melting of silicates and quenching them to glassy material (Rose, 1979; Dorschner *et al.*, 1986).

Before the introduction of these physical methods laboratory astrophysicists mostly used chemical processes in order to make amorphous silicates. Including chemical ways we complete the above list:

(iv) gel reactions resulting in what was called 'protosilicates' (Day, 1974; Duley and McCullough, 1977);

(v) cathode sputtering of targets consisting of Mg–Si or Fe–Si alloys in oxidizing atmosphere (Day, 1979, 1981);

(vi) simultaneously evaporating Mg and SiO in a reducing atmosphere of hydrogen and noble gases, which leads to silicate smokes (Day and Donn, 1978; Nuth and Donn, 1982).

3.2. MEASUREMENT OF THE OPTICAL DATA

The fundamental task of laboratory experiments carried out in order to reproduce astronomical observations of interstellar dust is the derivation of refractive indices of AIIS or the direct measurements of extinction and scattering efficiencies of small particles of the needed size, which were produced already in the chemical process of AIIS formation, for example, smoke particles, or by trituration of bulk AIIS.

Optical constants (complex dielectric function $\varepsilon = \varepsilon_1 + i\varepsilon_2$, complex refractive indices $m = n + ik$) over a broad wavelength range – in the ideal case from EUV to FIR – have been measured only on few silicates which in some cases are not AIIS in the strict sense of our definition, but rather are AIS. Table II lists those measurements that have been

TABLE II

AIS with measured optical constants

Authors	AIS	Wavelength range (µm)
Huffman and Stapp (1971)	crystalline enstatite	0.09–0.5
Huffman and Stapp (1973) Steyer (1974)	crystalline olivine	0.04–300
Egan and Hilgeman (1975)	cryst. augite, enstatite, bytownite, diopside, Bruderheim meteorite	0.18–0.4
Lamy (1978)	rocks and glasses measured by Pollack *et al.* (1973) in visual and NIR	0.05–0.45
Stephens (1980)	amorphous olivine and enstatite	0.12–1
Perry *et al.* (1972)	Apollo 11, 12, and 14 lunar rocks	2.5–500
Pollack *et al.* (1973)	basalt, andesite, basaltic glass, obsidian	0.4–6
Steyer *et al.* (1974)	amorphous quartz	7–25
Penman (1976)	dunite, serpentinite, chloritite, Vigarano and Murchison meteorites	6–33
Krätschmer and Huffman (1979)	amorphous olivine	8–30
Day (1979, 1981)	amorphous enstatite, forsterite, ferrosilite and fayalite	7–33
Mooney and Knacke (1985)	serpentine, chlorite	2.5–50
Dorschner *et al.* (1986b)	amorphous bronzite	7–40

used in the literature on interstellar dust. Most of these data concern the important range of the middle infrared (MIR) from 7 to 40 µm.

Optical data of MIR have mostly been determined on thin sections of compact material or thin films on infrared-transparent carriers. According to an often used standard measuring procedure n and k can be derived from reflectance and transmittance of a slab of sufficiently transparent material at near-normal radiation incidence. Optical data of more opaque material are suitably derived from reflectance at the (polished) surface at two oblique angles of incidence or from reflectance measurements at near-normal incidence, which have to be combined with the phase shift of the reflected light obtained by Kramers–Kronig analysis (cf. Bohren and Huffman, 1984). Methods coming out with reflectance measurements alone must, on principle, be used if optical data of irradiated silicates are derived. Because the penetration depth of the ions is only as large as about 1 μm (cf. Krätschmer and Huffman, 1979), the lattice is sufficiently disordered in an extremely thin layer at the surface only.

Basing on the measured optical data, cross-sections and scattering functions for small particles can be calculated in terms of the Mie theory for spherical and cylindrical particles of the needed size. Cross-sections for AIIS particles can, however, be determined experimentally. Such direct measurements are important because they concern particles much less regularly shaped than the ideal particles of the Mie theory.

Laboratory measurements of particle cross-sections have been carried out on sub-micrometre-sized smoke particles and microdroplets of glasses that were put as well-dispersed mono-layer of particles on an IR-transparent carrier (KBr, polyethylene) necessary for the spectrometer measurements. Particle radii must additionally be determined by electron microscopy. In the simplest case, from the transmission curves, mass absorption coefficients (MAC) representing the size- and shape-averaged absorption behaviour of the dust grains on the carrier are derived.

MAC can also be measured if the experimentally produced particulate is embedded in KBr pellets used in the commercial spectrometers. Mostly, the particles are manu-factured by trituration of bulk AIIS. In order to get particulates containing only submicrometre-sized grains (interstellar particles should be, at most, of this size), usually suspension techniques are used. For the details of this preparatory procedure, see Dorschner et al. (1977).

If the particles are very small compared with the wavelength (as it is the case if the MIR spectrum of particles smaller than 1 μm is measured), then the scattering cross-section can be neglected and the extinction is almost completely raised by the absorption. In this case, the extinction cross-section is sufficiently accurate given by the small-particle approximation of the Mie theory and the MAC become independent of the particle size.

It becomes even possible approximately to derive complex indices of refraction for such particulates. The measured MAC curve, say, in the spectral region 7–40 μm, can be represented in the usual way by a superposition of Lorentzian profiles. On the other hand, the mentioned small-particle approximation of the absorption cross-section contains the complex dielectric function (relative permittivity) only as the imaginary part of the term $\varepsilon - 1/\varepsilon + 2$. This term itself is, however, given by the well-known Clausius–Mosotti relation

$$\frac{\varepsilon - 1}{\varepsilon + 2} = \frac{\varepsilon_\chi - 1}{\varepsilon_\chi + 2} + \sum_j \frac{K_j}{\omega_j^2 - \omega^2 + i\gamma_j\omega} \ .$$

Thus, if ε_χ, the short-wavelength limit of ε, is given and if the parameters K_j, ω_j, γ_j of the Lorentzian profiles are known, then n and k can be derived. This method has been used by Dorschner *et al.* (1986b) in order to derive refractive indices from the MAC curve for amorphous bronzite grains (cf. Section 3.3).

Measuring MAC curves on the base of KBr embedding techniques renders it necessary to correct the influence of the embedding medium that causes some minor changes in the band profiles. These KBr effects were studied in greater detail by Dorschner *et al.* (1978), Schmidt (1981), and Henning (1986).

Cross-sections and scattering functions for particles of an arbitrarily chosen shape have also been determined by microwave analogy experiments. The pioneering work to establish this method was done by Greenberg *et al.* (1961) and by Giese and Siedentopf (1962). In connection with these microwave measurements Greenberg coined the term 'laboratory astrophysics', which was meanwhile extended to the whole field of laboratory experiments. The first twenty years of laboratory astrophysics with respect to dust research have been reviewed by Gürtler and Dorschner (1980). The main application field of microwave analogy measurements with centimetre-sized artificial 'grains' has been the study of scattering properties of irregularly shaped interplanetary particles (Zerrull *et al.*, 1980) up to complex particle arrangements, so-called bird's nests, which have been introduced in order to simulate the structure of cometary solids (Gustafson, 1980).

Meanwhile, interplanetary dust particles captured in the earth atmosphere can be individually investigated by mineralogical methods. Sandford and Walker (1985) showed that most of their particles consisted of crystalline olivine, pyroxene, or layer-lattice silicates. So, it is not clear if optical data of the material of these particles are useful for the investigation of the interstellar dust. We discussed this point already in Section 2.1.

3.3. EXPERIMENTAL STUDIES OF MIR SILICATE BANDS

The central task of the astronomical dust research is to understand nature and origin of cosmic particulates. A first absolutely necessary step in this direction is the identification of the interstellar silicate. For this aim, laboratory experiments of AIDS providing the dust researchers with realistic optical data for particles of the suitable size at the most informative wavelength range from 7 to 40 μm are indispensible. It is the comparison of such laboratory data with the observations, to whom we mainly owe our present knowledge on interstellar silicate dust. Absolute MAC, at least at certain distinguished wavelengths, for example, the absorption peaks of the 10 and 20 μm bands, are also needed if the amount of silicate dust is to be reliably estimated.

Normally, from the observations only relative MAC curves can be derived. For many purposes, for example, for modelling MIR source spectra by radiation transfer calculations, it suffices to know the MAC curve in a normalized manner and to render the absolute value at the normalization wavelength in a suitable form the model parameter.

Observationally-based relative MAC can be derived with only few assumptions from the spectra of sources that are optically thin in the silicate bands. In these sources the 10 and 20 μm bands are observed in emission. To gain such data from sources showing silicate self-absorption or, generally, deep absorption bands requires expensive radiative transfer calculations that contain numerous model assumptions, so that the results are conclusive only to a limited extent.

Henning *et al.* (1983), Pégourié and Papoular (1986), and Rowan-Robinson (1986) published such data that were derived from optically thin shells around late-type stars. These results show, that this circumstellar silicate must be distinguished from the silicate observed in molecular sources, but, apparently, also from silicate dust surrounding T Tauri stars. The only case, where relative MAC of molecular-cloud dust can be derived in a reliable manner is the Orion Trapezium nebula, which shows the 10 μm band in emission. For many calculations the Trapezium profile has been used as a kind of standard. It is, however, by no means clear whether Trapezium nebula dust is representative for the interstellar silicate with respect to structure, composition, and size distribution or not. In any case, in agreement with the 10 μm absorption bands of BN objects and the emission bands of T Tauri stars and related sources the 'Orion silicate' shows the 10 μm absorption peak at a wavelength clearly smaller than 9.9 to 10.0 μm where the M-star silicate has its absorption peak. In Table III the spectral properties of the two well-distinguished silicates, molecular-cloud silicate (MCS) and late-type star silicate (LSS), are listed. In our context, we prefer phenomenological designations of these silicates to such labelling that contains chemical or mineralogical terminology anticipating some kind of identification. Gürtler and Henning (1986), who aimed at identification, used the terms pyroxene-type silicate and olivine-type silicate for MCS and LSS, respectively.

For characterizing spectral properties the following parameters of the MAC curves have been used in Table III: peak absorption wavelengths (λ_c) of the 10 and 20 μm bands, full width of half maximum (FWHM), wavelength of minimum absorption in the trough between the bands (λ_t) and relative MAC (k_{rel}) at λ_t and λ_c (20 μm). The normalization of k_{rel} is such that $k_{rel} = 1$ for λ_c (10 μm). For the sake of comparison, the 'astronomical silicate' proposed by Draine and Lee (1984) has been added to Table III. This silicate was postulated by fitting observational, experimental, and theoretical curves in order to get a self-consistent run of representative optical data from FUV to FIR for practical use.

It is interesting that the observed 10 μm profile of the comets, which shows a very broad, plateau-like peak region, can be represented by a mixture of MCS and LSS, at least as a first approximation. For this reason, we do not postulate a special 'cometary silicate'.

Corresponding to the two groups of observed silicates the laboratory work also reflects such a division. In Table IV we have listed important AIIS produced and studied by the experimentalists in the last decade. The parameters characterizing the special properties are the same as in Table III. Additionally, the absolute value of MAC at the 10 μm peak has been enclosed. As in Table III the 'astronomical silicate' of Draine and Lee (1984) has been added. Table IV points strikingly to the fact, that the difference between the observationally well distinguished silicates MCS and LSS could, indeed, be based on different chemical composition. This is the reason, why Gürtler and Henning (1986) used the corresponding mineralogical terms. MCS can be tentatively

TABLE III

MIR silicate dust spectrum observed in optically thin sources

Sources	Authors	10 μm band		Trough		20 μm band		
		λ_c (μm)	FWHM (μm)	λ_t (μm)	k_{rel}	λ_c (μm)	FWHM (μm)	k_{rel}
Orion Trapezium nebula	Forrest et al. (1975)	9.75	3.5					
T Tauri stars	Cohen (1980)	9.6	2.4					
M-type stars	Henning et al. (1983)	9.9	3.4	14.4	0.32	19.7	8	0.54 [a]
	Pégourié and Papoular (1985)	10.0	2.5	14.2	0.17	18.0	11	0.34 [a]
	Rowan-Robinson (1986)	10.0	2.4	13.0	0.30	18.5		0.55
'Astronomical silicate'	Draine and Lee (1984)	9.5	2.9	14.0	0.21	18.0		0.35

[a] Fixed with regard to experiments.

identified with Mg-dominated silicates of pyroxene composition, whereas LSS can be attributed to Mg-dominated olivine.

A very interesting new AIIS candidate is glassy bronzite (Dorschner *et al.*, 1986). In Figure 3 relative MAC normalized in the described way are compared with other experimental curves. Our bronzite measurements show clearly that the ratio between the MAC values at the band peaks is remarkably higher than it is usually assumed in literature. The low value $k_{rel}(20\ \mu m) = 0.35$ of the 'astronomical silicate' is probably influenced by the extremely low value of Day's (1979) measurements that are also characterized by a very deep trough (cf. Figure 3). As the ratio $k_{rel}(20\ \mu m) = 0.55$ derived from IRAS low-resolution spectra for M-type stars (Rowan-Robinson, 1986) shows it is justified to suggest a larger value than Draine and Lee (1984) give.

A striking point of Table IV is the good agreement of the absolute MAC at the 10 μm peaks of the different AIIS. We, therefore, recommend to use the value 3000 cm^2 g^{-1} for the interstellar particles. Much less clear is the situation around the 20 μm band

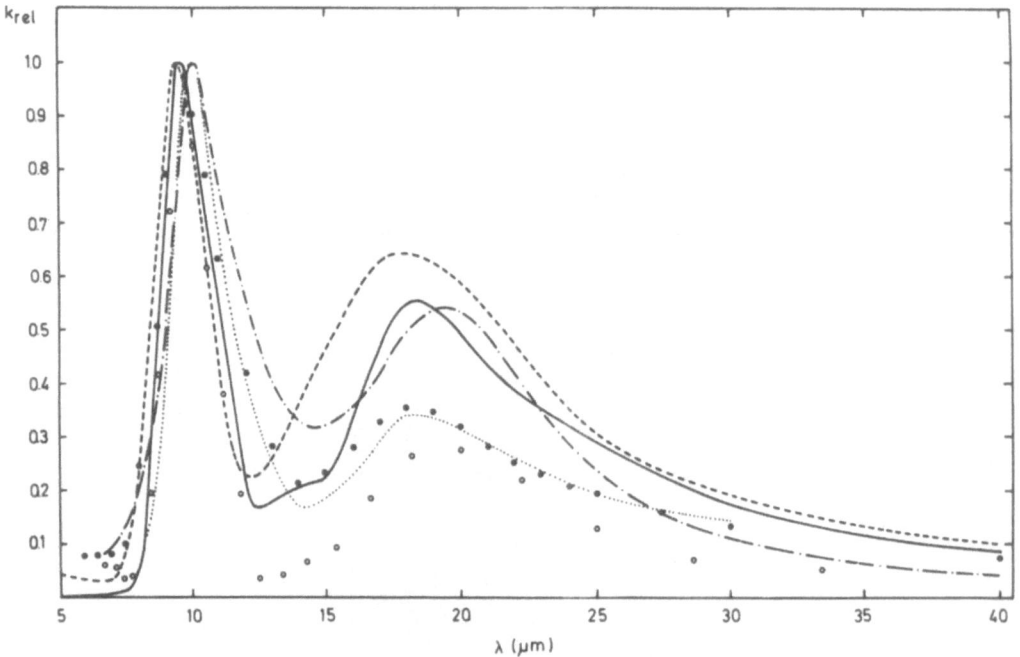

Fig. 3.   Normalized MAC curve of amorphous bronzite by Dorschner *et al.* (1986a, b; solid line) compared with those of particles (radius 0.1 μm) of amorphous MgSiO$_3$ by Day (1979; open circles), of the silicate used by Pearce and Evans (1984; dashed line) and of the 'astronomical silicate' by Draine and Lee (1984; filled circles). For the sake of comparison observationally-based curves by Henning *et al.* (1983; dashed-dotted line) and Pégourié and Papoular (1985; dotted line) have been included.

where the experimental results considerably differ from each other. In this wavelength range the observational data, too, are not in very good agreement. The situation could, however, clarify if more IRAS spectra are investigated.

TABLE IV
MIR silicate dust spectrum observed in optically thin sources

Silicate	Authors	λ_c (μm)	FWHM (μm)	MAC (cm² g⁻¹)	λ_t (μm)	k_{rel}	λ_c (μm)	FWHM (μm)	k_{rel}
		10 μm band			Trough		20 μm band		
Ca-protosilicate	Dorschner et al. (1977)	9.75	2.3	2940			21.5	≈5	0.39
Mg-protosilicate	Dorschner et al. (1977)	9.9	1.9	2880			21.0	≈10	0.48
Fe-protosilicate	Dorschner et al. (1977)	10.4	2.3	2760			21.5	≈15	0.43
Amorphous Mg_2SiO_4	Day (1979)	10.0	2.2	2300	13.4	0.05	19.5	11.5	0.36
Laser-vaporized olivine	Stephens and Russell (1979)	9.75	2.1						
Irradiated olivine	Krätschmer and Huffman (1979)	9.8	2.0	3900	13.5	0.08	17.2	≈5	0.35
Amorphous $MgSiO_3$	Day (1979)	9.5	2.0	2920	12.7	0.03	19.0	8.5	0.28
Laser-vaporized enstatite	Stephens and Russell (1979)	9.1	1.8						
Amorphous bronzite	Dorschner et al. (1986a, b)	9.5	2.0	3000	12.5	0.17	18.5	11.1	0.56
'Astronomical silicate'	Draine and Lee (1984)	9.5	2.9	3020	14.0	0.21	18.0	≈13	0.35

4. Summary and Conclusions

More than 2000 well-known IR sources with strong silicate emission or absorption bands prove that silicate particles are an important constituent of the interstellar dust in a wider sense, that means, inclusive of circumstellar condensates, which should be in the case of cool oxygen-rich stars mainly silicate particles, and dust grains in molecular clouds and in the surroundings of extremely young objects.

It is necessary to distinguish, at least, between three different silicates: (i) silicate clusters that primarily condense in cool oxygen-dominated circumstellar shells and outflows; (ii) the chemically processed particles that finally leave the circumstellar space (LSS); (iii) silicates occurring in dense interstellar regions, e.g., in molecular clouds (MCS).

From the cosmic abundance of elements, from the observed depletion of elements in the interstellar gas and from calculations of condensation processes leading to the formation of silicate particles chemical constraints result. They let us expect that mainly Mg silicate particles in which Mg ions are partly substituted by Fe ions and which contain traces of Al, Ca, Ni, and Na should occur.

For the investigation of the interstellar silicate dust, laboratory experiments are indispensable, especially what concerns the investigation of physical structure and chemical composition of these particles. Moreover, for various model calculations knowledge of realistic optical data of the particles' material is necessary, the determination of which presupposes very good laboratory simulation of the interstellar silicates. In this paper, a review of the experimental methods and results reached in preparing such astronomically important dust silicates (AIIS) has been given and optical properties of important AIIS have been presented and confronted with the observations. Laboratory experiments have yielded the following results of fundamental importance:

(i) The lattices of the silicate particles causing the observed MIR bands must be strongly disordered. The glassy state seems to be a good approximation. Additionally to structural disorder also chemical disorder is to be expected. Experiments with nonstochiometric Mg silicates that form as smoke particles in the result of chemical reactions simulating the conditions in circumstellar shells point to this.

(ii) The observed dust emission or absorption bands at 10 and 20 μm can be best represented by particles with diameters in the order of magnitude 0.1 μm, which consist of amorphous olivine-type and pyroxene-type silicates. O-type silicate particles having their 10 μm absorption peak near 10.0 μm are good candidates for identifying late-type star silicates (LSS). P-type silicate particles having their 10 μm peak at about 9.5 μm could explain molecular cloud silicates (MCS).

(iii) The peak mass absorption coefficient (MAC) for the 10 μm band amounts to $3000 \text{ cm}^2 \text{ g}^{-1}$ and the ratio MAC(20 μm)/MAC(10 μm) amounts to about 0.5.

In spite of large progress with the understanding of the nature of interstellar silicate dust, a lot of open questions in connection with laboratory experiments with AIIS remain. In our opinion, the most important ones are:

(i) Are there layer-lattice silicate particles in the interstellar matter? To approach this problem experimentally a great obstacle exists: Methods for producing disordered lattices, which have been described in Section 3.1, transform phyllosilicates in other silicates that do not contain water and that are no longer layer-lattice silicates.

(ii) How can we get silicates that are 'dirty' enough to have a relatively large imaginary part of the NIR refractive index in order to absorb sufficiently in NIR?

(iii) How do silicates contribute to FIR dust absorption? With which power of λ do the FIR cross-sections go? Are there absorption bands?

(iv) Which role play very small silicate particles (diameter $\ll 0.1$ μm) in connection with the interstellar extinction curve in the UV (220 nm hump) and in the FUV (steep rise for $\lambda^{-1} > 6$ μm^{-1})?

From this catalogue of open questions we recommend the experimentalists to devote themselves to the following problems:

(i) Preparing experimentally gained, consistent MAC curves which could be fitted in order to cover the whole wavelength range from FUV to FIR. These experimentally-based data should replace the 'astronomical silicate' data by Draine and Lee (1984) that are a mixture composed of observational, experimental, and theoretical data.

(ii) Finding experimental ways to study disordered layer-lattice silicates in the MIR without expelling bound water.

(iii) Investigation of the properties of very small silicate particles with regard to their possible role as cores of core-mantle grains, as possible cause of the MIR shoulder found in IRAS observations and, above all, as cause of or contributor to the UV extinction.

Of course, parallel to laboratory simulation, the observational base of dust research has to be broadened. The detailed investigation of the already existing IRAS low-resolution spectra offers the most obvious possibility.

References

Beichman, C. A., Neugebauer, G., Habing, H. J, Clegg, P. E., and Chester, T. J.: 1985, *Infrared Astronomical Satellite Catalogs and Atlases*, Explanatory Supplement.

Bohren, C. F. and Huffman, D. R.: 1984, *Absorption and Scattering of Light by Small Particles*, John Wiley and Sons, New York.

Cohen, M.: 1980, *Monthly Notices Roy. Astron. Soc.* **191**, 499.

Day, K. L.: 1974, *Astrophys. J.* **192**, L15.

Day, K. L.: 1979, *Astrophys. J.* **234**, 158.

Day, K. L.: 1981, *Astrophys. J.* **246**, 110.

Day, K. L. and Donn, B.: 1978, *Astrophys. J.* **222**, L45.

Donn, B. and Nuth, J. A.: 1985, *Astrophys. J.* **288**, 187.

Dorschner, J.: 1968, *Astron. Nachr.* **290**, 171.

Dorschner, J.: 1970, *Astron. Nachr.* **292**, 107.

Dorschner, J.: 1971, *Astron. Nachr.* **293**, 53.

Dorschner, J., Friedemann, C., and Gürtler, J.: 1977, *Astrophys. Space Sci.* **48**, 305.

Dorschner, J., Friedemann, C., and Gürtler, J.: 1978, *Astron. Nachr.* **299**, 269.

Dorschner, J., Friedemann, C., and Gürtler, J., and Duley, W. W.: 1980, *Astrophys. Space Sci.* **68**, 159.

Dorschner, J., Friedemann, C., Gürtler, J., Henning, Th., and Wagner, H.: 1986a, *Monthly Notices Roy. Astron. Soc.* **218**, 37P.

Dorschner, J., Friedemann, C., Gürtler, J., Henning, Th.: 1986b, in preparation.

Draine, B. T. and Lee, H. M.: 1984, *Astrophys. J.* **285**, 89.

Duley, W. W. and McCullough, J. D.: 1977, *Astrophys. J.* **211**, L145.
Duley, W. W., Millar, T. J., and Williams, D. A.: 1979, *Astrophys. Space Sci.* **65**, 69.
Egan, W. G. and Hilgeman, T.: 1975, *Astron. J.* **80**, 587.
Fadeyev, Yu. A. and Henning, Th.: 1986, in preparation.
Field, G. B.: 1974, *Astrophys. J.* **187**, 453.
Forrest, W. F., Gillett, F. C., and Stein, W. A.: 1975, *Astrophys. J.* **195**, 423.
Friedemann, C., Gürtler, J., and Dorschner, J.: 1979, *Astrophys. Space Sci.* **60**, 297.
Gaffey, M. J. and McCord, T. B.: 1979, in T. Gehrels (ed.), *Asteroids*, University of Arizona Press, Tucson.
Gail, H.-P. and Sedlmayr, E.: 1986, *Astron. Astrophys.*, in press.
Gerastchenko, A. N.: 1975, *Izv. Glav. Astron. Obs. Pulkovo*, No. 193, p. 65.
Giese, R.-H. and Siedentopf, H.: 1962, *Z. Naturforsch.* **17a**, 817.
Gillett, F. C., Low, F., and Stein, W. A.: 1968, *Astrophys. J.* **154**, 677.
Greenberg, J. M. and Chlewicki, G.: 1983, *Astrophys. J.* **272**, 563.
Greenberg, J. M., Pedersen, N. E., and Pedersen, J. C.: 1961, *J. Appl. Phys.* **32**, 233.
Gürtler, J. and Dorschner, J.: 1980, *Sterne* **56**, 300.
Gürtler, J. and Henning, Th.: 1986, *Astrophys. Space Sci.* **128**, 163.
Gustafson, B.: 1980, *Rep. Obs. Lund*, No. 17.
Hackwell, J. A., Gehrz, R. D., and Woolf, N. J.: 1970, *Nature* **227**, 822.
Henning, Th.: 1983, *Astrophys. Space Sci.* **97**, 405.
Henning, Th.: 1986, *Acta Universitatis Carolinae – Mathematica et Physica* (in press).
Henning, Th. and Svatoš, J.: 1985, *Astron. Nachr.* **307**, 1.
Henning, Th., Gürtler, J., and Dorschner, J.: 1983, *Astrophys. Space Sci.* **94**, 333.
Huffman, D. R.: 1970, *Nature* **225**, 833.
Huffman, D. R. and Stapp, J. L.: 1971, *Nature Phys. Sci.* **229**, 45.
Huffman, D. R. and Stapp, J. L.: 1973, in J. M. Greenberg and H. C. van de Hulst (eds.), 'Interstellar Dust and Related Topics', *IAU Symp.* **52**, 297.
Knacke, R. F. and Krätschmer, W.: 1980, *Astron. Astrophys.* **92**, 281.
Knacke, R. F., Puetter, R. C., Erickson, E., and McCorkle, S.: 1985, *Astron. J.* **90**, 1828.
Koike, C. and Hasegawa, H.: 1982, *Astrophys. Space Sci.* **88**, 89.
Krätschmer, W. and Huffman, D. R.: 1979, *Astrophys. Space Sci.* **61**, 195.
Lamy, P. L.: 1978, *Icarus* **34**, 68.
Larimer, J. W. and Anders, E.: 1967, *Geochim. Cosmochim. Acta* **31**, 1239.
Lebofsky, L. A.: 1980, *Astron. J.* **85**, 573.
Low, F. J. and Swamy, K. S. K.: 1970, *Nature* **227**, 1333.
Maas, R. W., Ney, E. P., and Woolf, N. J.: 1970, *Astrophys. J.* **160**, L101.
MacLean, S., Duley, W. W., and Millar, T. J.: 1982, *Astrophys. J.* **256**, L61.
Manning, P. G.: 1970, *Nature* **226**, 829.
Mooney, T. and Knacke, R. F.: 1985, *Icarus* **64**, 493.
Nuth, J. A. and Donn, B.: 1982, *Astrophys. J.* **257**, L103.
Palme, H., Suess, H. E., and Zeh, H. D.: 1981, in K. Schaifers and H. H. Voigt (eds.), *Landolt-Börnstein*, New Series, Group VI, Vol. 2a, p. 257.
Pearce, G. and Evans, A.: 1984, *Astron. Astrophys.* **136**, 306.
Pégourié, B. and Papoular, R.: 1985, *Astron. Astrophys.* **142**, 451.
Penman, J. M.: 1976, *Monthly Notices Ry. Astron. Soc.* **175**, 149.
Perry, C. H., Agrawal, D. K., Anastassakis, E., Lowndes, R. P., Rastogi, A., and Tornberg, N. E.: 1972, *Moon* **4**, 315.
Pollack, J. B., Toon, O. B., and Khare, B. N.: 1973, *Icarus* **19**, 372.
Rose, L. A.: 1979, *Astrophys. Space Sci.* **65**, 47.
Rowan-Robinson, M.: 1986, *Monthly Notices Roy. Astron. Soc.* **219**, 737.
Sandford, S. A. and Walker, R. M.: 1985, *Astrophys. J.* **291**, 838.
Savage, B. D. and Bohlin, R. C.: 1979, *Astrophys. J.* **229**, 136.
Schmidt, R.: 1981, *Astron. Nachr.* **302**, 235.
Seki, J. and Yamamoto, T.: 1980, *Astrophys. Space Sci.* **72**, 79.
Stein, W. A. and Gillett, F. C.: 1969, *Astrophys. J.* **155**, L179.
Stephens, J. R.: 1980, *Astrophys. J.* **237**, 450.

Stephens, J. R. and Russell, R. W.: 1979, *Astrophys. J.* **228**, 780.
Steyer, T. R.: 1974, Thesis, University of Arizona, Tucson.
Steyer, T. R., Day, K. L., and Huffman, D. R.: 1974, *Appl. Opt.* **13**, 1586.
Whittet, D. C. B.: 1984, *Monthly Notices Roy. Astron. Soc.* **210**, 479.
Woolf, N. J. and Ney, E. P.: 1969, *Astrophys. J.* **155**, L181.
Wong, J. and Angell, C. A. (eds.): 1976, *Glass Structure by Spectroscopy*, Marcel Dekker, Inc., New York, Basel.
Zaikowski, A., Knacke, R. F., and Porco, C. C.: 1975, *Astrophys. Space Sci.* **35**, 97.
Zerrull, R. H., Giese, R. H., Schwill, S., and Weiss, K.: 1980, in D. Schuerman (ed.), *Light Scattering by Irregularly Shaped Particles*, Plenum Publ. Corp., New York.

Discussion

H. J. Habing: I remember that about ten years ago P. J. Bedijn turned to make models for the Becklin–Neugebauer object and one big problem he met, which was also met by Jones and Merrill, is that silicates do not have enough absorption at about 2 μm and he invented the dirty silicate which was first an *ad hoc* assumption, but apparently everybody did it so at that time. Now, what happens to dirty silicates, are they there or not there? Have you any comment?

J. Dorschner: As a hypothesis, they, of course, yet exist, but I think nobody can exactly say what dirty silicates are. (To J. M. Greenberg) You must know it better.

J. M. Greenberg: No, no, no! It is the same question I always had: What produces these fortified imaginary parts? It is difficult to get them.

Th. Henning: There is no problem in the case of BN objects because the central energy sources radiate mainly in the ultraviolet region, but there is a problem for M stars!

H. J. Habing: Because they have to absorb.

Th. Henning: Yes, they have to absorb in the near infrared or visual part.

J. M. Greenberg: This was an excellent review and I appreciate what you have done there. When one talks about glassy or amorphous silicates, how do you get polarization by particles that are amorphous, if the particles are spherical more or less? Yet we know silicates produce polarization with the 10-μm feature.

J. Dorschner: I have no solution to this problem if the particles are indeed spherical.

W. Krätschmer: We usually have the problem of clumping of particles if we make the substrates. It is very hard to calculate from the extinction the cross-sections. How did you manage this problem?

J. Dorschner: In a strict sense, we did not solve this problem. What we did was to take electron-microscope pictures and we inspected the clumps formed and counted the particles in the clump of a certain size. We hoped that also in the pellets necessary for our IR spectrometry the same kind of clumping was present. But it is by no means clear if the things are comparable.

E. Krügel: The Draine and Lee data and the changes introduced compared with the previous ones are that they are a factor of three or so higher in the NIR and in the very long-wavelength range a factor of three lower. I should like to hear your opinion on the overall error estimate that one should have for the numbers that are published.

Th. Henning: I think, there are some difficulties to obtain such long wavelength range cross-sections or optical constants especially in two fields, in the NIR and the FIR. If you read the paper by Draine and Lee you will find that they select the data in the NIR in a rather arbitrary manner. The same is also true in the FIR where they took a λ^{-2} dependence. Rowan-Robinson, however, told us, that there must be a λ^{-1} dependence. So, there may be some changes. Another point is the ratio between the 20-μm peak and the 10-μm peak which is in accordance with experimental work. The value of 0.3 is taken from the experimental data by Day, we found a much larger value: 0.5. This gives a feeling how large the uncertainty is. We shall try to derive optical constants for our amorphous bronzite in order to make a detailed comparison possible.

P. G. Mezger: When I first used cross-sections up to 100 μm calculated by J. Mathis for the optical constants of graphite I was quite amazed by the fact how well it fitted. That was the first curve we published in 1980 and then it was E. Dwek who pointed out to us that apparently K. L. Day in his first measurements of amorphous silicates had made a mistake because he solved the material in water. That is why he got cross-sections that were too high. He was aware of it, and corrected it himself, but did not point it so clearly that everybody could get it so. Up to now the old Day values were quoted.

In the FIR the cross-sections have to be proportional to λ^{-2}. This is rather clear from model computations which did R. Chini and E. Krügel. There is no way to fit the observed spectrum between 350 and 1200 μm by anything which goes with λ^{-1}.

J. M. Greenberg: If λ^{-1} or λ^{-2} is really critical, because if it goes with λ^{-1} the particles are much cooler than if it goes like λ^{-2}.

P. G. Mezger: It looks rather convincing at least between 40 and 1000 μm and has to go like λ^{-2}.

J. M. Greenberg: That is very fine and I would not find that at all unpleasant.

P. G. Mezger: I find it pleasant that you do not find it unpleasant.

J. Dorschner: For the sake of modelling certain objects like BN objects we need a kind of standard opacity of dust over the whole wavelength range. The situation resembles that in the early model calculations on stellar interior where also the question of standard opacity was present. For the study of the nature of the interstellar dust, however, we must know the things a little bit more exact than such a standard mixture or standard curve can give it.

P. G. Mezger: That is completely correct. Of course, the deeper question is what these particles are made of. Only what irritates me is that the people get so terribly involved what is happening in the NIR and not what is happening in the FIR.

PROBLEMS WITH THE INTERPRETATION OF THE 220 nm INTERSTELLAR FEATURE*

C. FRIEDEMANN and J. GÜRTLER

Universitäts-Sternwarte Jena, G.D.R.

(Received 23 June, 1986)

Abstract. The bump in the ultraviolet part of the interstellar extinction curve provides a great challenge in the modelling of interstellar dust. Its shape can be well approximated by a classical dispersion profile with a total halfwidth of 48 nm centred at 217 nm. Apart from few slightly deviating cases the parameters of the band seem to be surprisingly constant in the solar neighbourhood.

The equivalent width W of the 217 nm band shows a very tight correlation with the colour excess $E(B - V)$. Studies of correlations with the strength of diffuse interstellar bands gave no conclusive results as to the nature of the band.

The most common interpretation of the 217 nm feature as originating from small graphite grians meets several difficulties. No final decision on the carrier can be made at present.

1. Introduction

At present the interstellar extinction law is known down to about 100 nm in the ultraviolet spectral region. The following general features may be discerned: – The linear rise in the visual region. Between 3–4 μm^{-1} the curve seems to reach a saturation.

– Centred at 4.6 μm^{-1} the flat part of the extinction curve is superposed by a broad 'bump'. Its properties make it very similar to an absorption band.

– The bump is followed by a flat minimum in the region 5–7 μm^{-1} and by a steep rise that continues to the limits of present-day observations, e.g., to 10 μm^{-1}.

The interpretation of the bump has been an unresolved problem until now. The commonly accepted interpretation for the increasing extinction toward shorter wavelengths is a bimodal grain-size distribution. Larger particles with radii of 0.15 μm^{-1} (amorphous silicates) cause the visual extinction, whereas smaller particles with radii of 0.05 μm^{-1} produce the rise of extinction in the ultraviolet spectral region.

Following Greenberg's (1973) suggestion the extinction curve can be conveniently divided into three shares:

– a sharp rise at the shortest wavelengths produced by very small grains,

– a plateau being the flat continuation of the visual extinction curve and attributed to larger grains, and

– an absorption band at about 220 nm wavelength, whose origin is unknown up to now.

Here, the basic question arises if the total extinction curve can be represented by chemically uniform particles obeying a suited size distribution or if particles of different

* Paper presented at a Workshop on 'The Role of Dust in Dense Regions of Interstellar Matter', held at Georgenthal, G.D.R., in March 1986.

chemical composition must be necessarily taken into account. At present only the existence of silicate particles in the interstellar medium has been revealed by infrared observations. A number of absorption and emission features in the near infrared provided some hints on the chemical compositions of the grain mantles. Investigations of the 220 nm feature may give some additional insight into the nature of the particles.

2. Extraction of Parameters of the 220 nm Feature from Observations

The basic properties of the 220 nm feature which have to be extracted from the observations are the central wavelength position, the width, and the strength as well as the variations of these parameters from star to star if present.

Dorschner (1973) first pointed to the fact that the feature can be approximated by a Lorentzian profile. Savage (1975) and all subsequent investigators confirmed this finding. As a consequence of this result Gürtler *et al.* (1982) approximated the extinction curve in the ultraviolet spectral region by an analytical expression consisting of three terms:

- one portion arising from the Lorentzian profile, and
- two others describing the 'continuous' extinction in the ultraviolet

$$e_\lambda = \alpha(\lambda^{-1} - \lambda_0^{-1})^n + \beta + \gamma \; \frac{\Gamma^2\lambda^2}{4\pi^2(\lambda^2 - \lambda_c^2)^2 + \Gamma^2\lambda^2} \; . \tag{1}$$

In this expression λ_c and Γ denote the wavelength position and the damping constant of the Lorentzian profile and λ_0 is the wavelength position where the share of the continuous UV extinction starts. The free parameters α, β, γ, and n were calculated from the observational data by a least-squares fit. The observed extinction A_λ and the visual extinction A_V are connected with e_λ via

$$e_\lambda = 0.4(A_\lambda - A_V) . \tag{2}$$

Seaton (1979) used a similar formula in which apart from the Lorentzian profile the continuous extinction is expressed by a power series of the second order. Massa and Fitzpatrick (1986) added to Seaton's formula a third-order term.

For statistical purposes we must be interested in the strength of the extinction hump. Different investigators have defined different quantities as a measure of its strength. Widely used is the extinction at the band's centre relative to the extinction at another wavelength outside the bump.

If a dispersion profile is fitted to the observed bump by a least-squares fit the result is a central depth that is based on the profile on the *whole* and not on few selected points. Thus, the strength of the band derived in this way is more reliable than other measures because the description of the continuous extinction is less arbitrary. Using Equation (1) we calculated equivalent widths W for a large number of stars observed with TD-1, OAO-2, IUE, and ANS satellites (Gürtler *et al.*, 1982; Friedemann *et al.*, 1983; Friedemann and Röder, 1987).

3. Statistical Properties of the 220 nm Band

Figures 1–3 show the correlations between the parameters α, β, and W of the extinction curve and colour excess $E(B - V)$. The weakest correlation exists between α and $E(B - V)$. An obvious interpretation is that different grain populations are responsible for visual and far UV extinctions. Expectedly the correlation between β and $E(B - V)$ is the strongest because parameter β by its definition is intimately related to $E(B - V)$. The correlation between the equivalent width W and $E(B - V)$ is relatively tight and points to a relationship between the 220 nm band and the population of dust particles causing the visual extinction rather than to a connection with grains responsible for the far UV extinction. This finding is in accordance with correlation analysis between the parameters α, β, and γ. Results show a tight correlation between β and γ ($r = 0.94$), whereas both β and γ are more weakly correlated with α ($r = 0.85$). Concordant results were reached by several authors (e.g., Danks, 1980; Meyer and Savage, 1981; Morgan et al., 1982; Franco et al., 1985).

The apparent relationship of the 220 nm dust feature to the diffuse interstellar lines (DIL's) in the optical wavelength region suggests the search for correlations between

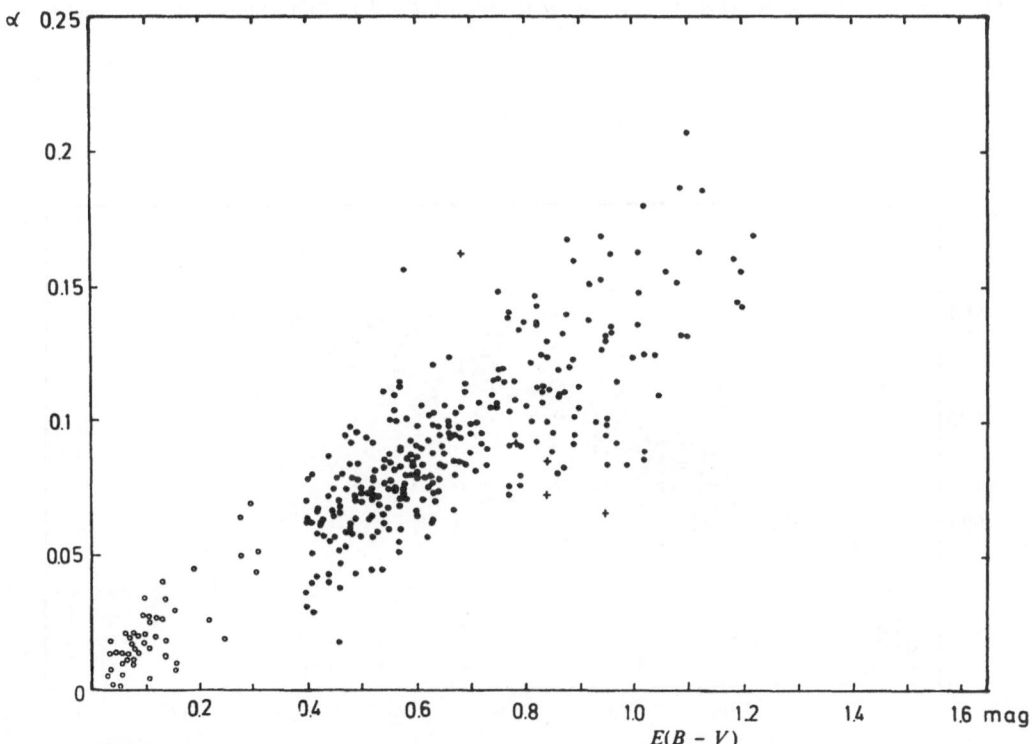

Fig. 1. Relationship between parameter α and colour excess $E(B - V)$ for a sample of stars observed by ANS (filled circles), TD-1 (open circles), and IUE satellites (crosses). The diagram contains only stars for which the observations fulfill certain criteria of accuracy.

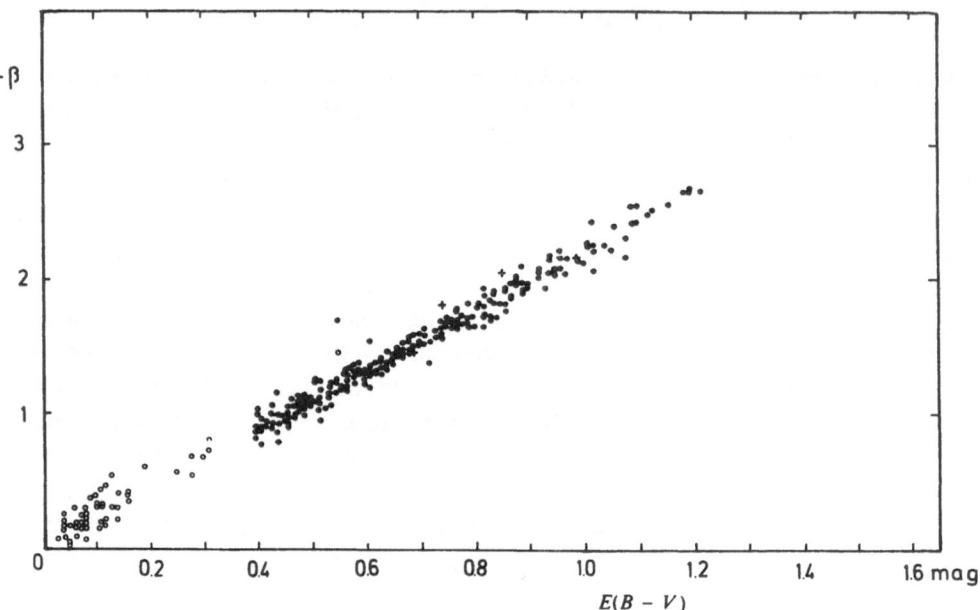

Fig. 2.   Relationship between parameter β and colour excess $E(B - V)$ for the same sample of stars as in
Figure 1.

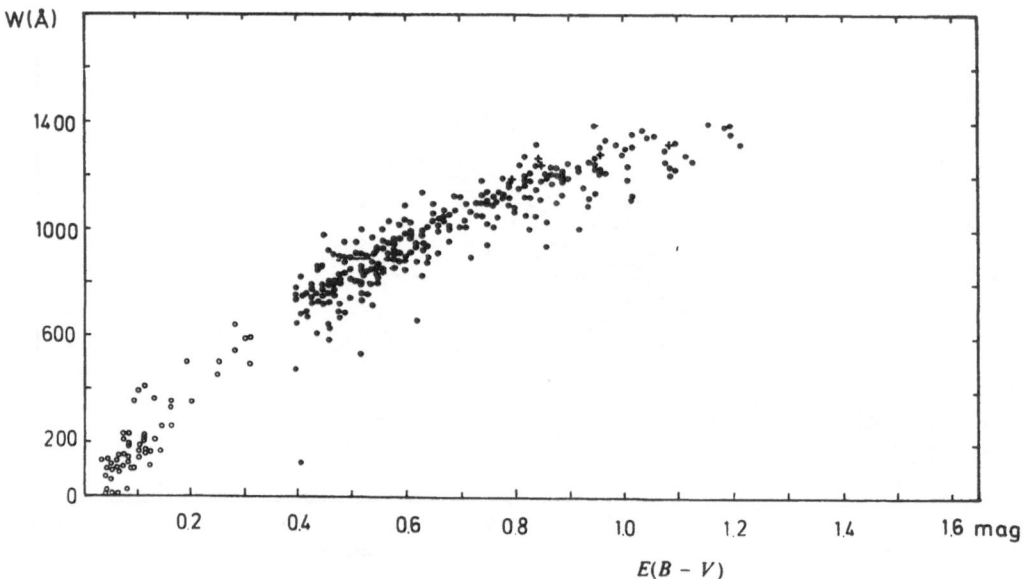

Fig. 3.   Relationship between equivalent width W of the 220 nm band and colour excess $E(B - V)$ for the
same sample of stars as in Figure 1.

TABLE I

Correlations between the strength of the 220 nm band and other interstellar features

W	E_{bump}	FUV vs $E(B - V)$ (125 nm)	A_c (443 nm)	DIL			References
				W (578 nm)	W (579.7 nm)	W (628.4 nm)	
×		~ 0.96	0.87	0.89	> 1		Dorschner et al. (1977)
	×	~ +	+				Danks (1980)
×		~ 0.974	0.998	0.834		0.634	Wu et al. (1981)
×		~	+ [a]	-		-	Seab and Snow (1984)
	×	~	-	-		-	Seab and Snow (1984)
×		~	0.73				Witt et al. (1983)
	×	~	- 0.53				Witt et al. (1983)

Notes:
+ : Correlation exists, but no correlation coefficient is given.
- : No correlation between normalized parameters exists.
[a] Normalized and unnormalized correlation.

them. Several authors performed statistical investigations to this end. Table I summarizes the results. These studies generally show good correlations between the strengths of the 220 nm feature and the DIL's. The case is most convincing for the 443 nm band mainly because the observed sample of stars is relatively large. The interpretation of the correlation is, however, not conclusive concerning the physical interpretation because the strength of the 220 nm band as well as that of the DIL's correlate with $E(B - V)$. To overcome these difficulties Witt et al. (1983) studied correlations between the strengths of the 443 nm and the 220 nm bands, normalizing their strengths by dividing them by $E(B - V)$. A weak correlation between the normalized strengths was found. Seab and Snow (1984) related the normalized strengths of the 220 nm bump with the deviations of the strengths of some DIL's from their mean relations with $E(B - V)$. Whereas a good correlation was found for the 443 nm band no correlations seem to exist for the DIL's at 578 and 628.4 nm.

Recently, Duley (1985) studied the correlation between the strength of the 220 nm band and the column density of interstellar hydrogen, where the strength of the UV band was modified by taking the depletion of several elements into account. Improvements of correlations between hydrogen density and equivalent width W was found in the case of silicon and magnesium, whereas a significant deterioration in the case of carbon occurred. This may indicate that carbon is no main constituent for the carrier of the 220 nm band. This conclusion agrees well with an earlier result by Millar (1979).

The correlation analyses have given conflicting results. Partly this is caused by the methods used and by the observational data available at present. A remarkable scatter of the data points is apparent, as is especially well illustrated in Figures 1 and 3. Investigations of the origin of this scatter revealed that it is only partly caused by observational errors and uncertainties in the derivation of the extinction curves. There remains a significant scatter due to intrinsic variations in the extinction law (see, e.g., Meyer and Savage, 1981; Gürtler *et al.*, 1982; Friedemann *et al.*, 1983; Massa *et al.*, 1983; Friedemann and Röder, 1986). Seab *et al.* (1981) pointed out four kinds of extinction anomalies:

- weak 220 nm features,
- low far UV extinction,
- a shortward shifted 220 nm band coupled with high far UV extinction, and
- strong 220 nm features.

A possible way to solve the open question of what parameters of the interstellar features correlate with each other could be the study of extreme cases.

4. Intrinsic Properties of the 220 nm Band

Investigations revealed three main properties of the band which must be accounted for in all attempts for its identification:

- The shape of the band is always Lorentzian, independently of its relative strength.
- The total halfwidth seems to be rather constant from star to star. Gürtler *et al.* (1982) give a mean of 48 nm for it in excellent agreement with the result reached by Savage (1975) and with the mean for the three open clusters NGC 3293, Trumpler 14, NGC 6231 as well as the Cepheus OB 3 association derived by Massa and Fitzpatrick (1986). On the other hand for members of the cluster NGC 2244 the latter authors found the deviating value of 42 nm.
- According to Gürtler *et al.* (1982) the mean position of the central wavelength is $\bar{\lambda}_c$ = 217 nm for 551 TD-1 stars. This value agrees quite well with $\bar{\lambda}_c$ = 217.5 $\pm$ 0.1 nm derived by Massa and Fitzpatrick (1986).

A few stars seem to deviate from the mean position by a larger amount than it would be expected by the observational errors alone. The extremes are 210 nm (HD 50083, $E(B - V) = 0.30$; HD 6273, $E(B - V) = 0.14$) and 227 nm (HD 45321, $E(B - V) = 0.06$; HD 69080, $E(B - V) = 0.08$).

The distribution function of the individual wavelength positions is slightly asymmetric toward shorter wavelengths. This is in accordance with the finding of Seab *et al.* (1981) who point to some stars (HD 169454, o Persei, Persei ζ, Ophiuchi ζ) for which the band's centre is shifted to shorter wavelengths, too.

Investigations of the diffuse galactic light by Witt and Lilley (1973) showed a significant decrease of the albedo in the band region. On the contrary Mathis *et al.* (1981) found no such a minimum in the direction toward the Orion nebula. These results must not contradict each other because in the Orion nebula the strength of the 220 nm band is anomalously weak. The finding of Witt and Lilley (1973) is commonly interpreted as evidence for the 220 nm feature being a true absorption band.

5. Where does the 220 nm Feature Originates

There exists a tight correlation between the strength of the 220 nm band and the colour excess $E(B - V)$. Apart from a few local deviations the strength of the 220 nm band shows a remarkably small variation. It indicates that the carrier of the absorption feature must be common, well mixed, and relatively stable against modifications of its properties by different physical conditions in the interstellar environment.

The correlations between the strengths of some diffuse interstellar lines and the 220 nm band support the assumption of its interstellar origin, too.

In an earlier investigation Krełowski and Strobel (1979) claimed a partial circumstellar origin of the 220 nm band based on a luminosity dependence of the band strength. However, the luminosity effect turns out to be spurious if a larger sample of stars is used.

Direct evidence against a circumstellar origin of the 220 nm band came from IUE observations of ε Aurigae (Boehm et al., 1984) and the two RV Tauri stars AC Herculis and R Scuti (Baird and Cardelli, 1985). The 220 nm feature present in the spectra of these variables shows no obvious phase-dependent variations. Thus, if there are hints on influences on the relative strength of the band at all then there seem to be connections with the interstellar environment rather than the circumstellar space. As discussed in Section 3 there are correlations with elemental depletion. Additionally, there is a tendency that stars with anomalously weak 220 nm bands are associated with dense interstellar clouds (see Table II).

TABLE II

Anomalies of the extinction curve

1. Weak 220 nm hump			
HD 29647	Taurus		Snow and Seab (1980)
$9^1, 9^2$ Ori	Orion	dense clouds	Bohlin and Savage (1981)
σ Sco	Scorpius		Savage (1975)
Stars in SMC			Prévot et al. (1984)
Stars in LMC			Nandy et al. (1981)
			Koorneef and Code (1981)
2. Low far UV extinction			
$9^1, 9^2$ Ori			Bless and Savage (1972)
σ Sco			Savage (1975)
ζ Oph			Seab et al. (1981)
3. Shortward shifted 220 nm band coupled with high UV extinction			
HD 147419 (WN)			Willis and Stickland (1981)
Stars in LMC			Nandy et al. (1981)
			Koorneef and Code (1981)
4. Strong 220 nm hump			
HD 192163 (WN6)			Willis and Wilson (1975)
HD 156385 (WC7)			Willis and Wilson (1977)

6. Attempts of the Identification of the 220 nm Band

In the past the theoretical efforts for an identification of the 220 nm band focussed on graphite particles as the carrier of the band. Most attractive of this hypothesis is that small particles ($a < 0.03$ µm) of that material exhibit such a band. Therefore, graphite is an essential component in various models for interstellar dust. However, the observed Lorentzian profile of the 220 nm absorption band and its nearly invariant parameters impose severe constraints on the distribution function of the particle radii. Furthermore, the conclusiveness of the model calculations is restricted from a theoretical point of view. As Draine and Lee (1984) pointed out, the rigorous application of the classical Mie theory to optical anisotropic particles is not allowed.

Another problem arises from the dust model proposed by Mathis *et al.* (1977). It consists of a mixture of silicate and graphite grains. Their contributions to the total extinction in the far UV are comparable. As pointed out by Greenberg and Chłewicki (1983) this composition of dust implies a correlation between colour excess $E(B - V)$ and the extinction in the far UV. The lack of a tight correlation indicates, however, that the amount of graphite presumed in the dust model of Mathis *et al.* (1977) must be reduced drastically.

Another objection to the graphite hypothesis was raised by Czyzak *et al.* (1982). They investigated the conditions under which graphite grains originate in stellar atmospheres and found all available facts to show that the production of graphite is a two-step process: first condensation into some sort of disordered carbon grains and second application of an appropriate heat-treatment to convert the disordered carbon into graphite. There is no evidence that such a process takes place in the atmospheres of carbon stars and circumstellar shells. These theoretical expectations are in agreement with Draine's (1984) finding that in the circumstellar shell around the carbon star IRS + 10°216 the graphite absorption band at 11.52 μm is lacking.

Experimental work on amorphous carbon grains was done by several authors and led to contradicting results as to the 220 nm band. Measurements of the optical constants by Duley (1984) display no band in the 220 nm region. Different from this, laboratory investigations by Stephens (1980), Borghesi *et al.* (1985), and Maggipinto *et al.* (1985) showed a broad feature which centres at 250 nm. Also the substance Sakata *et al.* (1983) synthesized in the laboratory and called quenched-carbonaceous composite (QCC) exhibits a band in the 220 nm region. Obviously, the strength of this band depends on the way of producing the amorphous carbon and the wavelength position on the particle radii (Borghesi *et al.*, 1985). In the light of the laboratory data available at present, it seems difficult to reproduce the astronomical observations by amorphous carbon. Common to all this laboratory investigations is, however, a sharp rise of the extinction in the far UV.

Since the discovery of the 220 nm feature by Stecher (1965) a number of other possible identifications have been suggested. Among the more extensively discussed hypotheses is the proposal by Duley (1976) that small magnesium oxide grains may explain the 220 nm feature. At present there exist no laboratory measurements of the absorption efficiency of the MgO particles in the UV. The strongest experimental evidence for this mechanism came from the study of fluorescence radiation produced by UV incident upon magnesium oxide grains. This radiation reaches a maximum at 220 nm. Duley's model predicts several additional absorption features in the UV. However, careful searches (see Massa *et al.*, 1983; Seab and Snow, 1985) revealed no such bands. From the lack of the feature at 160 nm, Massa *et al.* (1983) estimated that the total amount of magnesium bound in crystalline magnesium oxide is smaller than 3% of the available magnesium.

Huffman and Stapp (1971) suggested that silicate grains, especially brown enstatite (Mg, Fe)SiO$_3$, may be responsible for the 220 nm band. According to their measurements crystalline enstatite has an absorption edge near 220 nm. Mie calculations based on the optical data of Huffman and Stapp revealed that the wavelength position and the shape of the band-like feature is extremely sensitive to the grain size. Thus, silicates seem not to be a serious identification for the 220 nm feature.

Subsequent measurements by Egan and Hilgeman (1975) did not verify the data of Huffman and Stapp, but they found other silicate material to have the desired properties. Vapour-condensed amorphous silicates studied by Stephens (1980) do not show an absorption peak in the 220 nm region.

Large organic molecules with conjugated double and triple bonds as chains and rings were proposed by several authors as possible candidates for the explanation of the 220 nm band (see, e.g., Donn, 1968; Hoyle and Wickramasinghe, 1977a, b; Webster, 1978). Laboratory work provided evidence that polycyclic aromatic hydrocarbons (PAH) are likely carriers for the DIL's (see, e.g., van der Zwet and Allamandola, 1985; Crawford *et al.*, 1985). PAH's are interesting in connection with the identification of the 220 nm band, too. But as van der Zwet and Allamandola (1985) pointed out, the PAH's may be the carrier of both the DIL's and the 220 nm feature only if the oscillator strength in the visible is much less than in the UV near 220 nm.

At present the identification of the 220 nm feature is an unresolved problem. None of the proposed hypotheses explain the main observed properties of the band, namely its virtually invariant Lorentzian profile – which demands its origin in very small particles – as well as the tight connection to visual extinction.

References

Baird, S. R. and Cardelli, J. A.: 1985, *Astrophys. J.* **290**, 689.
Bless, R. C. and Savage, B. D.: 1972, *Astrophys. J.* **171**, 293.
Boehm, C., Ferluga, S., and Hack, M.: 1984, *Astron. Astrophys.* **130**, 419.
Bohlin, R. C. and Savage, B. D.: 1981, *Astrophys. J.* **249**, 109.
Borghesi, A., Bussoletti, E., and Colangeli, L.: 1985, *Astron. Astrophys.* **142**, 225.
Crawford, M. K., Tielens, A. G. G. M., and Allamandola, L. J.: 1985, *Astrophys. J.* **293**, L45.
Czyzak, J. S., Hirth, J. P., and Tabak, R. G.: 1982, *Vistas Astron.* **25**, 327.
Danks, A. C.: 1980, *Publ. Astron. Soc. Pacific* **92**, 52.
Donn, B.: 1968, *Astrophys. J.* **152**, L129.
Dorschner, J.: 1973, *Astrophys. Space Sci.* **25**, 405.
Draine, B. T.: 1984, *Astrophys. J.* **277**, L71.
Draine, B. T. and Lee, H. M.: 1984, *Astrophys. J.* **285**, 89.
Duley, W. W.: 1976, *Astrophys. Space Sci.* **45**, 253.
Duley, W. W.: 1984, *Astrophys. J.* **287**, 694.
Duley, W. W.: 1985, *Astrophys. Space Sci.* **112**, 321.
Egan, W. G. and Hilgeman, T.: 1975, *Astron. J.* **80**, 587.
Franco, M. L., Magazzù, A., and Stalio, R.: 1985, *Astron. Astrophys.* **147**, 191.

Friedemann, C. and Röder, U.-K.: 1987, *Astron. Nachr.* **308**.

Friedemann, C., Gürtler, J., Schielicke, R., and Dorschner, J.: 1983, *Astron. Nachr.* **304**, 237.

Greenberg, J. M.: 1973, in J. M. Greenberg and H. C. van de Hulst (eds.), 'Interstellar Dust and Related Topics', *IAU Symp.* **52**, 3.

Greenberg, J. M. and Chłewicki, G.: 1983, *Astrophys. J.* **272**, 563.

Gürtler, J., Schielicke, R., Dorschner, J., and Friedemann, C.: 1982, *Astron. Nachr.* **303**, 105.

Hoyle, F. and Wickramasinghe, N. C.: 1977a, *Nature* **266**, 241.

Hoyle, F. and Wickramasinghe, N. C.: 1977b, *Nature* **270**, 323.

Huffman, D. R. and Stapp, J. L.: 1971, *Nature Phys. Sci.* **229**, 45.

Koorneef, J. and Code, A. D.: 1981, *Astrophys. J.* **247**, 860.

Krełowski, J. and Strobel, A.: 1979, *Acta Astron.* **29**, 627.

Maggipinto, G., Minafra, A., and Titto, F.: 1985, *Astrophys. Space Sci.* **108**, 101.

Massa, D. M. and Fitzpatrick, E. L.: 1986, *Astrophys. J. Suppl. Ser.* **60**, 305.

Massa, D., Savage, B. D., and Fitzpatrick, E. L.: 1983, *Astrophys. J.* **266**, 662.

Mathis, J. S., Perinotto, M., Patriarchi, P., and Schiffer III, F. H.: 1981, *Astrophys. J.* **249**, 99.

Mathis, J. S., Rumpl, W., and Nordsiek, K. H.: 1977, *Astrophys. J.* **217**, 425.

Meyer, D. M. and Savage, B. D.: 1981, *Astrophys. J.* **248**, 545.

Millar, T. J.: 1979, *Monthly Notices Roy. Astron. Soc.* **188**, 507.

Morgan, D. H., MacLachlan, A., and Nandy, K.: 1982, *Monthly Notices Roy. Astron. Soc.* **198**, 779.

Nandy, K., Morgan, D. H., Willis, A. J., Wilson, R., and Gondhalekar, P. M.: 1981, *Monthly Notices Roy. Astron. Soc.* **196**, 955.

Prévot, M. L., Lequeux, J., Maurice, E., Prévot, L., and Rocca-Volmerange, B.: 1984, *Astron. Astrophys.* **132**, 389.

Sakata, A., Wada, S., Okutsu, Y., Shintani, H., and Nakada, Y.: 1983, *Nature* **301**, 493.

Savage, B. D.: 1975, *Astrophys. J.* **199**, 92.

Seab, C. G. and Snow, T. P.: 1984, *Astrophys. J.* **277**, 200.

Seab, C. G. and Snow, T. P.: 1985, *Astrophys. J.* **295**, 485.

Seab, C. G., Snow, T. P., and Joseph, C. L.: 1981, *Astrophys. J.* **246**, 788.

Seaton, M. J.: 1979, *Monthly Notices Roy. Astron. Soc.* **187**, 73P.

Snow, T. P. and Seab, C. G.: 1980, *Astrophys. J.* **242**, L83.

Stecher, T. P.: 1965, *Astrophys. J.* **142**, 1683.

Stephens, J. R.: 1980, *Astrophys. J.* **237**, 450.

Van der Zwet, A. P. and Allamandola, L. J.: 1985, *Astron. Astrophys.* **146**, 76.

Willis, A. J. and Stickland, D. J.: 1981, *Monthly Notices Roy. Astron. Soc.* **197**, 1P.

Willis, A. J. and Wilson, R.: 1975, *Astron. Astrophys.* **44**, 205.

Willis, A. J. and Wilson, R.: 1977, *Astron. Astrophys.* **59**, 133.

Witt, A. N., Bohlin, R. C., and Stecher, T. P.: 1983, *Astrophys. J.* **267**, L47.

Witt, A. N. and Lilley, C. F.: 1973, *Astron. Astrophys.* **25**, 397.

Wu, C. C., York, D. G., and Snow, T. P.: 1981, *Astron. J.* **86**, 755.

Discussion

J. M. Greenberg: In a recent paper with Chłewicki we have been able to determine from about six very carefully observed diffuse bands that they are not correlated with each other, so it is difficult to talk of a narrow correlation of diffuse bands with 220 nm hump because they itself have a problem with each other. There are several that form groups. One group may be correlated with $E(B - V)$ and with each other and the others are absolutely uncorrelated. Although I do not think we want to go into details, there is at least the question that the correlation of the 220 nm hump with the diffuse bands seems to be really not confirmed. PAHs may be candidates for diffuse bands, but I do not believe that they are really suggested as candidates for the 220 nm hump. There was, however, a proposal by de Groot from our group who found that irradiated ring molecules in matrices show evidence for the production of vacuum UV radiation. We did not yet totally identify the unstable compounds, but they contain eight carbon atoms. That is what we know. They have also some IR features. Depending on which molecule we started we get somewhat different features, but they all peak at 220 nm. An interesting thing about that is they have all a very, very strong absorption per carbon atom, probably it may be five times as strong as the absorption per carbon atom of graphite. To me that is very efficient. You need only about 5 o 10% of carbon to produce the 220 nm hump. I can guarantee that diffuse bands are not made by particles of any sort unless the particles are really subtile particles. The molecules, yes!

C. Friedemann: I think, we should stress the following point: We are sure that the correlation between the diffuse interstellar lines in the optical region and the UV is not a settled one.

J. M. Greenberg: I did say that they are not correlated!

C. Friedemann: We should point to the fact from our analysis that the strength of the UV band 220 nm is tighter correlated with visual extinction than to far UV and this would mean that the carrier is connected with larger particles.

J. M. Greenberg: Well, this is a very good point. Yes, it is correlated with large particles, there is no question about this. We did find the same correlation. So, how is it that something which is not a particle is correlated with something which is so different? Maybe these things come from the large grains, maybe that is the reason they correlate with them. I would like to dispose of that question that is really one of the deep questions. How do you get a correlation between two entirely fundamentally different kinds of species, molecules of some type on one hand, particles on the other?

P. G. Mezger: P. Cox and co-workers of our institute searched in vain for a correlation between the 12 μm IRAS flux and the UV bump. They did not find one.

J. M. Greenberg: I think, there are really enough reasons to believe that they are not correlated rather than correlated. The question of the diffuse bands, although they form a kind of family of the same sort, they are not correlated with each other. They exist together but they are not correlated. Correlation coefficients of 0.6 mean that there is not a good correlation.

H. J. Habing: There must be various families of molecules.

J. M. Greenberg: Yes, I agree. But I think, the stupid question is still the same. Goes is correlated with $E(B - V)$ just like the 220 nm hump correlates with $E(B - V)$? Have you got that nice correlation? I think, it is very curious.

C. Friedemann: There are some restrictions concerning the particle radius. It must be less than 0.2 μm otherwise you will get an asymmetric shape of the band. All the observations show the Lorentzian profile.

OBSERVATIONS OF THE VERY BROAD-BAND STRUCTURE BY COMBINED *uvby* AND *UBV* PHOTOMETRY*

H.-G. REIMANN

Universitäts-Sternwarte Jena, G.D.R.

(Received 27 June, 1986)

Abstract. We discuss the papers of various authors which have observed the very broad-band structure (VBS) in the spectra of highly-reddened stars using different methods. The centre of the VBS is situated in the range of the intermediate band *y*-filter of the Strömgren system and of the broad-band *V* filter of the *UBV* system. It is possible to use the colour excess ratio $E(b - y)/E(B - V)$ as a measure of the strength of the VBS. We found that this ratio varies between 0.7 and 0.8 for intermediately and highly-reddened stars with observed VBS. We calibrated the colour excess ratio in terms of the central depth of the VBS. From the calibration stars and 13 reddened young open clusters containing more than 10 B-type stars we derived a relation between the central depth of the VBS and the reddening $E(B - V)$ which is in good agreement with the results of van Breda and Whittet (1981) and supports the assumption that the VBS is a continuous phenomenon originating in solid grains.

1. Introduction

At this workshop our main interest is directed to clarify the role of dust in dense regions of interstellar matter, using infrared, radio, and ultraviolet spectral features as the principal diagnostic. Nevertheless, in the optical range of the spectrum there are also some interesting structures which may be expected to give important clues to the chemical composition and internal structure of the dust grains. In my talk I shall described a photometric method to measure the central depth of very broadband structure (VBS) in the interstellar extinction curve.

Normally in the visual range of the extinction curve we distinguish three kinds of structures.

(i) The diffuse bands of width from 0.5 to 3 nm which could be regarded as fine structure as Herbig (1975) has suggested.

(ii) Structure of width from 3 to 10 nm which has been identified by York (1971). For these features the term 'broad-band structure' was introduced.

(iii) The 'very broad-band structure', a term which is used to characterize features of width greater than or equal to 10 nm.

In addition we find in the 440 nm range the so-called 'knee', a change in the slope of the interstellar extinction curve. We examined in detail the ability of a photometric method to measure the strength of the VBS which has the form of a very broad shallow modulation in the 500–600 nm region – seen in emission – with a central depth amounting up to 0.06 mag per magnitude reddening $E(B - V)$.

---

* Paper presented at a Workshop on 'The Role of Dust in Dense Regions of Interstellar Matter', held at Georgenthal, G.D.R., in March 1986.

2. Discussion of Previous Studies of the VBS

The earliest studies of the interstellar reddening curve by Whitford (1958) only showed the absorption to be proportional to reciprocal wavelength with a change of slope in the 440 nm region. Whiteoak (1966) extended the work to fainter stars and first noticed the shallow modulation in the 400–700 nm region. Whiteoak compared 28 reddened O-stars with model atmospheres of Mihalas (1965). The theoretical models were fitted to the reddened stars on the basis of observed Balmer discontinuities. This is a procedure which can only be successful if the uncertainties in the used absolute calibration of the energy distributions of the stars are smaller than the observational errors and if the Balmer discontinuities from the models of a given spectral type agree very well with those from the observed stars. Unfortunately both conditions were not fulfilled in the procedure described by Whiteoak. Additional to the observational error of 0.02 mag there are systematical errors of the same order, caused by the use of the Oke (1964) calibration of stellar energy distributions, which is certainly in error.

Hayes *et al.* (1973) have further studied this structure on the basis of the same data as Whiteoak. In distinction to Whiteoak they derived the extinction curves for the individual stars by subtracting the mean energy distribution of seven observed comparison stars with small reddening and applied corrections to the reddening curves for the difference in Balmer discontinuity between the mean unreddened star and each reddened O-star. This procedure also seems to contain a source of systematic error, because most of the reddened stars have earlier spectral types than the mean unreddened comparison star. Because of the shallowness of the very broadband structure compared to the general slope of the extinction curve Hayes introduced the method to plot the VBS as residuals, after a straight line is subtracted from the measured extinction curves. His reference line passes through the points at $\lambda = 457$ and 880 nm. If one compares this picture with the common approximation of the interstellar reddening curve by two straight lines intersecting at $\lambda = 435$ nm, it seems to be evident that the change of slope in the interstellar reddening curve occurs with a higher probability in the vicinity of 456 nm. Only averaging over the curvature of the VBS leads to a break point at shorter wavelength. Hayes derived from his procedure an error for the central depth of the VBS for a single star of 0.02 mag.

Schild (1977) presented a new investigation of the interstellar reddening curve based upon new scanner observations in selected reddened open star clusters. His reduction procedure was quite different to the methods mentioned earlier. First all stars were dereddened with the reddening curve for the Perseus region as given by Radick (1973). This curve was modified for each cluster, by a trial-and-error fitting procedure, to bring the energy distributions of the cluster stars into agreement with the models by Kurucz *et al.* (1974). Unfortunately, there are a lot of Be-stars in the sample of Schild whereas the models from Kurucz give the energy distributions for normal B-stars. Additionally in case of Be-stars it is very difficult to separate the effects of intrinsic and foreground reddening. Averaging the VBS over the cluster members should, therefore, not reduce the systematic error which seems to be present in the Schild measurements. Critically analyzing his results Schild gives 0.04 mag as a typical error per wavelength interval. This is a high value compared to a central depth of the VBS of 0.06 mag per magnitude reddening $E(B - V)$, and is a further argument against using the Schild results for the calibration of a photometric measure of the VBS. The Schild results are comparable to the measurements by Hayes, because Schild also has subtracted a straight line from his data using the same normalization points as Hayes *et al.* (1973). From the results by Schild the VBS appears to be greatest in the direction of Perseus and smallest in the direction of NGC 6709 (Aquila). First he offered the idea that the c_1 and m_1 indices of the Strömgren system should show considerable scatter when heavily reddened stars are investigated. Viewed in another way the UBV system or the $uvby$ system could possibly be a powerful tool for the investigation of the VBS.

We tried to clarify in detail if a photometric detection of the VBS is possible.

3. Estimation of the Photometric Effect of the VBS

Our principal idea was to measure the strength of the VBS by a combination of a narrow band filter and a broadband filter, with the boundary condition not to introduce an additional photometric system. The necessity of measurements of highest accuracy seems to favour the use of colour indices, instead of brightness in the given spectral range. Because the y filter of the $uvby$ system and the V filter of the UBV system should be influenced by the VBS, whereas the b and B filter should be nearly uneffected by the structure, the colour excess ratio $E(b - y)/E(B - V)$ should be a measure of the strength of the VBS.

To examine a possible influence on the colour excess ratios in the UBV and $uvby$ system caused by the VBS, we have got an estimation of some photometric indices of both photometric systems by integration over filter transmission curves, stellar energy distributions, and two different interstellar extinction curves.

First we used the extinction curve for the χ Persei region according to Schild (1977) normalized to $E(B - V) = 1$ mag which is equivalent to a central depth of the VBS in the order of 0.06 mag. Secondly we chose an approximation of the extinction curve without any structure by two straight lines with the point of intersection at 457 nm. The equations of the two lines are, with $\kappa = 1/\lambda$

$$A(\kappa) = 2.26\kappa - 0.82 \quad \text{if} \quad \kappa < 2.19 \, \mu\text{m}^{-1} \tag{1}$$

and

$$A(\kappa) = 1.58\kappa + 0.61 \quad \text{if} \quad \kappa > 2.19 \, \mu\text{m}^{-1}. \tag{2}$$

Using these two extinction curves we found by integration that the indices m_1 and $b - y$ are affected by the VBS whereas the $U - B$, $B - V$, and c_1 indices are not influenced by this structure. This means that we can use the colour excess ratios $E_{m_1}/(B - V)$ or $E(b - y)/E(B - V)$ as *a* means to detect the VBS. The m_1 index contains the v-magnitude which is strongly influenced by the Hδ-line. Because it is difficult to separate luminosity, temperature, and reddening effects in one parameter we prefer the colour excess ratio $E(b - y)/E(B - V)$. A simple estimation relative to the photometric accuracy which is necessary to determine the colour excess ratio leads to

$$\Delta[E(b - y)/E(B - V)] = (E(B - V)\Delta E(b - y) + E(b - y)\Delta E(B - V))/E(B - V)^2.$$

For the estimation of the error we set $E(b - y)/E(B - V) = 0.743$ getting

$$\Delta[E(b - y)/E(B - V)] = (\Delta E(b - y) + 0.743\Delta E(B - V))/E(B - V). \quad (3)$$

If we assume realistic errors of the colour excess in both systems ($\Delta E(b - y) <$

0.005 mag and $\Delta E(B - V) < 0.01$ mag) we get

$$\Delta(E(b - y)/E(B - V)) = 0.012/E(B - V). \quad (4)$$

This is equivalent to an error in the colour excess ratio of 0.012 per magnitude reddening $E(B - V)$. From the integration we obtained for the colour excess ratio $E(b - y)/E(B - V) = 0.814$ in the case with VBS and $E(b - y)/E(B - V) = 0.700$ in the case without any structure.

Firstly this means, taken into account the above discussed errors, that the VBS is detectable with combined $uvby$ and UBV photometry.

Secondly, we cannot further regard the colour excess ratio $E(b - y)/E(B - V)$ to be constant, as it was done before in various investigations of open galactic clusters. $E(b - y)/E(B - V) = 0.743$ seems to be only a mean value for intermediate colour excess.

We carefully payed attention to other influences which could change the colour excess ratio, such as reddening, spectral type, or luminosity class. From the integration we found that the colour excess ratio $E(b - y)/E.(B - V)$ is nearly constant over the total range of B-stars, that means if we restrict the problem to B-stars we can neglect other influences.

4. Calibration of the Colour Excess Ratio

In a next step we calibrated the colour excess ratio $E(b - y)/E(B - V)$ in terms of the central depth of the VBS. We used the very accurate measurements of the VBS by van Breda and Whittet (1981). They examined in detail 22 stars – mainly reddened B-type supergiants – in the southern Milky Way. Van Breda and Whittet measured the strength of the VBS as the difference between reddened and comparison star. Two stars of their sample situated in the Rho Ophiuchi dark cloud show a form of VBS which is quite different from that in the general interstellar medium, an effect which is possibly connected with the particle size or chemical changes occurring on the grain surfaces. Van Breda and Whittet carefully corrected the VBS for the effect of all the principal diffuse interstellar bands and give a resulting error of 0.01 mag for the central depth of the VBS of a single star.

For all comparison stars and for 11 of the 22 VBS stars measured by van Breda and Whittet $uvby$ and UBV photometry was available. The $uvby$ data were taken from the catalogue of Hauck and Mermilliod (1979). For additional 3 stars we derived the *uvby* indices from our own photoelectric observations with the 90 cm telescope at Großschwabhausen observing station. Observing technique, colour equations, and reduction procedure are described in Reimann *et al.* (1984). The photometric data for this 3 stars are given in Table I.

TABLE I

uvby photometry of VBS stars

HD	V	$b - y$	m_1	c_1
169454	6.57	0.738	-0.182	0.005
170938	7.82	0.726	-0.167	0.002
172488	7.64	0.490	-0.103	-0.002

Because most of the stars with observed VBS are B-type supergiants, we need a calibration for supergiants in the *uvby* system, independent of the colour excess ratio $E(b - y)/E(B - V)$ in order to accurately determine the reddening. The calibration recently published by Zhang (1983) is, therefore, not usable in this connection. Zhang used a constant colour excess ratio $(E(b - y)/E(B - V) = 0.730)$ to get the unreddened colours of supergiants in the *ubvy* system on the basis of the reddening in the *UBV* system. We derived new reference lines for zero reddening for the supergiants only on the basis of the slope between unreddened B-type stars in both photometric systems. Therefore, we compared empirical calibration for Main-Sequence stars with models for various values of surface gravity from Kurucz (1979). We found no significant difference between the slope for Main-Sequence stars or supergiants in a $(b - y)_0$ versus $(B - V)_0$ diagram. For the slope we derived

$$\Delta(b - y)_0 = (0.435 \pm 0.007)\Delta(B - V)_0 . \tag{5}$$

With this relation we calculated the unreddened index $(b - y)_0$ of supergiants in the *uvby* system. A detailed description of our calibration method is given in Reimann (1986).

To avoid additional errors in our calibration of the colour excess ratio in terms of the VBS, resulting from the data handling procedure, we tried to copy the method of van Breda and Whittet so exactly as possible in a photometric way. At first we derived from our supergiant calibration the reddening for the comparison stars in the *uvby* system (using $E(b - y) = (b - y) - (b - y)_0$) and correct the corresponding $E(B - V)$ value for the colour excess ratio in the case without VBS $(E(b - y)/E(B - V) = 0.7)$ in accordance with our result from the integration. Than we derived the reddening of the VBS stars and subtracted the colour excess of the comparison stars. After this we calculated with the reduced colour excesses the ratios for the VBS stars. The data and the results are contained in Table II. Starting from the left are listed: HD number, $b - y$; $E(b - y)$, MK-type; Δm, central depth of the VBS (van Breda and Whittet, 1981); $E(b - y)_C$ of the comparison stars; $E(b - y)_R$, the reduced reddening of the VBS stars;

TABLE II

Photometric indices of VBS stars, reduction and resulting colour excess ratios as a measure of the central depth of the VBS

HD	$b - y$	$E(b - y)$	Sp. type	Δm	$E(b - y)_C$	$E(b - y)_R$	$B - V$	$E(B - V)$	$E(B - V)_C$	$E(B - V)_R$	$E(b - y)_R : E(B - V)_R$
92964	0.245	0.302	B2.5Ia	0.023	0.025	0.277	0.265	0.405	0.035	0.370	0.749
115842	0.286	0.378	B0.5Ia	0.034	0.065	0.313	0.295	0.505	0.093	0.412	0.760
142468	0.524	0.616	B0.5Ia	0.049	0.013	0.603	0.580	0.790	0.019	0.771	0.782
144969	0.777	0.869	B0.5Ia	0.047	0.111	0.758	0.930	1.140	0.159	0.981	0.773
147084	0.615	0.520	A5II	0.000	0.003	0.517	0.836	0.726	0.004	0.722	0.716
147889	0.649	0.749	B2V	0.000	0.040	0.709	0.855	1.095	0.057	1.038	0.683
148379	0.476	0.549	B1.5Ia	0.038	0.062	0.487	0.550	0.725	0.089	0.636	0.766
148688	0.309	0.389	B1Ia	0.030	0.056	0.333	0.331	0.521	0.080	0.441	0.755
152235	0.466	0.546	B1Ia	0.036	0.111	0.435	0.550	0.740	0.159	0.581	0.749
152236	0.439	0.512	B1.5Ia	0.042	0.056	0.456	0.510	0.685	0.080	0.605	0.754
156201	0.575	0.667	B0.5Ia	0.038	0.056	0.611	0.650	0.860	0.080	0.780	0.783
169454	0.738	0.818	B1Ia	0.038	0.056	0.762	0.900	1.080	0.080	1.000	0.762
170938	0.726	0.806	B1Ia	0.046	0.056	0.750	0.850	1.040	0.080	0.960	0.781
172488	0.490	0.610	B0.5II	0.042	0.056	0.554	0.530	0.805	0.080	0.725	0.764

$B - V$, $E(B - V)$, $E(B - V)_C$ of the comparison stars; $E(B - V)_R$, the reduced reddening of the VBS stars, and the colour excess ratio $E(b - y)_R/E(B - V)_R$. A least-squares
solution taking observational errors to lie entirely in the colour excess ratio gives:

$$\Delta m = 0.579(E(b - y)/E(B - V)) - 0.405 , \quad r = 0.89 . \tag{6}$$

The regression line together with the data are plotted in Figure 1.

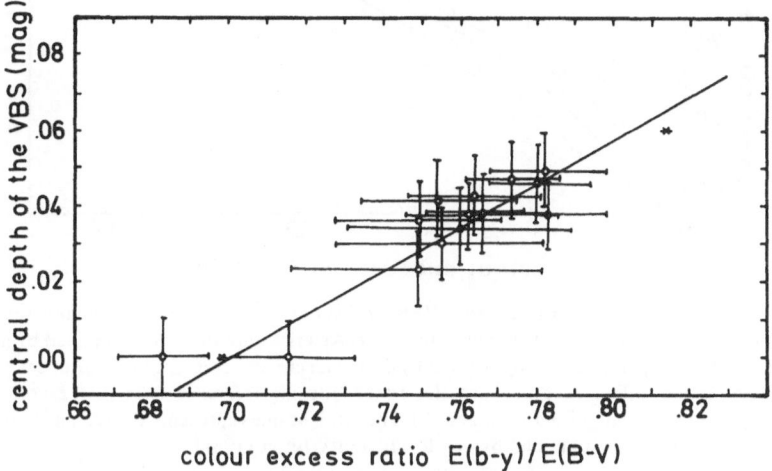

Fig. 1.   The calibration of the colour excess ratio $E(b - y)/E(B - V)$ in terms of the central depth of the
VBS. Open circles are single-star points plotted from the data of Table II. Asterisks mark the positions of
the values obtained from the integration. Error bars are calculated from Equation (4). The solid line
represents the least-squares solution given by Equation (6).

5. Discussion

As a first result we are able to present a relation between the central depth of the VBS
and the colour excess $E(B - V)$. The relation, shown in Figure 2, is in a good agreement
with that, found by van Breda and Whittet (1981), assuming their least-squares fit which
passes through the origin. Our relation based on the determination of the colour excess
ratios for open clusters containing more than 10 B-type stars. The data were taken from
the catalogue of Hauck and Mermilliod (1979) and reduced with the calibration
explained above. For the highly reddened open cluster NGC 1502 we obtained from our
own observations (Reimann and Pfau, 1986) the relation

$$E(b - y)/E(B - V) = 0.770 \pm 0.006 ,$$

equivalent to a central depth of the VBS of 0.04 mag. The position of this cluster is
marked in Figure 2. The fact of a systematic deviation from the relation given by
van Breda and Whittet towards smaller values of $E(B - V)$ seems to be a poor
photometric effect. It can be explained if we assume a systematic difference δ between

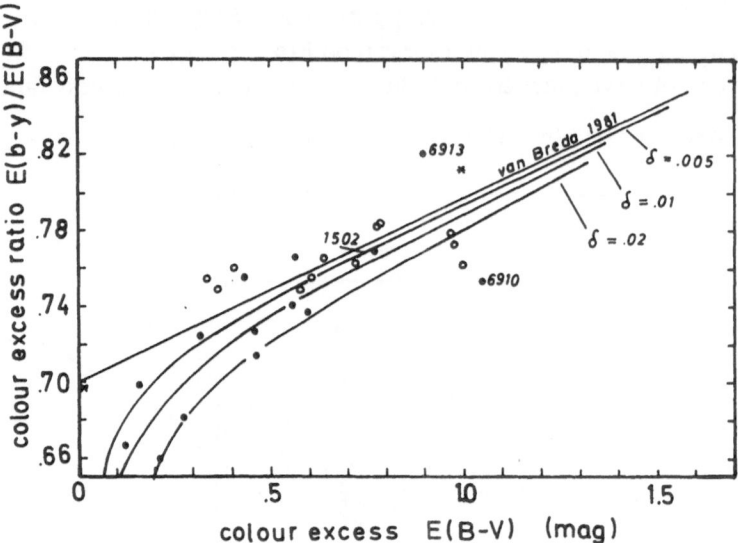

Fig. 2. The reddening dependence of the VBS measured in terms of the colour excess ratio $E(b - y)/E(B - V)$. Open circles are the calibraiton stars. Asterisks are the values obtained by integration. The mean values for 13 open clusters are denoted by dots. The curves represent the theoretical dependence assuming a systematic difference δ between the zero-reddening reference lines in the *uvby* and *UBV* photometric systems, calculated from Equation (7). The straight line represents the reddening dependence obtained by van Breda and Whittet (1981).

the zero-reddening reference lines in the *uvby* and *UBV* systems. If we set

$$R = E(b - y)/E(B - V)$$

and

$$R_0 = ((b - y) - (b - y)_0)/((B - V) - (B - V)_0 + \delta),$$

we get

$$R = R_0/(1 + \delta/E(B - V)). \tag{7}$$

Equation (7) is plotted in Figure 2 with different values of δ. The best fit for the observations we obtained for $\delta = 0.01$ mag.

Because the VBS is not easy to delineate even when the best choice of unreddened comparison stars has been made, once calibrated the measurement of the colour excess ratio is more straightforward in the sense that no comparison star is needed, but it is necessary to pay attention to specific photometric effects.

The correlation found between the VBS strength estimates and the colour excess $E(B - V)$ seems to confirm that this feature is connected with the grains, but the relative great scatter also at greater values of $E(B - V)$ indicates that there exist local differences in the strength of the VBS. These variations could possibly caused by different grain sizes, as it seems to be indicated from the observations by van Breda and Whittet. In the near future we intend to start an observing program for young open clusters to study this problem in detail. Whereas it is not possible to detect some fine structure within the VBS, we think this method is a good way to get an overview about the general occurrence of the VBS in the Milky Way.

References

Hauck, B. and Mermilliod, M.: 1979, *Astron. Astrophys. Suppl.* **40**, 1.

Hayes, D. S., Mavko, G. E., Radick, R. R., Rex, K. H., and Greenberg, J. M.: 1973, in J. M. Greenberg and H. C. van de Hulst (eds.), 'Interstellar Dust and Related Topics', *IAU Symp.* **52**, 83.

Herbig, G. H.: 1975, *Astrophys. J.* **196**, 129.

Kurucz, R. L.: 1978, *Astrophys. J. Suppl.* **40**, 1.

Kurucz, R. L., Peytremann, E., and Avrett, E. H.: 1974, *Blanketed Model Atmospheres for Early-Type Stars*, U.S. GPO, Washington, D.C.

Mihalas, D.: 1965, *Astrophys. J. Suppl.* **9**, 321.

Oke, J. B.: 1964, *Astrophys. J.* **140**, 689.

Radick, R. R.: 1973, 'An Application of Classical Dispersion Theory to the Interstellar Extinction Problem', Thesis, Rennselaer Polytech. Inst., New York.

Reimann, H.-G.: 1986, in preparation.

Reimann, H.-G. and Pfau, W.: 1986, in preparation.

Reimann, H.-G., Pfau, W., and Stecklum, B.: 1984, *Astron. Nachr.* **305**, 227.

Schild, R. E.: 1977, *Astron. J.* **82**, 337.

Van Breda, I. G. and Whittet, D. C. B.: 1981, *Monthly Notices Roy. Astron. Soc.* **195**, 79.

Whiteoak, J. B.: 1966, *Astrophys. J.* **144**, 305.

Whitford, A. E.: 1958, *Astron. J.* **63**, 201.

York, D. G.: 1971, *Astrophys. J.* **166**, 65.

Zhang, E.-H.: 1983, *Astron. J.* **88**, 825.

Discussion

E. Krügel: Is it necessary that you know the spectra of the standard stars and the stars under investigation in the wavelength range where you measure with an accuracy of several or one hundredths of a magnitude? Does your method allow to measure stars with different reddenings in a manner that the influence of the shape of the spectrum cancels out?

H.-G. Reimann: I think, from the photometric measurements it is impossible to identify any fine structure in the wavelength range of the VBS. But this is not needed, because in our case a comparison star is not necessary. Viewed in another way, this means in the range of B-type stars our method is independent of the spectral type. The only one, which has to be ensured is that the zero-reddening line has to agree very well between both photometric systems. In the case of *uvby* and *UBV* photometry a small systematic difference in the range of 0.005 to 0.01 mag seems to be present. If one takes this into account there are no problems concerning the relation between the strength of the VBS and the colour excess ratio.

P. G. Mezger: Could it be that some of the problems which arise when we look for correlations come from the assumptions that in the interstellar matter with the exception of molecular clouds the dust particles are very well mixed with the gas, that they have all the same composition and that the abundance gradients in the Galaxy do not influence the particle characteristics, but only their total number?

J. M. Greenberg: This is a very general question which we should discuss tomorrow in the afternoon. My comment to the VBS is that it is really an emission feature, not an absorption feature! This has to be kept in mind. It may be associated with a kind of fluorescence, for example.

CARBON MOLECULES AS POSSIBLE CARRIER OF THE DIFFUSE INTERSTELLAR BANDS*

W. KRÄTSCHMER

Max-Planck-Institut für Kernphysik, Heidelberg, F.R.G.

(Received 27 June, 1986)

Abstract. Douglas (1977) proposed linear carbon molecules as one of the carriers of the diffuse interstellar bands. In particular, he suggested that either the species C_5, C_7, or C_9 should produce the most intense interstellar 443 nm band. We have performed laboratory experiments to investigate whether the basic assumption of this hypothesis is fulfilled, namely whether a species of carbon molecules exhibits a strong absorption in the vicinity of 443 nm. For this purpose, we studied the UV–VIS spectra of larger carbon molecules applying the matrix isolation technique. We found that in fact a carbon molecule with such an absorption does exist. A rather preliminary interpretation of our data suggests that this band is produced by the linear molecule C_7. Because the laboratory spectra are distorted by matrix-effects, a conclusive comparison with the interstellar absorptions is not yet possible.

1. Introduction

The spectrum of the diffuse interstellar bands consists of about 40 absorptions of which the strongest feature is located at 443 nm while all the other known lines are situated at longer wavelengths (for details, see, e.g., Herbig, 1975). Even though these bands have challenged many investigators to speculate about possible carriers in the form of grains or molecules (see, e.g., Huffman, 1977), so far no satisfying explanation for their origin could be given. Among the speculations with regard to molecular carriers, the hypothesis of Douglas (1977) appears to be the most appealing, at least from the viewpont of a laboratory spectroscopist, since it is rather specific and thus can be checked by experiments. Douglas's hypothesis rests on two assumptions, namely (a) that a carbon chain molecule C_n with n either 5, 7, or 9 exists having a strong absorption at 443 nm, and (b) that this molecule is one of the carriers of the diffuse interstellar bands. The first crucial test of the Douglas hypothesis is thus to see whether any carbon molecule produces a feature at around 443 nm.

Larger carbon molecules have been studied spectroscopically from the UV to the IR domain by applying the matrix isolation technique (Weltner *et al.*, 1964; Weltner and McLeod, 1964; Kok Wei Chang and Graham, 1982; Thompson *et al.*, 1971; Graham *et al.*, 1976). Absorption features in the VIS, which probably belong to larger carbon molecules were reported by Wdowiak (1980) and Krätschmer *et al.* (1985). A stronger absorption at 447 nm (in an argon matrix) has in fact been detected and in a preliminary analysis of the spectra assigned to the linear C_7 molecule (Krätschmer *et al.*, 1985). In the following, our recent results on the problem are presented.

---

* Paper presented at a Workshop on 'The Role of Dust in Dense Regions of Interstellar Matter', held at Georgenthal, G.D.R., in March 1986.

2. Experimental Procedures

The experimental set-up has already been described (Krätschmer *et al.*, 1985) and thus merely the principle of the approach will be presented here. Carbon vapour is produced by resistive heating of carbon rods in vacuum. At the same time, a stream of argon gas is introduced into the vacuum system. The carbon vapour is deposited onto a cooled sapphire window (held at about 10 K) on which simultaneously the flow of argon condenses. The molecules of carbon vapour are thus trapped in an inert matrix of solid argon. We usually adjusted the deposition rates such that carbon concentrations between 0.1 and 1 mol% with respect to the matrix were achieved. Transmission spectra of the matrix deposited on the cold window were recorded between 200 and 850 nm with 0.1–0.2 nm resolution. In order to produce larger carbon molecules, the matrices were thermally annealed such that the lighter molecules C, C_2, and C_3 which are the main species in carbon vapour could diffuse through the matrix and react with each other, e.g., according to $C_m + C_n = C_{m+n}$. Thus the spectra of the matrices were recorded at increasing degrees of thermal annealing until the argon ice started to sublime (at about 35–40 K). From the growth and decay of features during annealing informations on the structure and composition of the relevant carbon molecules in the matrix may be obtained. More recently we used a similar experimental set-up to study the IR spectra of the carbon molecules between 4000 and 650 cm^{-1}.

3. Results and Discussion

The spectra of carbon molecules obtained immediately after matrix-deposition at 10 K and after increasing degrees of thermal annealing are shown in Figure 1. In the initial 10 K spectrum, the main absorption is known to originate from the $\Sigma_g \rightarrow \Pi_u$ transition of C_3 at 410 nm (Weltner *et al.*, 1964). Sub-features between 410 and 380 nm belong to the same transition. The other absorption band, exhibiting distinct vibrational sub-structure and centered at 247 nm we have tentatively assigned to the allowed $\Sigma_g \rightarrow \Sigma_u$ transition in the linear C_4 molecule (other details in the spectra and the arguments leading to the assignments of features given here can be found in Krätschmer *et al.*, 1985). As an allowed transition, the 247 nm band should be strong; from our spectral data we, in fact, estimated an oscillator strength of about 0.3 for this band. Upon annealing, the C_3 band decreases and the 247 nm feature increases in strength. Other features start to grow as well, namely bands centered at 311, 348, 394, 447 nm and at still longer wavelengths. A close inspection of Figure 1 shows a weak band at about 238 nm decaying at the shoulder of the increasing 247 nm band. This band certainly belongs to C_2 (Milligan *et al.*, 1967). Thus the light species C (which has no spectral features in this wavelength domain), C_2 and C_3 are feeding the build-up of the larger carbon molecules.

As far as we could study the growth rates of the different features, it appears that they, at least up to the 447 nm band, are uncorrelated, indicating that each band originates from a different molecule. An exception is the weak 348 absorption which grows rather

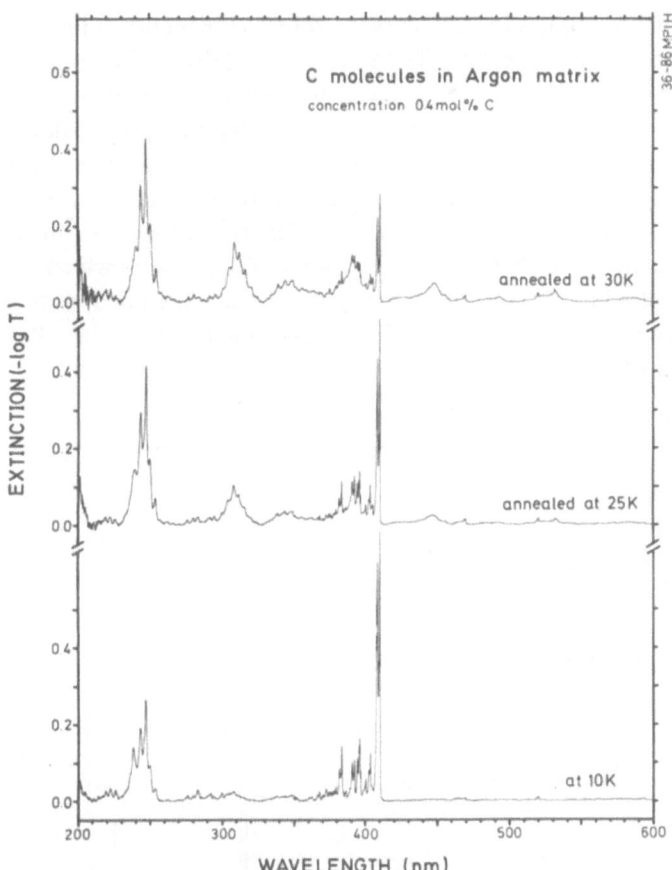

Fig. 1. The UV–VIS spectra of carbon molecules matrix-isolated in solid argon. A smooth background continuum, mainly produced by scattering of the matrix has been subtracted from the spectra. After deposition (bottom spectrum) the matrix has been thermally annealed at the temperature indicated. While the C_3 band at 410 nm decreases upon annealing, other bands are growing in strength. These bands belong to larger carbon molecules which are produced by chemical reactions between the lighter carbon species within the matrix. The absorption at 447 nm originates from a specific larger carbon molecule, probably C_7.

similarly to the adjacent 311 nm band. We suspect that both bands belong to the same, or to two rather similar molecules. We, furthermore, observed that with increasing degrees of annealing, the distinctly growing features are located at longer wavelengths while the bands at shorter wavelengths tend to decay. This suggests that the wavelength position of the main absorption increases with the size of the molecule.

The tentative assignments of the features is based on the assumption that the absorptions observed all belong to the same type, namely the $\Sigma_g \rightarrow \Sigma_u$ transition from the ground state to the adjacent π orbital of the carbon chain molecules C_4 (247 nm band) up to C_7 (447 nm band). If this assignment is correct, the first part of the Douglas hypothesis would be confirmed. However, our recent IR data indicate that the situation is more complex than we so far assumed.

We performed IR studies to gain more insight into the nature of the carbon molecules produced upon matrix annealing. Thompson *et al.* (1971) already performed similar experiments and we essentially reach the same conclusion in our work: the IR spectra are hard to interpret in terms of linear molecules alone; it appears that a considerable fraction of the larger carbon molecules exhibits nonlinear structures. To what extent the nonlinear species contribute to the UV-VIS spectra remains to be investigated.

Even though future research may suggest that the above assignments have to be revised, our studies show that a specific carbon molecule exists which absorbs strongly in the vicinity of 443 nm. Thus in a more general sense the initial part of the Douglas hypothesis seems to be correct, although the carrier molecule may be different from what Douglas assumed.

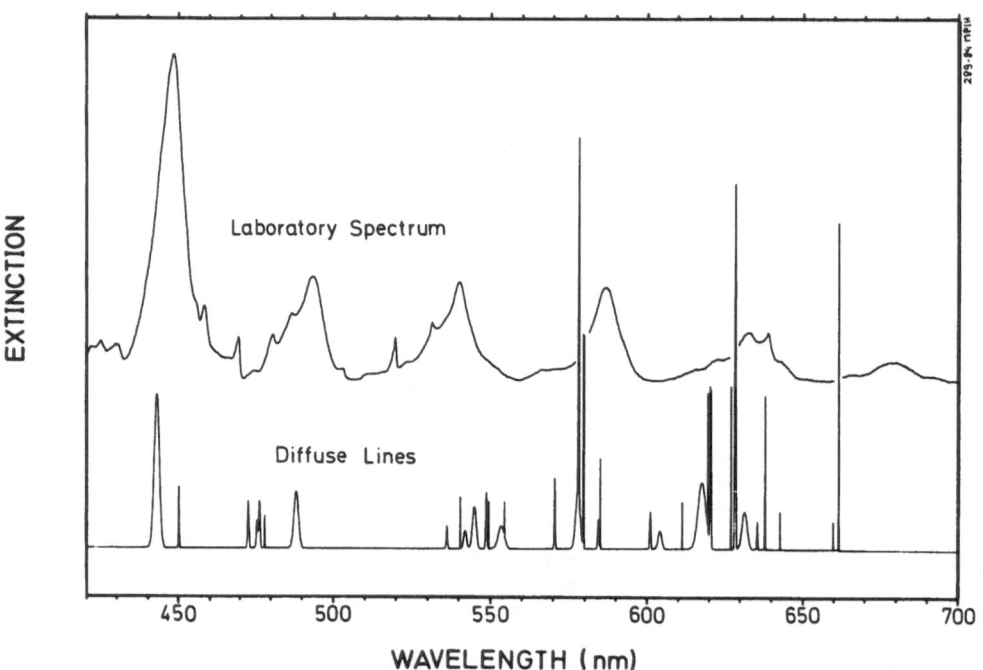

Fig. 2.   Comparison between a laboratory spectrum of larger carbon molecules, trapped in an argon matrix with the spectrum of the diffuse interstellar bands. The laboratory-spectrum is distorted by matrix-effects, which tend to shift and to broaden the absorptions. Even though the general pattern of features in both spectra appears to be quite similar, a more conclusive test of the Douglas hypothesis is required, i.e., by spectroscopy of free carbon molecules.

In order to check how far the carbon molecules can explain the diffuse interstellar band absorptions, a comparison of both spectra is shown in Figure 2. The laboratory spectrum exhibits the absorptions of larger carbon molecules in an argon matrix longwards of 420 nm. It is known that besides other distortions, the bands in a matrix are shifted and broadened by molecule-matrix interactions (see, e.g., Weltner *et al.*,

1964). To get an idea how large the matrix-shifts in argon are, one may compare the position of the C_3 band in argon (410 nm) with that of the free molecule (405 nm) yielding a shift of -5 nm for this particular band. If a similar matrix-shift is assumed for the 447 nm absorption, the band of the free molecule should be located at about 443 nm, i.e., rather close to the position of the strongest diffuse interstellar feature.

Disregarding the broadening, it appears that the pattern of bands present in both spectra is remarkably similar at least in the domain between 440 and 560 nm. The two narrower lines at 470 and 520 nm in the laboratory spectrum probably belong to the linear C_4 molecule (Graham *et al.*, 1976). According to Douglas (1977) this and the other molecules lighter than the carrier of the 443 nm band should photodissociate under interstellar conditions and thus not significantly contribute to the interstellar absorption. The strong bands in the laboratory spectrum located shortwards of 447 nm (and which are not shown in Figure 2) thus should not occur in the interstellar medium since these are very probably produced by lighter species.

The distortions introduced by matrix effects do not yet allow us to check whether the idea of Douglas is, in fact, correct. We believe that it is certainly worthwhile to continue the research on carbon molecules, e.g., try to obtain spectra in less distorting matrix environments. However, the ultimate test of the Douglas hypothesis requires the spectroscopy of free large carbon molecules, a difficult experimental problem.

Acknowledgements

The author thanks N. Sorg and K. Nachtigall for their support in performing the measurements and D. R. Huffman for his stimulating interest in this work.

References

Douglas, A. E.: 1977, *Nature* **269**, 130.
Graham, W. R. M., Dismuke, K. I., and Weltner, W.: 1976, *Astrophys. J.* **204**, 301.
Herbig, E.: 1975, *Astrophys. J.* **196**, 129.
Huffman, D. R.: 1977, *Adv. Phys.* **26**, 129.
Kok Wei Chang and Graham, W. R. M.: 1982, *J. Chem. Phys.* **77**, 4300.
Krätschmer, W., Sorg, N., and Huffman, D. R.: 1985, *Surface Sci.* **156**, 814.
Milligan, D. F., Jacox, M. E., and Abouaf-Marguin, L.: 1967, *J. Chem. Phys.* **46**, 4562.
Thompson, K. R., DeKock, R. L., and Weltner, W.: 1971, *J. Am. Chem. Soc.* **93**, 4688.
Weltner, W. and Mcleoad, D.: 1964, *J. Chem. Phys.* **40**, 1305.
Weltner, W., Walsh, P. N., and Angell, C. L.: 1964, *J. Chem. Phys.* **40**, 1299.
Wdowiak, T. J.: 1980, *Astrophys. J.* **241**, L55.

Discussion

H. J. Habing: Would you see also lines from C_9 and C_{11}, etc.?

W. Krätschmer: In the laboratory spectra, the absorptions longwards of 450 nm probably originate from carbon molecules larger than C_7. The spectral features of the large species may contribute to the diffuse interstellar lines as well.

C. Friedemann: In the laboratory measurements shown you labeled the temperature during the experiments. It is impressive to me that the conditions under which these substances are formed depend very strongly on temperature.

W. Krätschmer: Upon thermal annealing, the lighter molecules can diffuse through the matrix and form larger species by chemical reactions with already existing molecules. Thus the yield of large carbon molecules depends on annealing temperature, i.e., the thermal history to which the matrix-sample is subjected.

W. Friedemann: Does this mean that for the cosmic environment this temperature must have a fixed value?

W. Krätschmer: We did not intend to simulate the formation of larger interstellar carbon molecules. If these species are actually existing in space they are probably formed in processes rather different from what we did in the laboratory. Our aim was to produce carbon molecules by some means to study their spectral features. The method described here is probably the most simple experimental approach. Obviously, larger members of different carbon molecules are formed in this procedure. The main problem is to identify individual molecules by their spectral features. Since not very much is known on the larger carbon molecules, the identification work is highly uncertain.

W. Pfau: At least, we can learn from your laboratory experiments that the relative strength of different lines is dependent on the formation process. Maybe, we should look for the ratios between the strengths of different interstellar lines. Herbig has done this already and he found much better correlations between certain groups of diffuse lines than between all of them. With the improved observational material of today one should study this again.

W. Krätschmer: Such investigation would be rather interesting. I would like to remark that changes in the strengths and positions of lines occur when the molecule is trapped within a matrix. This increases the difficulties to identify possible molecular carriers of the diffuse interstellar lines by matrix-isolation-spectroscopy.

E. Krügel: Are there relations to the well-known chain molecules that happened to be observed in molecular clouds?

W. Krätschmer: The chain molecules observed in molecular clouds are of the type HC_nN, whereas this experiment refers to pure C_n molecules. For larger n, the optical spectra of both types of molecules are not known. Rather than in molecular clouds, the diffuse interstellar lines originate in an environment which is less shielded against destructive UV-photons.

Th. Henning: Do you expect infrared emission lines from these chain molecules?

W. Krätschmer: We studied the infrared absorptions of the matrix-isolated carbon molecules and found no coincidences to the well-known unidentified interstellar IR-emissions. The original idea of Douglas was that the molecules C_n with n either 5, 7, or 9 should be large enough to survive in the interstellar UV flux. The unidentified interstellar IR emissions are associated with regions in which the UV fluxes are rather strong. In order to survive such conditions and to convert the UV energy absorbed into IR emission, molecules significantly larger than C_9 may be required. The emission of the chain molecules considered here should predominantly occur in the 7 to 4.5 µm domain (stretching vibrations) and beyond 100 µm (bending vibrations).

J. Dorschner: In the spectra of some comets there are bands of molecules with C–C bonds. Are there coincidences with your experiments? Are there any correlations?

W. Krätschmer: The only coincidences are C_2 and C_3. I do not know about heavier carbon molecules in comets.

J. M. Greenberg: As far as a possible connection of interstellar and cometary molecules is concerned: C_2 and C_3 seem not to be parent molecules, they are daughter molecules.

W. Krätschmer: This is true. As far as I know, there were also unsuccessful attempts reported to detect spectral features in comets at the positions of the diffuse interstellar lines.

H. J. Habing: This morning it was mentioned that interstellar molecules never have carbon-double-bonds. Don't these molecules have double bonds?

W. Krätschmer: The odd n C_n molecules in fact should have double bonds, the even n ($n > 2$) species should exhibit alternating single-triple-bonds. I think there is no reason to generally exclude the occurrence of carbon-double-bonds in interstellar molecules.

ASTROPHYSICAL INFLUENCES ON THE DIFFUSE
INTERSTELLAR LINES*

W. PFAU

Jena University Observatory, Jena, G.D.R.

(Received 23 June, 1986)

Abstract. After general remarks on the diffuse interstellar lines (DIL) the necessity to select the objects under investigation upon astrophysical criteria is stressed. The dependence on galactic longitude of the relative strength of the DIL at $\lambda = 4430$ Å is given. A possible connection between the strength of DIL originating in stellar aggregates and radiation density is shown and discussed.

The Diffuse Interstellar Lines (DIL) pose the by far oldest unsolved problem of astronomical spectroscopy. The first observations are dated earlier than 1920 and their interstellar origin was unambiguously proven in 1936 by Merrill and collaborators at Mount Wilson Observatory. Nowadays some 40 DIL in the optical spectral region longward of 440 nm are known.

The identification problem points to a complex nature of the absorbers, which, in addition, may show traces of a different history and influences from their environments. A considerable part of the difficulties is purely observational, however, and arises from the weakness of the majority of the features in the spectra of mostly very faint stars. It is easy to speak of central depths of the DIL or even equivalent widths as figures in catalogues, but only those who did reduce such observational material by themselves know about the uncertainties inherent in the data.

One possible way to tackle the problem is correlation analysis. From this side it seems to be very difficult to draw conclusions other than the statement that the absorber is interstellar in origin. The reason is that all the interstellar quantities characterizing the gaseous as well the dusty components are more or less correlating with each other. There is a clear correlation of DIL strength with various colour excesses, e.g., $E(B - V)$. But it is well-known that even if regional effects are taken into account a considerable cosmic scatter remains. From a statistical treatment I could show that the scatter is not normally distributed but rather there is a preference of line strengths being too weak relative to colour excess (1984, unpublished). Obviously the DIL respond to astrophysical conditions otherwise than the colour excesses.

In order to get insight into the nature of the absorber one should restrict correlation analysis to certain types of objects or special regions in the sky. Former investigations of this kind were:

---

* Paper presented at a Workshop on 'The Role of Dust in Dense Regions of Interstellar Matter', held at Georgenthal, G.D.R., in March 1986.

Astrophysics and Space Science **128** (1986) 101–109.
© 1986 by *D. Reidel Publishing Company*

– Stoeckly and Dressler (1964) and Smith *et al.* (1977) investigated DIL absorbed by clouds of different space velocities. The authors are contradictory in their results.

– Zimmermann (1982) and Kumar *et al.* (1982) both treated regions of different H_2 column densities. For the weak data base neither of the papers offers unambiguous conclusions.

– Despite considerable extinction Snow and Wallerstein (1972) do not find DIL originating in circumstellar regions.

– From the work of Snow and Cohen (1974) and others it seems to be certain that dust concentrated in dense clouds produces relatively weak DIL. One example in this respect is the dark cloud near ρ Ophiuchi.

The main problem with this type of treatment is that by far the largest part of the body of data on DIL is very inhomogeneous as to the methods of observation and reduction and for purely observational reasons had been collected according to brightness of the stars and not upon astrophysical significance.

At Jena we have introduced the concept of, as we say, pathological cases. That means that we concentrate interest on cases extremely deviating from the normal behaviour and expect to get insight into the nature of the absorber from a discussion of a broad spectrum of properties in the direction of such objects. We hope that once we can point to some type of special condition prevalent in the direction to weak DIL stars that may explain the exceptional behaviour just there.

From a statistical treatment of a homogeneous sample of data I redetermined the dependence on galactic longitude of the ratio

$$RC = \frac{\text{Strength of DIL 4430}}{E(B - V)}.$$

The material consists of 506 early-type stars for which the Danish group of Rudkjöbing determined a photoelectric index for the strength of DIL 4430 (Baerentzen *et al.*, 1967). In Figure 1 it can be seen that the sector from 60 to 100° galactic longitude is exceptional in the sense of in the mean relatively weak DIL with considerable scatter. This is the direction to Cygnus along the Orion spiral arm. There the reddening law is exceptional, too. This is illustrated by the paper from Lucke (1980), where the slope of the reddening law between 345 and 400 nm is evaluated from photoelectric measurements in the Geneva photometric system. Drawn versus galactic longitude the reciprocal slope shows the same trend as our relative strength of the DIL (Figure 2). The ordinates in the figure mean that along the Orion spiral arm the preponderance of weak DIL is connected with a steeply rising reddening law in the ultraviolet wavelength region near 350 nm. After Deutschmann *et al.* (1976) the same is true for $E(16 - V)/E(B - V)$. This is a statistical behaviour, however, and it would be interesting to see to what extent such a correlation exists for every single star or stellar aggregate in the region. From ANS data Kiszkurno *et al.* (1984) find a small colour excess ratio $E(22-33)/E(B - V)$ in the direction of galactic longitude 80°. That means weak absorption in the 220 nm bump and is the expression of a correlation between the 220 nm bump and the DIL strength as it had been proven to exist by Dorschner *et al.* (1977a, b).

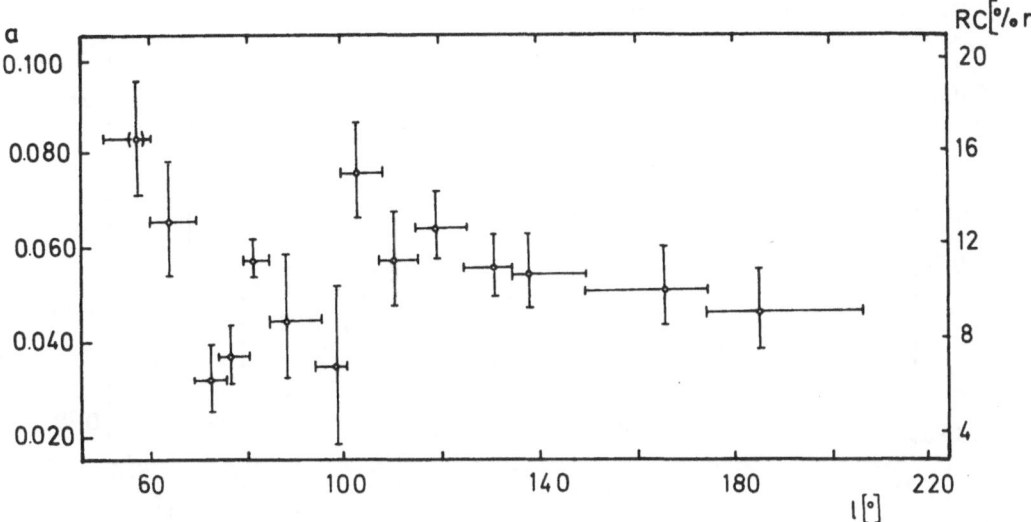

Fig. 1.   Dependence on galactic longitude of RC = strength of DIL 4430/$E(B - V)$. The right scale gives RC as central depth in percent of continuum intensity per magnitude of reddening. The left scale refers to the photoelectric index characterizing the DIL strength and is dimensionless. The data are averaged over intervals in galactic longitude as indicated by the horizontal bars. The vertical bars give the r.m.s. deviation. In most intervals the number of stars is larger than 20, at mean longitude 57°4 there are only seven stars and the data point is considered uncertain.

One of the astrophysical influences in the following discussed in some detail is the interstellar radiation density. The radiation field can in principle influence the absorptivity of the interstellar material in the DIL directly by the impinging photons or via the particle temperature. Changes of the internal structure, surface conditions, and diameter distribution of particles or excitation, ionization, or dissociation processes altering the chemical status of the absorbers are feasible. In this respect the particle temperature determined from infrared observations could also be indicative of a radiative influence.

Regions of high-radiation density are the spatial concentrations of early-type stars in the form of young clusters and OB associations. Aggregates of this type are suited to study a possible connection between radiation field and strength of DIL when most of the reddening along the line-of-sight is due to dust concentrated within the aggregate and thus being under influence of the radiation.

In order to discuss a possible relationship between DIL and radiation density the relative strength of the DIL is formed as

$$RC = \frac{A_{c,\,obs} - A_{c,f}}{E(B - V)_{obs} - E(B - V)_f} ,$$

where the index f marks the foreground contribution determined from surrounding stars as $E(B - V)_f$ and $A_{c,f} = a(l)E(B - V)_f + const.$ The latter relation is the longitude dependence determined earlier.

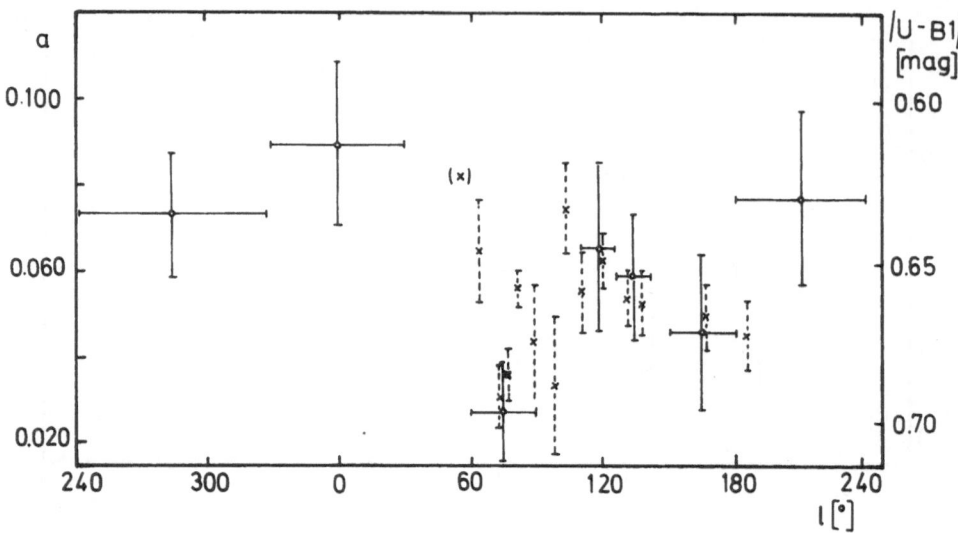

Fig. 2. Comparison of the dependence on galactic longitude of the relative strength of DIL 4430 expressed as the dimensionless quantity *a* (crosses) with the slope of the reddening law between λ345 nm and λ400 nm after Lucke (1980) (open circles). Vertical error bars and horizontal bars indicating the sampling interval in longitude (for the Lucke data only) are drawn.

The radiation densities within the aggregates result from the contributions of embedded and surrounding stars and from the mean interstellar radiation field. The latter one can be neglected because in all cases it is smaller by at least two orders of magnitude. As long as we can only speculate about a physical process influencing the DIL absorber we have in principle to take into account all the radiation incident on the grains, either directly from the stars or after interaction with gas and dust. As a first approximation it is reasonable to assume the hypothetic process affecting the absorbers as proportional to the grain absorption cross-sections and to the stellar radiation flux. In the vicinity of early-type stars both are large near the wavelength of 220 nm and below 50.4 nm. 220 nm marks the position of the well-known absorption bump in the reddening law and the importance of the short-wavelength radiation for dust heating had been demonstrated by Mezger *et al.* (1982) from the correlation of IR excesses and the He II abundance in H II regions. In the investigation presented here the radiation density for both these wavelengths had been considered.

The radiation density resulting at a certain point from a single star at distance *r* can be written as

$$u_\lambda = \frac{4}{c} \, \pi F_\lambda \, \frac{R_*^2}{4r^2} \, \exp\left[-\tau_\lambda(r)\right].$$

In the case of several stars their positions relative to the point, the spatial distribution of dust density, and the radiation transport are of importance. Furthermore, the influence of radiation on every point within the dust column along the line-of-sight should be taken into account. For simplification a model was constructed under the assumption that a number of n stars are evenly distributed within a spherical volume of radius R_{cl} and of constant dust density. The radiation transport is treated with purely forward-scattering particles, which is reasonable if $2\pi a \gg \lambda$. Then we can put the optical depth τ_λ as equal to the optical absorption depth $\tau_{\lambda, \, abs}$. If, finally, the radiation density in the centre of the cloud is taken as representative we get

$$u_\lambda = \frac{1}{c} \; \bar{w}_\lambda \sum_{i=1}^{n} (\pi F_\lambda)_i R^2_{*\,i}$$

with

$$\bar{w}_\lambda = \frac{3}{\tau_\lambda R^2_{cl}} \, [1 - \exp(-\tau_\lambda)] \, .$$

For the reason given above these equations are to be used for $\lambda = 220$ nm as well as for $\lambda = 50$ nm. From the mean interstellar extinction curve by Savage and Mathis (1979) we find $\tau(220) = 8.9E(B - V)$ and from the compilation by Mezger et al. (1982) that $\tau(220) \approx \tau(50)$. We, therefore, put

$$\tau_{abs}(50) = \tau_{abs}(220) = 8.9E(B - V) \, .$$

The model is tied to our real stellar aggregates with the number and types of stars and the total reddening present there.

Altogether 14 stellar clusters with observed DIL strengths are available to look for a possible connection between the DIL data and radiation density. These clusters were discussed as to their dimensions, stellar content, and the amount of extinction produced within or in the immediate vicinity. The result is shown in Figure 3 for $\lambda = 220$ nm. It is quite similar to that for $\lambda = 50$ nm. The ordinate RC is a measure of the number of DIL absorbers per 'ordinary' dust particle producing $E(B - V)$ within the aggregate. The point farthest to the left stands for four clusters with internal reddening $E(B - V)_{cl} < 0.2$ mag. Here most of the dust is under the influence of the mean interstellar radiation field only and accordingly the point is drawn at $u(200 \text{ nm}) = 3 \times 10^{-23} \text{ J cm}^{-3} \text{ nm}^{-1}$ as given by Witt and Johnson (1973). In the diagram the points remain at about the same strength of absorbers with the radiation density rising up to $5 \times 10^{-21} \text{ J cm}^{-3} \text{ nm}^{-1}$. Then the absorbers seem to disappear.

The cluster IC 348 is crucial in respect of the behaviour of DIL strength with rising radiation density. As in case of the Orion Trapezium the DIL are very weak or absent. This is typical for the whole association Per OB2 of which IC 348 is a member. Owing to a relatively large galactic latitude most of the members of the Per OB2 association show foreground reddening only and normal DIL. IC 348, however, is physically connected with interstellar matter, shows signs of recent star formation, and is at the border of the dark cloud Barnard No. 5 for which star formation is also reported

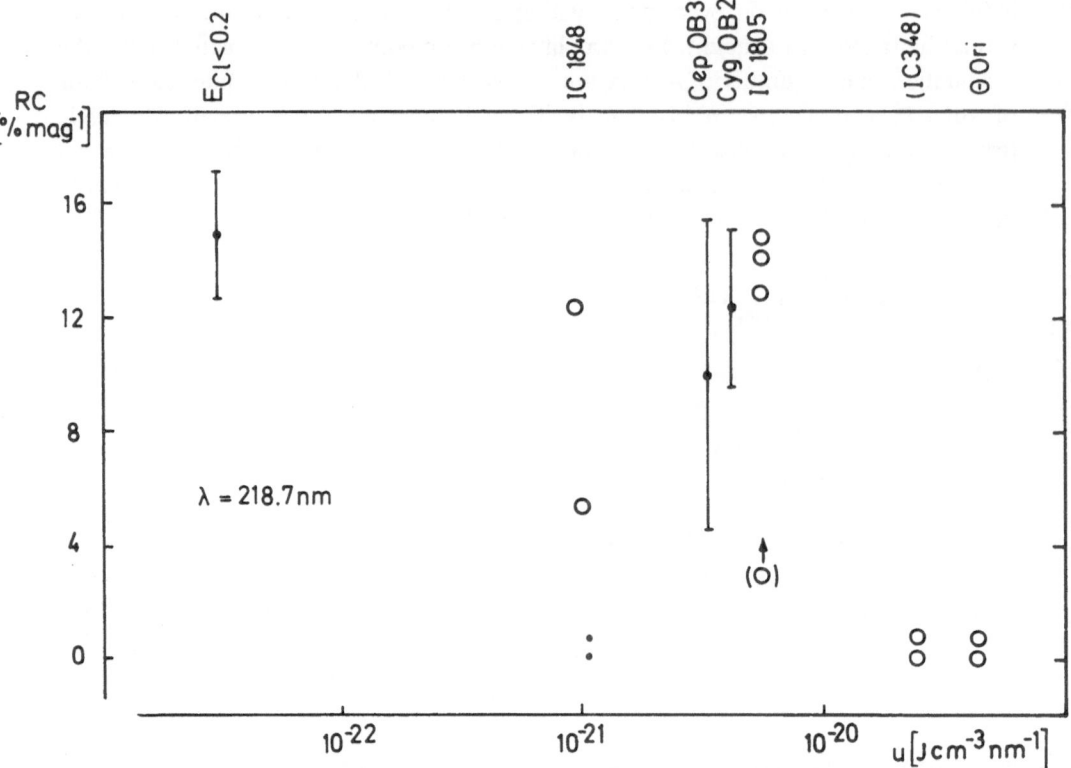

Fig. 3.   Relative strength of DIL 4430 for stellar aggregates drawn versus radiation density at λ218.7 nm.
Open circles are for single stars in clusters, points are mean values (with error bars) for two associations
and for a group of four clusters showing mainly foreground reddening.

(Beichman *et al.*, 1984). In the diagram of Figure 3 IC 348 is shown at
$u = 2.5 \times 10^{-20}$ J cm^{-3} nm^{-1}. From the visible stars alone one gets a radiation density
of $u = 1 \times 10^{-21}$ J cm^{-3} nm^{-1} only (small filled circles in Figure 3). For the cluster
radiofrequency observations at 318 Mpc are available. If interpreted as thermal free-free
radiation from the H II region these observations can be used to determine the
Lyman-continuum flux and thus the stellar content. The procedure follows Churchwell
and Walmsley (1973) who give for the excitation parameter

$$z = n_e^{2/3} R_{cl} = 13.3 \, \nu_{GHz}^{0.03} \, T_4^{0.12} (d_{kpc}^2 S)^{1/3}$$

(T_4 kinetic temperature in 10^4 K, d_{kpc} distance, and S flux density in Jy), and for the
Lyman-continuum flux in photons per second

$$L_{UV} \geq 1.23 \times 10^{56} \beta_2 z^3$$

(β_2 recombination coefficient for $n \geq 2$). In our case we get $L_{UV} \geq 1 \times 10^{46}$ phot s^{-1}.

That demands the presence of additional stars not yet detected and equivalent to one star of spectral type B1V or earlier. The earliest type observed is B5. The addition of one B1V-type rises the radiation density within the cluster considerably. In Figure 3 the cluster is drawn with the radiation density derived from the radiofrequency observations. In the literature radiofrequency observations can be found for IC 1805 and IC 1848, too. Here the radiation density inferred from the visible stars is in accordance with the one based on the radio method.

It can be stated that IC 348 and the Orion Trapezium are two regions which are both lacking in DIL and similar in respect to the radiation density and the number of particles exposed to excessive radiation. Pertaining data are collected in Table I.

TABLE I

Special data for IC 348 and the Orion Trapezium

	IC 348	Orion Trapezium
Cluster radius, R_{cl} (pc)	0.5	1.7
Colour excess, $E(B - V)_{cl}$ (mag)	0.34	0.22
Radiation density at λ220 nm,		
u (J cm^{-3} nm^{-1})	2.5×10^{20}	4.6×10^{-20}
Column densities of particles		
from $E(B - V)_{cl}$, N (cm^{-2})	3.9×10^8	2.5×10^8

As another object showing weak DIL the Herbig Be-type star HD 200775 was investigated. It is embedded in an isolated moderate dark cloud and illuminates the reflection nebula NGC 7023. The total reddening of HD 200775 amounts to $E(B - V) \approx 0.45$ mag, at least half of which is due to surrounding material (Pfau *et al.*, 1986). Without going into detail it is only reported here that in the vicinity of the star we find a number of dust grains per square centimetre comparable to those quoted in Table I at a temperature above the one a particle in free-interstellar space attains. The radiation density is also comparable to those within the two clusters IC 348 and the Orion Trapezium.

From the accordance of data in those examples alone we cannot conclude that it is necessarily the radiation density which is responsible for the weakness of the DIL. Both IC 348 and the Orion Trapezium are indicative of recent star formation and HD 200775 is also a member of a very young aggregate. The clusters may be examples of blister structure in star forming regions. The clue to the weak DIL may, therefore, lie in processes connected with star formation and the history of the interstellar matter present there. In view of our highly evolved observational techniques it is recommended to design programmes to investigate other such interesting regions in detail.

References

Baerentzen, J., Gammelgaard, P., Hilberg, T., Joergensen, K. F., Kristensen, H., Nissen, P. E., and Rudkjöbing, M.: 1967, *J. Obs.* **50**, 83.

Beichman, C. A., Jennings, R. E., Emerson, J. P., Baud, B., Harris, S., Rowan-Robinson, M., Aumann, H.
    H., Gautier, T. N., Gillett, F. C., Habing, H. J., Marsden, P. L., Neugebauer, G., and Young, E.: 1984,
    Astrophys. J. **278**, L45.
Churchwell, E. and Walmsley, C. M.: 1973, *Astron. Astrophys.* **23**, 117.
Deutschman, W. A., Davis, R. J., and Schild, R. E.: 1976, *Astrophys. J. Suppl. Ser.* **30**, 97.
Dorschner, J., Friedemann, C., and Gürtler, J.: 1977a, *Astrophys. Space Sci.* **46**, 357.
Dorschner, J., Friedemann, C., and Gürtler, J.: 1977b, *Astron. Astrophys.* **58**, 201.
Kiszkurno, E., Kolos, R., Krelowski, J., and Strobel, A.: 1984, *Astron. Astrophys.* **135**, 337.
Kumar, C. K., Federman, S. R., and Vanden Bout, P. A.: 1982, *Astrophys. J.* **261**, L51.
Lucke, P. B.: 1980, *Astron. Astrophys.* **90**, 350.
Mezger, P. G., Mathis, J. S., and Panagia, N.: 1982, *Astron. Astrophys.* **105**, 372.
Pfau, W., Piirola, V., and Reimann, H. G.: 1986, *Astron. Astrophys.*, submitted.
Savage, B. D. and Mathis, J. S.: 1979, *Ann. Rev. Astron. Astrophys.* **17**, 73.
Smith, W. M., Snow, T. P., and York, D. G.: 1977, *Astrophys. J.* **218**, 124.
Snow, T. P. and Cohen, J. G.: 1974, *Astrophys. J.* **194**, 313.
Snow, T. P. and Wallerstein, G.: 1972, *Publ. Astron. Soc. Pacific* **84**, 492.
Stoeckly, R. and Dressler, K.: 1964, *Astrophys. J.* **139**, 240.
Witt, A. N. and Johnson, M. W.: 1973, *Astrophys. J.* **181**, 363.
Zimmermann, H.: 1982, *Astrophys. Space Sci.* **84**, 505.

Discussion

P. G. Mezger: I expect that a more sensitive indicator for an increased radiation density
is the grain temperature. If you increase the intensity of the radiation field by a factor
of 100 the grain temperature should go up by a factor of two over what the grain
temperature is in the general radiation field. My intuitive conclusion is: The strength of
the DIL is not connected with star formation, it is connected with the higher temperature
of the grains. The absorber may reside in the mantle of a grain which is evaporated if
the grain temperature grows by a factor of two or so. The idea I wanted to suggest is
that you have taken a rather circumstantial and sophisticated estimation of the radiation
density. You might have taken the grain temperature itself as most important indicator
of the intensity of the radiation field.

W. Pfau: Well, I completely agree with you. The assumption that the temperature is
somehow influencing the particles was really the working hypothesis behind the
investigation. But when I started it there was nor reliable indicator of grain temperature
available for a greater number of interesting regions. Today, with the IRAS data the
situation has completely changed.

J. Palouš: There exists this Gould's Belt among young stars. It would be interesting to
see if the radiation density and DIL are somehow connected with their plane. There
would exist some difference between Gould's Belt plane and the galactic central plane
in this respect.

W. Pfau: I think it is not yet clear whether the more evenly distributed interstellar dust
also shows this tilted structure which you find within part of the early-type stars.

J. M. Greenberg: It is important to look for all these correlations. There are correlations between diffuse band strengths and polarization. An interesting thing is that you have shown here a correlation with galactic longitude. If you look along the Cygnus direction, generally the polarization is weaker and the diffuse bands are weaker, too.

W. Pfau: This region behaves exceptional in several respects. This is of great importance, but it poses severe problems to correlation analysis.

SUBMM/FAR-INFRARED OBSERVATIONS OF COLD AND WARM DUST CLOUDS*

P. G. MEZGER

Max-Planck-Institut für Radioastronomie, Bonn, F.R.G.

(Received 27 June, 1986)

Abstract. Basically, the dust emission from complex regions in interstellar space can be discriminated in emission from cold dust, which is heated by the general interstellar radiation field and in emission from warm dust, which is heated by OB stars. This is demonstrated in the context of the submm/FIR from the galactic disk and center region.

1. Introduction

Diffuse dust emission from our and from external spiral galaxies is observed in the wavelength range from ~ 3 to ~ 1000 μm. At longer wavelengths free-free and synchrotron emission begin to dominate the observed spectra. At shorter wavelengths direct stellar radiation from M and K stars is the principal emission source. The radiation mechanism for dust emission is simple: dust absorbs radiation at shorter wavelengths and reradiates the absorbed energy as quasi-black-body radiation. The spectrum of the dust emission is determined by both dust temperature and the wavelengths dependence of dust opacity at FIR wavelengths. We use the Mathis *et al.* (1977, hereafter to as MRN) dust model as extended to wavelengths beyond 10 μm by Mezger *et al.* (1982), but with optical constants as revised by Draine and Lee (1984). These dust cross-sections per H-atom and corresponding mass absorption coefficients are shown in Figure 1.

For $\lambda \sim 40$ μm these dust cross sections can be approximated by

$$\sigma_\lambda \simeq 7 \times 10^{-21} \lambda_{\mu m}^{-2} \quad cm^2/H\text{-}atom , \tag{2}$$

with $\lambda_{\mu m}$ the wavelength in μm. The corresponding dust emission spectrum is given by a modified Planck spectrum of the form $\nu^2 B_\nu (T_d)$, which integrated over all frequencies yields an intensity $B \propto T_d^6$. Hence, the dust temperature depends on the integrated mean intensity of the ISRF according to

$$T_d^6 \sim \frac{\langle \sigma_{opt, UV} \rangle}{\langle \sigma_{IFR} \rangle} \int_0^x 4\pi J_\lambda \, d\lambda , \tag{1}$$

with $\langle \ldots \rangle$ indicating appropriate average dust cross-sections at optical/UV wave-

---

* Invited paper presented at a Workshop on 'The Role of Dust in Dense Regions of Interstellar Matter', held at Georgenthal, G.D.R., in March 1986.

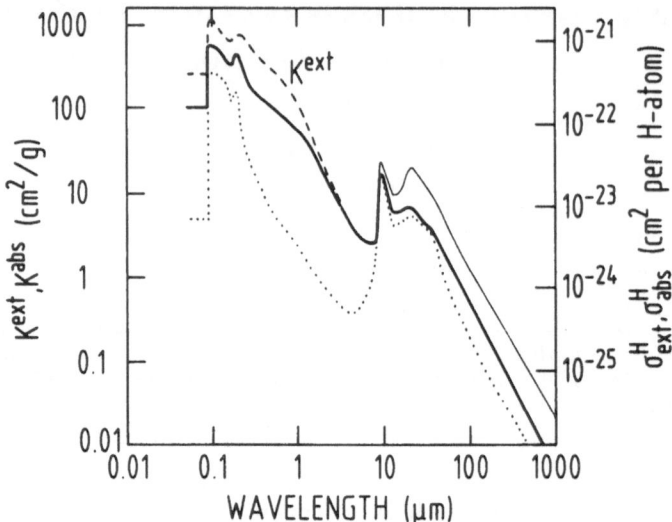

Fig. 1.   The mass absorption (solid curve) and mass extinction (dashed curve) coefficient of dust (in cm²
per g of interstellar matter) as a function of wavelength for the solar vicinity. The extinction curve for silicate
grains is indicated by dots. These curves are based upon the optical constants as given by Draine and Lee
(1984). Shown as thin solid curve for wavelengths $\lambda \gtrsim 5$ μm are the absorption cross sections used earlier.
The scale on the right-hand ordinate gives the corresponding dust cross sections in cm²/H-atom.

lengths (where grains absorb) and FIR wavelengths (where grains emit), respectively.
Since $\langle \sigma_{opt, UV} \rangle \propto a^2$ and $\langle \sigma_{FIR} \rangle \propto a^3$ it follows that for a given radiation density
$T_d^6 \propto a^{-1}$, the grain radius.

In the following we deal with cold dust ($\sim 10-14$ K) which is associated with diffuse
atomic hydrogen and with molecular hydrogen contained in dense clouds, respectively.
This dust is heated by the general interstellar radiation field (ISRF; see Mathis *et al.*,
1983, hereafter referred to as Paper I). We are further dealing with a warm dust
component (30–50 K), which requires for its heating much higher radiation densities
than the ISRF can provide, but which can occur in the vicinity of luminous stars. In
our Galaxy O- and early B-stars have been identified as prime sources of warm dust
emission. Only these stars are luminous and numerous enough to account for the
observed warm dust luminosity of nearly $10^{10} L_\odot$.

2. Observations

At submm wavelengths $\lambda \gtrsim 350$ μm observations are possible in a number of atmos-
pheric windows from high altitude sites ($\gtrsim 3000$ m). One of the best high altitude sites
is the 4200 m high Mauna Kea on the island of Hawaii. At FIR wavelengths
($\lambda \sim 350-20$ μm) observations can only be made with telescopes above the tropopause,
which – depending on geographical latitude and season – is located at altitudes ranging
from 11–14 km above sea level. Appropriate carriers for FIR telescopes are balloons,

astroplanes (such as the NASA Kuiper Airborne Observatory) and satellites (such as the US/Dutch/UK IRAS satellite).

At wavelengths $\lambda \lesssim 150-200$ μm photoconductors are used as detectors. At longer wavelengths Ga doped Germanium detectors, cooled to temperatures as low as 0.3 K, are used for observations of the broadband dust emission. Observations of warm dust emission have been pioneered by Frank Low (University of Arizona) and his students. Observations of cold dust emission have been pioneered by Roger Hildebrand (University of Chicago) and his students.

3. Dust Emission from Spiral Galaxies

Figure 2 shows the dust emission spectrum from the inner part ($R \lesssim 8$ kpc) of our Galaxy. This spectrum is taken from Cox *et al.* (1986; hereafter referred to as Paper II). Note that in this diagram the quantity $\lambda I_\lambda = \nu I_\nu$ is plotted, so that energies per frequency interval can immediately be compared. This galactic spectrum attains its maximum around ~ 100 μm and has a secondary maximum at mid-IR (MIR) wavelengths around ~ 10 μm. This secondary maximum contains between 10 and 15% of the total IR luminosity of the Galaxy and is due to 'hot dust' (several 100 K) emission, which appears to be primarily due to very small dust particles (probably polycyclic aromatic hydrogen = PAH particles as suggested by Léger and Puget, 1984; see also Puget *et al.*, 1985 and Krügel, this issue), which are temporarily heated to temperatures as high as ~ 600 K and which emit most of the absorbed energy via a 'forest of lines' in the MIR. The second contribution to the galactic MIR emission, shown in Figure 2 as dash-dotted curve, was suggested in Paper II as being due to dust in circumstellar shells of M and K giants (notably in so-called OH/IR stars). This contribution was probably over-estimated, since as a result of IRAS observations the number of these stars in the galactic disk has to be revised downward (see Habing, this issue).

In Figure 2 the FIR part of the galactic spectrum ($\lambda \gtrsim 25$ μm) is modelled with the three dust components discussed already in Section 1. Of these only the contribution by cold dust associated with atomic hydrogen is well determined, because we believe to know all three parameters fairly accurately, which enter into the model computation: (i) The distribution of atomic H from $\lambda 21$ cm line observations. (ii) The dust cross sections (see Figure 1 and text). And (iii) the ISRF (see Paper I). The contributions from warm dust (shown in Figure 2 as dotted curve) and very cold dust (shown by crosses) are needed to fit the observed spectrum. Corresponding dust luminosities and associated gas masses as derived in Paper II are given in Table I. Values given in brackets related to very cold dust as derived in Paper II; a 1.7 times larger mass of molecular hydrogen, however, appears to be more realistic (see the thorough discussion by Puget, 1983). To maintain the same total luminosity of very cold dust with this increased mass of H_2 its temperature has to be decreased from 14 to 13 K. The total visual extinction between $R = 10$ and 2 kpc then increases from 11 mag (Paper II) to be about 13–14 mag. Both lower dust temperature and higher visual extinction appear to fit recent observations better.

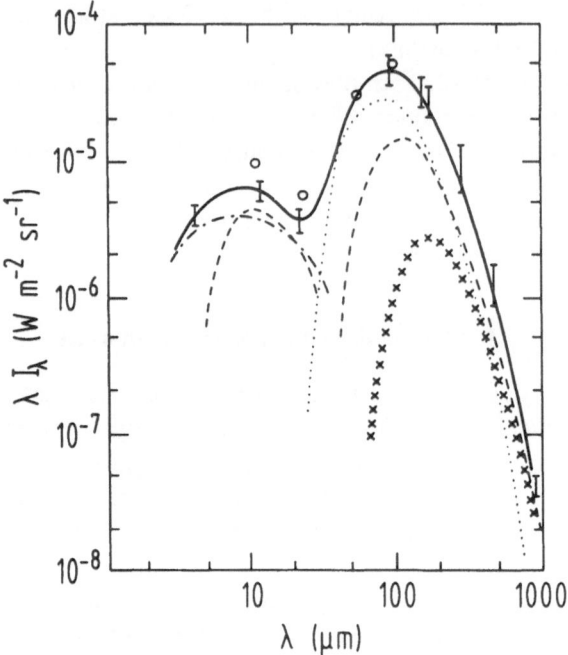

Fig. 2.   Spectrum of the dust emission between 4 and 900 μm from the inner part ($R \lesssim 8$ kpc) of our Galaxy averaged over galactic longitudes 3–35° and latitude $|b| < 1°$. The bars are error estimates of the observed spectrum as compiled by Pajot *et al.* (1986) with modifications as discussed in Paper II. The open circles are preliminary IRAS results. Shown are the following components which fit the observed spectrum: (i) *cold dust* (15–25 K) associated with atomic hydrogen (dashed curve) and *very cold dust* ($\langle T \rangle \lesssim 14$ K) (crosses) associated with quiescent molecular clouds. In both cases, dust grains are heated by the general ISRF; (ii) *warm dust* (30–50 K) heated by O and B stars (dotted curve). The heavy solid line hows the superposition of the computed spectra. Note that the spectrum in the middle IR (the 'MIR shoulder') is the result of the sum of two contributions: the *very small grains* mixed with atomic hydrogen and heated by the general ISRF, and normal grains heated to 250–450 K by *M giants with heavy mass loss*. The spectra of these two components are shown as dashed curve and dash-dotted curves, respectively.

TABLE I

Luminosities of and gas masses associated with the three dust components, which contribute to the FIR emission from the galactic disk within $R \lesssim 8$ kpc

Component	T_d/K	$L_{IR}/L_\odot$	Heated by	Assoc. with	$M_H/m_\odot$
Very cold dust	13	5E8	ISRF	molec. H_2	1.5E9
	(14	5E8			9E8)
Cold dust	15–20	4.4E9	ISRF	atomic H	6E8
Warm dust	30–50	7.3E9	OB stars	$H^+ (\sim \frac{2}{3})$; $H_2(\sim \frac{1}{3})$	4E7

Extended model computations show that O- and early B-stars are the most likely sources of heating of the warm dust (Paper II). We estimate that these stars contribute in the ratio 2 : 1. A comparison of hydrogen masses associated with very cold and warm dust shows that the latter accounts for only a few percent of the total mass of molecular hydrogen within the solar circle. Warm dust clouds thus constitute only a small fraction of all dust clouds in the galactic disk. Since gas and dust are thermally coupled only at relatively high gas densities ($n_H \gtrsim 10^6$ cm^{-3}), warm dust clouds must not necessarily be identical with warm molecular clouds, where 'warm' in this case refers to the kinetic gas temperature of the clouds as derived from molecular line observations.

While the total luminosity of very cold dust (or cold dust clouds, respectively) amounts to only $\sim 10\%$ of the luminosity of cold dust associated with diffuse atomic hydrogen, emission from very cold dust dominates the spectrum at $\lambda 1200$ µm. At this wavelength even the densest of the Giant Molecular Clouds (GMCs) are optically thin and the flux density is $S_{250} \propto T_d$ (note that subscripts in flux densities refer to frequencies in GHz). This fact allows pretty reliable total mass estimates of atomic and molecular hydrogen, even if the temperatures of cold and very cold dust are not too well known.

Observed dust emission spectra of external spiral galaxies are pretty similar to that of our Galaxy. The two-component structure of the FIR part of the spectrum is seen even more clearly. By fitting modified Planck spectra to the observed FIR spectra we derive dust temperature ranging from 13–17 K for cold dust and from 27–49 K for warm dust (Chini *et al.*, 1986; see also Chini, this issue). Here cold and very cold dust are lumped together in one spectral feature, whose colour temperature represents a mean value between the actual temperature of cold and very cold dust. One might argue about the relation between colour temperature derived by fitting a modified Planck spectrum and the physical dust temperature, since the MRN dust model has a size distribution $f(a) \propto a^{-3.5}$ and we have noted earlier that $T_d \propto a^{-1/6}$. This relation has been discussed in Mezger *et al.* (1982) and it was found, that the colour temperature agrees in most practical cases pretty well with the corresponding mean dust temperature.

4. Observations of the Sagittarius Giant Molecular Cloud

The interstellar matter in the Galactic Center region is a complex of ionized and neutral gas which extends to distances of about $|l| \lesssim 1°$ in longitude. The dynamical center of the Galaxy, located inside Sgr A, is determined by the gravitational center of an extended cluster of M giants. Close to this gravitational center appears to be located a supermassive object of $\sim 1.5 \times 10^6 \, m_\odot$, which is encompassed by the H II region Sgr A West, of size $\Delta l \times \Delta b \sim 82'' \times 66''$ and free-free flux density $S_{22} \simeq 27$ Jy (Mezger and Wink, 1986; and references quoted therein). Towards negative galactic latitudes extends a belt of Giant Molecular Clouds (GMCs) over $|l| \lesssim 0°2$, the most conspicuous of which are labelled (according to their positions in l, b) M$-0.13-0.08$ and M$-0.02-0.07$, or (according to their radial velocities as derived from molecular line observations) the '20 km s^{-1}' and '40–50 km s^{-1}' clouds, respectively. These clouds have been extensively mapped in various molecular lines; they were also the first cold clouds mapped in dust continuum emission of $\lambda_{eff} = 540$ µm by Hildebrand *et al.* (1978). Subsequently, the physical state of these clouds and their position relative to the galactic center has been investigated using various ammonia emission lines at $\lambda 1.3$ cm (Güsten *et al.*, 1981), H_2CO $\lambda 6.2$ cm and $\lambda 2.1$ cm absorption lines (Güsten and Henkel, 1983; and references quoted therein) and multiple rotation transitions of HC_3N at cm and mm wavelengths (Walmsley *et al.*, 1986), respectively. Most recently the dust emission of these clouds

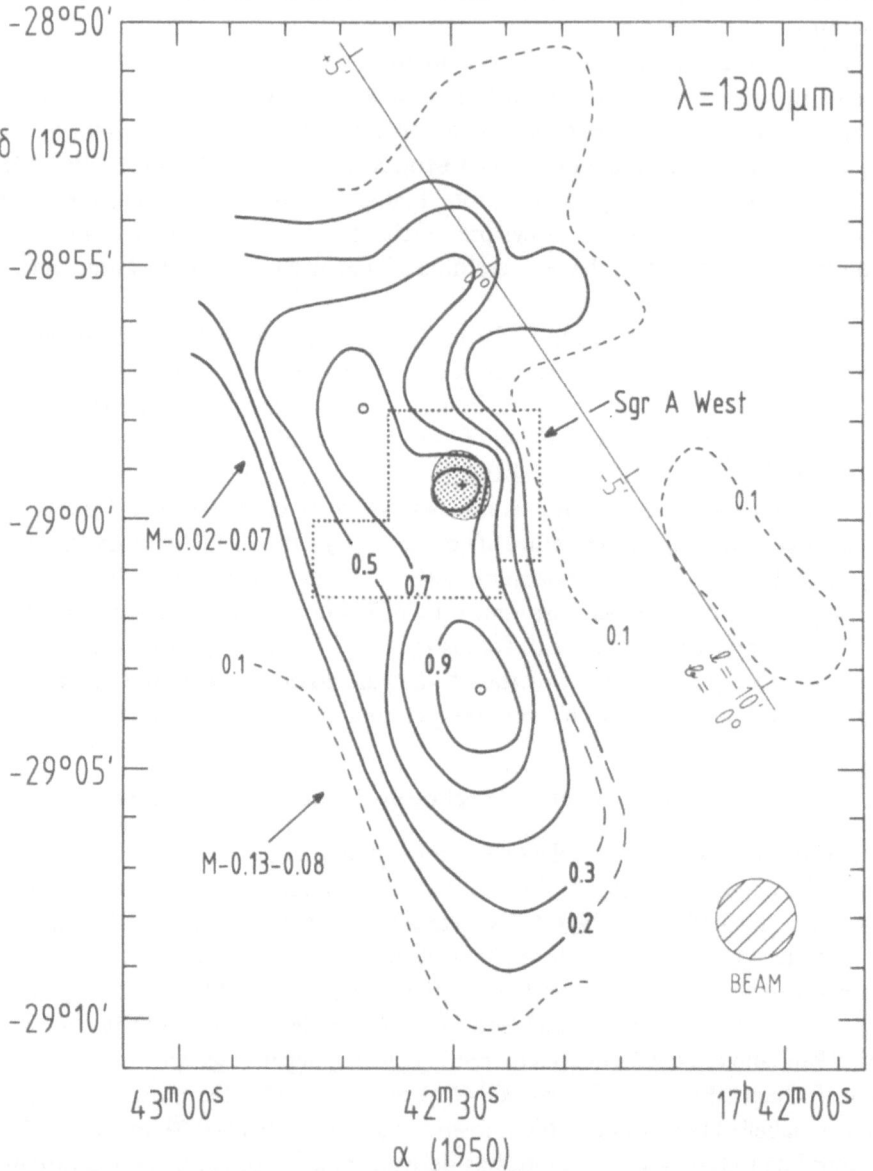

Fig. 3. $\lambda 1300$ μm map of the GMCs M−0.13−0.08, M−0.02−0.07 and the H II region Sgr A West. Open dots indicate the positions of dust emission peaks. The stippled area indicates the solid angle subtended by Sgr A West, the cross refers to the position of the compact non-thermal source Sgr A*. Contours are in units of 25.8 Jy in a 90″ beam, the peak flux density observed with the telescope centered on Sgr A*. Sampling intervals are $\Delta \delta = 2'$, $\Delta \alpha = 1'$ (Mezger *et al.*, 1986).

at $\lambda 1300\ \mu m$ has been mapped using the MPIfR ³He cooled germanium bolometer in the 3-m Infrared Telescope Facility (IRTF) on Mauna Kea. Details may be found in Mezger *et al.* (1986).

Figure 3 shows the contour map representation of the observed dust emission. This map deviates from that obtained by Hildebrand *et al.* (1978) in that it shows strong free-free emission at the position of the H II region Sgr A West. At 540 μm, where Hildebrand *et al.* made their observations, the contribution of free-free emission (which scales with frequency as $S_{ff} \propto v^{-0.1}$) becomes negligible as compared to the dust emission, whose flux density scales with $S_d \propto v^3-v^4$, depending on dust temperature.

Figure 4 shows an overlay of a $\lambda 55\ \mu m$ contour map obtained by Dent *et al.* (1982). This emission represents the distribution of warm dust and, hence, of regions of recent OB star formation. No warm dust emission is observed in the southern part of the GMCs. In the direction of the galactic center we see along the line-of-sight a mixture

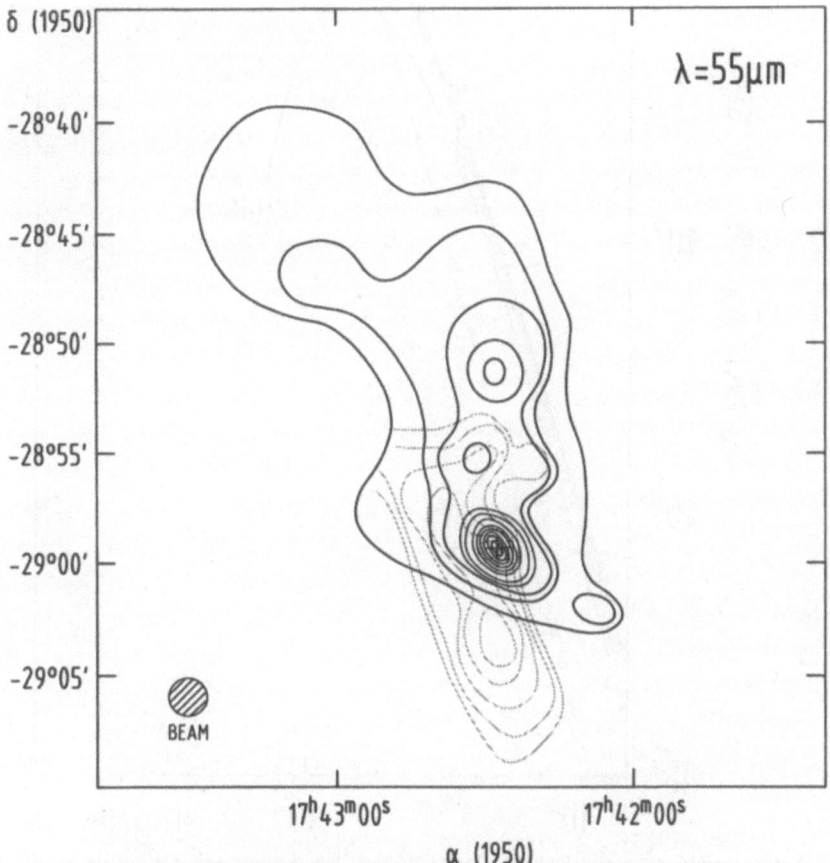

Fig. 4.   Overlay of the $\lambda 1300\ \mu m$ contour map shown in Figure 3 (dashed contours) on the $\lambda 55\ \mu m$ contour map of Dent *et al.* (1982). Contours are in units of 1000 Jy in a 60″ beam. The $\lambda 55\ \mu m$ map has been slightly shifted in α so that the positions of Sgr A* in the two maps coincide.

of both free-free emission and emission from warm and cold dust. Figure 5 shows the emission spectrum within the inner 30″ surrounding the galactic center. It can be decomposed into contributions from warm and cold dust. The cold dust represents a

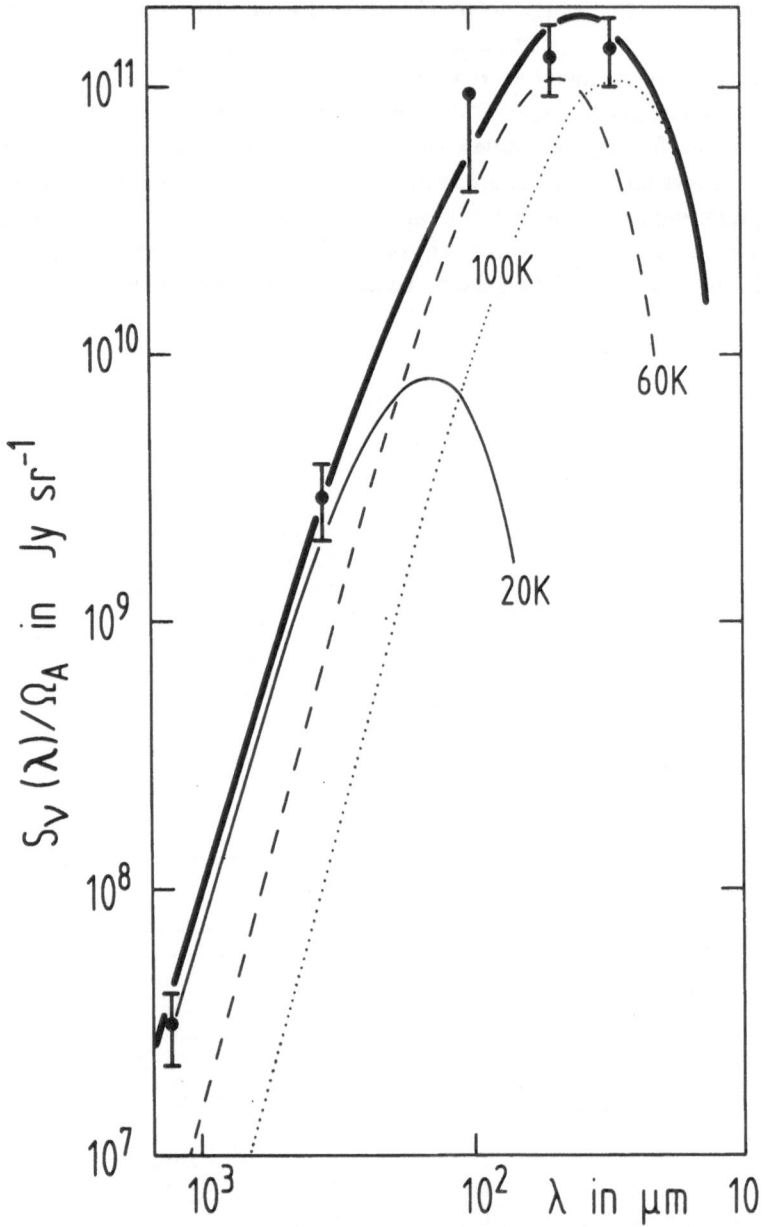

Fig. 5.   Surface brightness of the dust emission at five wavelengths derived for the inner 30″ of Sgr A West (heavy solid curve). The thin curves relate to a model fit with dust at three different temperatures (as labelled) and with optical depths, expressed (with increasing dust temperatures) in visual magnitudes of 140, 7.4, and 0.6 mag.

dust column density which corresponds to a visual extinction of $A_v \sim 140-170$ mg. If this part of the GMC were located in front of Sgr A West the galactic center would be unobservable for wavelengths $\lambda \lesssim 40\,\mu m$.

Figure 6 shows an overlay of the $\lambda 1300\,\mu m$ dust emission contours on a contour map representing the $\lambda 1.3$ cm NH$_3$ emission. There is an excellent correlation between molecular line and cold dust emission apart from Sgr A West, whose free-free emission is superimposed on the cold dust emission. However, while the cold dust has an average

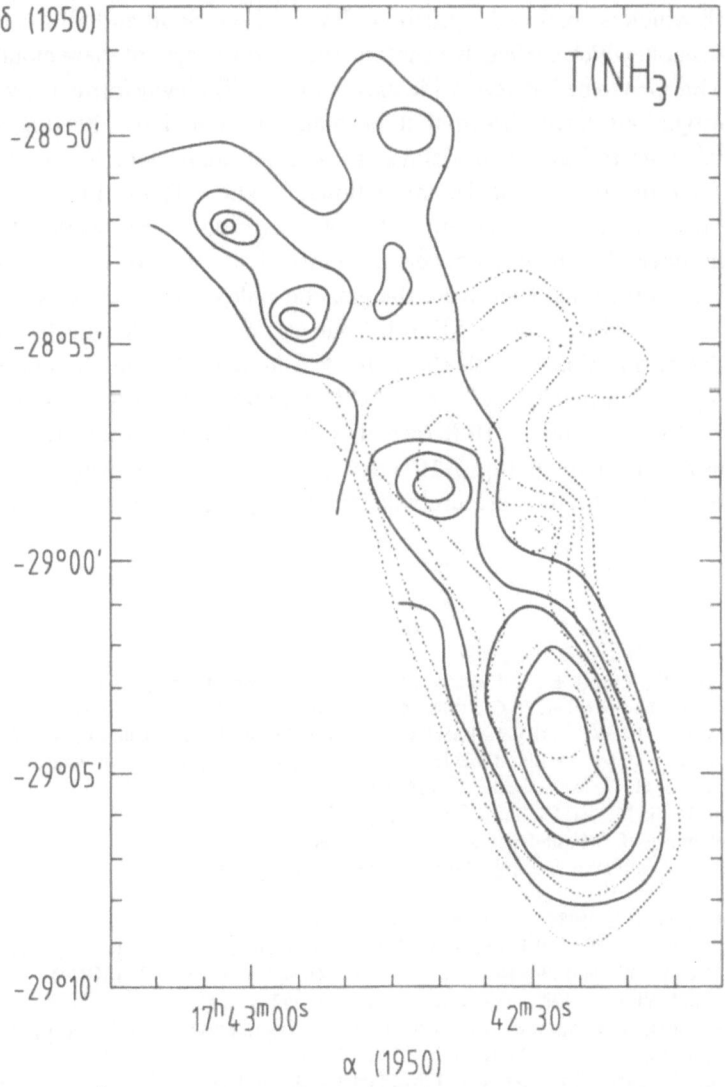

Fig. 6. Overlay of he $\lambda 1300\,\mu m$ contour map shown in Figure 1 (shaded contours) on a contour map of the NH$_3$ (1, 1) transition by Güsten *et al.* (1981). Contours are in units of brightness temperature in a 40″ beam.

temperature of ~ 20 K, the kinetic gas temperature, as derived from various molecular transitions, exhibits temperatures well in excess of 60 K. Sources of heating of both dust and gas in these clouds are not yet well understood

5. The Heating of Dust in Cold GMCs

'Cold molecular clouds' in our definition means that no massive stars are formed inside these clouds, while lower mass ($m \lesssim 3\, m_{\odot}$) star formation may occur throughout the cloud. There are three basic heating sources for the dust contained in cold clouds: (i) The ISRF which is absorbed in the outer layers of the cloud and thus is transformed into FIR emission, which can easily penetrate the densest cores of these clouds. (ii) Field stars, which happen to be located inside these clouds. (iii) Newly formed low-mass stars.

Our investigations have shown that normally the two latter heating sources are negligible. We refer to Paper I for details. Two major changes have occurred since we published the results of our model computations: (i) The FIR opacities of silicate grains have been decreased by about an order of magnitude (see Section 1). (ii) IRAS observations have shown that noticeable diffuse dust emission at FIR wavelengths comes from all directions, even from the galactic poles. This increases the integrated (over the sky) intensity of the IRSF in the galactic disk at FIR wavelengths by more than an order of magnitude. Both effects tend to increase the dust temperature inside GMCs. It appears now that even in the densest cores of molecular clouds the dust temperature rarely falls below 10 K (see also Puget, 1983). In those rare cases where gas densities get high enough ($n_{\mathrm{H}} \gtrsim 10^6$ cm^{-3}) that gas and dust will become thermally coupled, the gas kinetic temperature will rapidly approach the dust temperature.

References

Chini, R., Kreysa, E., Krügel, E., and Mezger, P. G.: 1986, *Astron. Astrophys.* (submitted).

Cox, P., Krügel, E., and Mezger, P. G.: 1986, *Astron. Astrophys.* **155**, 380 (Paper II).

Dent, W. A., Werner, M. W., Gatley, I., Becklin, E. E., Hildebrand, R. H., Keene, J., and Withcomb, S. E.: 1982, in R. Riegler and R. D. Blandford (eds.), *The Galactic Center*, AIP Conf. Proc., p. 33.

Draine, B. and Lee, H.: 1984, *Astrophys. J.* **285**, 89.

Güsten, R. and Henkel, C.: 1983, *Astron. Astrophys.* **125**, 136.

Güsten, R., Walmsley, C. M., and Pauls, T.: 1981, *Astron. Astrophys.* **103**, 197.

Hildebrand, R. H., Whitcomb, S. E., Winston, R., Steining, R. F., Harper, D. A., and Moseley, S. H.: 1978, *Astrophys. J.* **219**, L101.

Léger, A. and Puget, J.-L.: 1984, *Astron. Astrophys.* **137**, L5.

Mathis, J. S., Mezger, P. G., and Panagia, N.: 1983, *Astron. Astrophys.* **128**, 212 (Paper I).

Mathis, J. S., Rumpl, W., and Nordsieck, K. H.: 1977, *Astrophys. J.* **217**, 425 (MRN).

Mezger, P. G. and Wink, J.: 1986, *Astron. Astrophys.* **157**, 252.

Mezger, P. G., Chini, R., Kreysa, E., and Gemünd, H. P.: 1986, *Astron. Astrophys.* (in press).

Mezger, P. G., Mathis, J. S., and Panagia, N.: 1982, *Astron. Astrophys.* **105**, 372.

Pajot, F., Boissé, F., Gispert, R., Lamurre, J. M., Puget, J.-L., and Serra, G.: 1986, *Astron. Astrophys.* **157**, 393.

Puget, J.-L.: 1983, in Lucas, Omont, and Stora (eds.), 'Birth and Infancy of Stars', *Proc. of XLI Les Houches Meeting*, North-Holland, Amsterdam, p. 77.

Puget, J.-L., Léger, A., and Boulanger, F.: 1985, *Astron. Astrophys.* **142**, L19.
Walmsley, C. M., Güsten, R., Angerhofer, P., Churchwell, E., and Mundly, L.: 1986, *Astron. Astrophys.* (in press).

Discussion

Th. Henning: Have you any evidence about the distribution of cold and warm clouds in our Galaxy?

P. G. Mezger: No, not on a large scale. You would have to compare observations of the dust emission at 300 µm, which have the same angular resolution as the IRAS data, with the 60 µm IRAS observations. This is not yet done because of a lack of submm data. There are balloon observations available between 150 and 300 µm made by a NASA group with an angular resolution of 10'. This is not good enough. At present one can compare maps of special regions like, e.g., the galactic centre, where measurements of high angular resolution and at different wavelengths are available. One then usually observes a mixture of cold and warm clouds. In most cases these warm clouds are centres of activities of small angular sizes which are embedded in the much more extended cold dust clouds.

W. Pfau: My question is in the same direction as the one by Th. Henning. In one of your diagrams you showed the IR spectrum of the Galaxy which was averaged over all galactic longitudes. Don't you expect some differences in the relative amounts and relative contributions of cold and warm dust in the directions toward the galactic centre and the anticentre, for instance?

P. G. Mezger: This average which I had shown to you was integrated over 30 deg in galactic longitude and 2 deg in latitude. So the spectrum relates to the inner part of the Galaxy which roughly corresponds to the emission from all dust inside 8 kpc.

J. Palouš: In case of observations of cold dust of 14 K in the FIR, is it possible to distinguish this radiation from 3 K background radiation?

P. G. Mezger: This background emission is homogeneous and, therefore, acts like a 3 K receiver noise. Sources which we observe are superimposed on this emission. On a smaller scale we encounter a similar situation when we observe clouds in the direction towards the galactic centre. From this direction there comes a very extended and uniform emission. We, on the other hand, are chopping with the chopper throw of 5', and thus measure the difference between the extended background and the source of some arc min extension, which we are observing.

J. Palouš: Some people try to distinguish inhomogeneities in the cosmos background radiation. Does the radiation from the clouds influence such efforts?

P. G. Mezger: Yes, you have to be very careful to avoid this type of confusion. For example the MIT group, which is interested in measurements of the cosmic background, surveyed the galactic plane at 400 µm and determined the contribution from dust emission. Or Kardashov and Sagdeev, in their survey of the sky at 8 mm, found the galactic plane standing out because of its free-free and dust emission. Obviously, you have to stay away from the galactic plane if you want to investigate the microwave background emission.

J. M. Greenberg: I have some comments on the temperature distribution of dust inside dense clouds of interstellar gas. Usually the calculation of UV radiation penetration rests on the assumption of an average extinction in the molecular cloud, where many small particles may dominate the extinction at UV wavelengths. There are good physical reasons to believe, however, that these small particles may tend to disappear in the molecular clouds (because they accrete ice mantles) and then the UV radiation could penetrate much deeper into the cloud as compared with normal calculations. This fact should be kept in mind that the UV radiation intensity in the model computations by Mathis *et al.* is a lower limit. Another source of UV radiation is produced by penetration of cosmic rays into the clouds. A third source of UV radiation are the stellar winds. These extra factors should be at least attempted to be incorporated into model computations of the radiation field in clouds.

P. G. Mezger: This is certainly correct. I do not think, however, that these somewhat secondary processes are terribly important for the heating of grains. But additional UV photons are certainly important for your theory of exploding ice mantles.

J. M. Greenberg: No, I disagree. If a good deal of radiation, which hits the grains, is in the form of UV photons then it would be significant for calculations of the grain temperature to know, how much energy absorbed is from the visible and from the UV.

P. G. Mezger: I think it is necessary to make more model computations but always to state clearly which assumptions were made. To compare the results with observations that's the only way to improve the situation.

J. Dorschner: Do you think that 10 K is the absolute lower limit of dust temperature or are there any places where the temperature is lower?

P. G. Mezger: We extended our calculations for clouds to depths as far as about 40 mag extinction in the visual, measured from the surface of the cloud to its centre. Using the revised dust cross sections for silicate particles the lowest temperatures would amount to 7 or 8 K. J.-L. Puget made analogous but less detailed computations and found even somewhat higher temperatures. It appears, in fact, that for some reasons the dust temperature inside dense clouds is rather uniform and amounts to values between, say 10 and 15 K. Part of the heating is due to the strong FIR radiation field. Other sources of heating can only be guessed.

Another problem of interest relates to the interaction of gas and dust particles through collisions. There exist calculations concerning the coupling of gas and dust, in which the 'accomodation coefficient' plays an important role, but unfortunately its value is not very well known. Computations made with the assumption of an accomodation factor of order unity show that with molecular densities of order 10^5 cm^{-3} a temperature equilibrium between dust and gas should be reached. Such an equilibrium would mean that the gas attains the temperature of dust (and not *vice versa*) because at the same temperature the cooling rate of dust is 100 times higher than that of gas. Now, in the case of the galactic centre cloud, the temperature of the dust appears to be rather uniform and about 20 K. The kinetic temperature of the gas is between 100 and 300 K, i.e., is very high. At present we do not yet know the source of heating for neither gas nor dust. The mean gas density is in the order of 10^4 hydrogen molecules per cm^3 but is locally certainly an order of magnitude higher. So apparently at these densities there exists not yet a temperature equilibrium and coupling between gas and dust. This may allow some guesses of the numerical value of the accomodation coefficient. Actually, there are only a few known cases (such as the BN object in OMC I) where actually gas and dust appear to have the same temperature. But then the gas densities are already as high as 10^6 to 10^7 molecules cm^{-3}.

W. Pfau: Is a certain minimum temperature of particles maintained by chemical reactions or similar processes?

P. G. Mezger: I do not know because we considered only radiative heating of the particles. Therefore, I cannot tell how, for example, grain mantle explosions suggested by J. M. Greenberg would contribute to the heating of the particles.

J. M. Greenberg: The total energy of these reactions is extremely small.

P. G. Mezger: And in general chemical reactions of molecules which meet on the surface of grains?

J. M. Greenberg: The only reaction of some importance is the hydrogen molecule formation.

I just wonder if there is not another component that is producing the higher temperature. We had done calculations with core-mantle particles. We get low temperatures. So it is difficult to understand why the temperature of the larger grains is higher. I think there may be two grain components: one which achieves the very low temperature which we expect; and another component, which is not yet included in the calculations and which gives the unexpected higher temperature.

P. G. Mezger: At least in the solar vicinity one assumes the interstellar radiation field (ISRF) in the optical and UV part to be reasonably well known. Observations of the dust emission of globules are compatible with a colour temperature of ~ 16 K, in agreement with model computations for this ISRF. However, this ISRF cannot explain mean dust temperatures of ~ 14 K observed in GMCs. What amazes me most is the fact that, with this dust temperature, the luminosity-to-mass ratio of interstellar matter expressed in solar units comes out to about 0.1. This looks to be amazingly high.

ON THE DUST AND GAS ASSOCIATED WITH
SHARPLESS 252*

L. HAIKALA

Max-Planck-Institut für Astronomie, Heidelberg, F.R.G.

and

Observatory and Astrophysics Laboratory, Helsinki, Finland

(Received 27 June, 1986)

Abstract. The stellar cluster and the massive molecular cloud associated with the H II region Sharpless 252 have been studied by means of multicolour polarization and molecular line measurements. The average wavelength of the maximum polarization λ_{max} and polarization efficiency for the cluster stars are similar to the values observed for the nearby field stars. Two local maxima lying only 2' apart were found in the molecular cloud core in CO, NH_3, and HCO^+. The excitation conditions and radial velocities associated with the maxima are different.

1. Introduction

The H II region Sharpless 252 (diameter $0°5$) lies in the galactic plane in the direction of the galactic anticentre at a distance of 2.2 kpc. The main ionizing source of the region is the O6.5V star HD 42088. Two stellar clusters (S 252main and S 252small) lie in the direction of the H II region (Pismis, 1969). The ages of these clusters are of the order of 1 (S 252main) and 6 million years (S 252small) and they are both associated with S 252 (Haikala, 1986). Six small size thermal radio sources (S 252A, B, C, E, and F) were detected by Felli *et al.* (1977) at 1415 and 4995 MHz. These sources coincide with small nebulosities seen projected on the diffuse and extended H II region. S 252C is the strong ionization front at the interface of the molecular cloud and the H II region. The molecular cloud (extension $1°0–1°5$ associated with S 252 was mapped by Lada and Wooden (1979) (here after LW) in ^{12}CO and ^{13}CO using 3' spatial resolution. LW found also an H_2O maser near S 252A.

2. *UBVRI* Polarimetry

The observations were made in November 1982 with the two-channel photometer-polarimeter (Proetel *et al.*, 1975) at the 1.23 m telescope in Calar Alto. Early-type stars, 37 in the direction of S 252 and 60 within two degrees of it, were observed in the approximate Johnson BV and *UBVRI* bands. These observations have been discussed in Haikala (1986) together with the photometric data.

The observed polarizations of field stars lie mostly between 1 and 3%. The

---

* Paper presented at a Workshop on 'The Role of Dust in Dense Regions of Interstellar Matter', held at Georgenthal, G.D.R., in March 1986.

Astrophysics and Space Science **128** (1986) 125–133.
© 1986 *by D. Reidel Publishing Company*

polarizations can be well fitted with the empirical interstellar wavelength dependence of polarization (Serkowski, 1973). The normalized field star polarizations are plotted in Figure 1(a) together with the 'Serkowski curve'. The average of the wavelengths where the polarization reaches its maximum λ_{max} for these stars is $0.54 \pm 0.02\mu$ (error of the mean, ten stars). This is a typical value for normal interstellar polarization. The polarization efficiency (the ratio of the polarization degree and visual extinction) is high in the general direction of the galactic anticentre (Hiltner, 1956). The present observations are in agreement with this (see Figure 1(b)). This implies that one is observing nearly perpendicular to the galactic magnetic field and that in this direction the direction of the field does not essentially change within 2 kpc from the Sun.

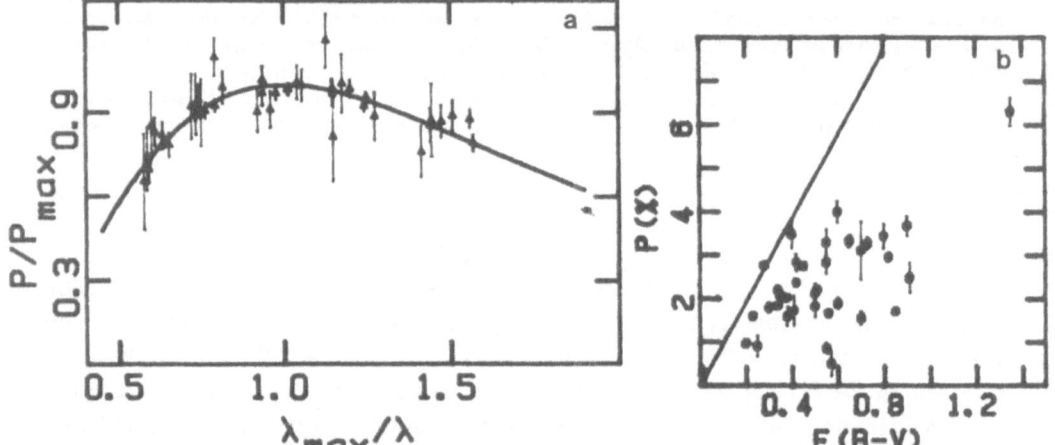

Fig. 1a–b.   (a) Normalised field star polarization. The continuous curve is 'Serkowski's law'. (b) Observed visual polarization and the colour excess of the field stars. The empirical upper limit $P = 3E(B - V)$ is also shown.

The observed polarizations of stars associated with S 252 range between 2 and 6%. Two exceptionally high polarizations, 8.9 and 12%, are observed for the probable ionization sources of S 252A and C, respectively.

The polarizations of the cluster stars do not show any significant deviations from the 'Serkowski's law' (Figure 2(a)). The average λ_{max} for these stars is $0.53 \pm 0.01\mu$ (nine stars) which is the same as for field stars. The observed polarization efficiency for the stars associated with S 252 is as high as for the field stars (Figure 2(b)).

It was shown by Serkowski et al. (1974) (hereafter referred to as SMF) that the three colour excess ratios $E(V - I)/E(V - R)$, $E(V - K)/E(V - R)$, and $E(V - K)/E(B - V)$ correlate well with the wavelength of maximum polarization. Using the relation $R = A_V/E(B - V) = 1.1 \times E(V - K)/E(B - V)$ given by Carrasco et al. (1973) and the observed correlation between λ_{max} and $E(V - K)/E(B - V)$ SMF calculate that R and λ_{max} are related by $R = 5.5\lambda_{max}$.

The colour excess ratio $E(B - V)/E(V - R)$ and λ_{max} can be calculated for eight stars

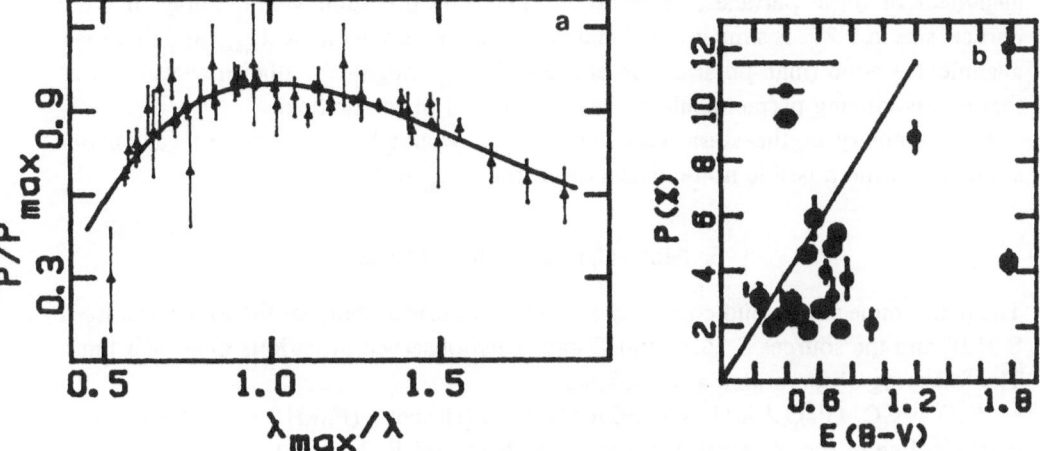

Fig. 2a–b.   As Figure 1 but for cluster stars. In (b) the symbol size is related to the probable error in the colour excess as shown in the upper left corner.

in S 252. These values are shown in Figure 3 (triangles) together with the data given in SMF (dots). The correlation between these two variables for stars in S 252 is by far not so strong as in the data by SMF. K photometry is available only for four stars in S 252 and so no proper comparison with SMF can be made.

Large λ_{max} values up to 0.8μ and poor polarization efficiencies have been reported for stars in the direction of the parts of dense molecular clouds associated with star formation (e.g.,Carrasco *et al.*, 1973; Vrba, 1977) and in dense parts of dark clouds (e.g., Vrba *et al.*, 1981). The polarization efficiency decreases and λ_{max} increases with growing extinction. Such effects are not observed for stars in S 252. High λ_{max} and poor polarization efficiency have been interpreted to be due to large particle sizes and poor

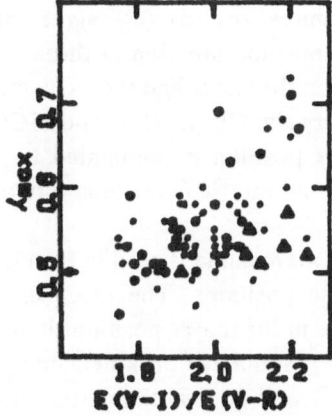

Fig. 3. $E(V - I)/E(V - R)$ and λ_{max}. Dots are data from SMF. Large symbols denote data which are more reliable according to SMF. Triangles are stars in S 252.

alignment of these particles, respectively. This interpretation would imply that the particle size in S 252 is similar to normal interstellar medium (same λ_{max}) and that their alignment is good (high-polarization efficiency). High-alignment efficiency also shows that one is looking perpendicular to a well-ordered magnetic field.

K photometry of the stars with known λ_{max} would be important to gain better statistics on the possible non-correlation of the λ_{max} and colour excesses.

3. Molecular Line Observations

The dense molecular cloud core near S 252A, the surroundings of the ionization front S 252F and the sources S 252C and E have been observed in various molecular lines. The following observations are available:

^{12}CO, ^{13}CO, $C^{18}O$ $J = (1 \rightarrow 0)$ FCRAO 14 m telescope (FWHP ~ 45″) S 252A, F (participating in the observations were P. Friberg and K. Mattila).

NH_3 $((J, K) = (1, 1), (2, 2))$ Effelsberg 100 m telescope, FWHP ~ 43″) S 252A, F (Ch. Henkel and K. Mattila).

^{12}CO, ^{13}CO, $C^{18}O$ $J = (1 \rightarrow 0)$ Onsala 20 m telescope (FWHP ~ 30″) S 252A, C, E, F (L. E. B. Johanson and K. Mattila).

HCO^+ $J = (1 \rightarrow 0)$ Metsähovi 14 m (FWHP ~ 1′), S 252A, F.

The beam sizes of all the telescopes are nearly the same which makes the interpretation of the data easier.

4. The Cloud Core

A region of 6′ by 4′ has been mapped using 45″ spacing in all the molecules. ^{12}CO and ^{13}CO integrated line intensity contour maps (Figure 4(a) and (b)) show a very dissimilar structure. The ^{12}CO intensity peaks at the H_2O maser position (marked with an asterisk) 45″ to the East from S 252A. ^{13}CO again has a maximum 2′ SE from the maser position. The emission of molecular species that require dense regions to be excited (NH_3, $C^{18}O$, and HCO^+, Figures 5(c), (d), (e)), agree with each other. In particular both the ^{12}CO and the ^{13}CO maxima are seen in these contour maps.

Individual spectra observed in the maser and the ^{13}CO maximum positions are shown in Figures 5(a) and (b), respectively. $C^{18}O$, NH_3, and HCO^+ temperatures have been multiplied by four. The maser position is dominated by one velocity component at 9.2 km s^{-1} in $C^{18}O$ and NH_3. At the ^{13}CO maximum two velocity components at 9.2 and 8.1 km s^{-1} are present.

The different nature of the two maxima is readily seen by comparing the NH_3 (1, 1) and (2, 2) line ratios in these positions. The (2, 2) emission is much stronger in comparison with the (1, 1) line in the maser position than in the ^{13}CO maximum. The (2, 2) line rotation temperature (estimate of the kinetic temperature) calculated from the (1, 1), (2, 2) line ratios are 19 K and 16 K at the maser and ^{13}CO maximum positions, respectively. The optical depth of the centre group of hyperfine components is 0.6 for both the positions. The ammonia data show that the molecular cloud has two components. The cold and dense cloud core lies at the ^{13}CO maximum position (velocity 8.1 km s^{-1}). A warm and less dense component is associated with the H_2O maser (velocity 9.2 km s^{-1}).

The two components and the change in the excitation temperature is also the reason for the different structure observed in ^{12}CO and ^{13}CO emission.

The ^{12}CO excitation temperature and the optical depth of the ^{13}CO line centre calculated from the observations are 35 K, 0.5 and 28 K, 0.7 for the maser and ^{13}CO

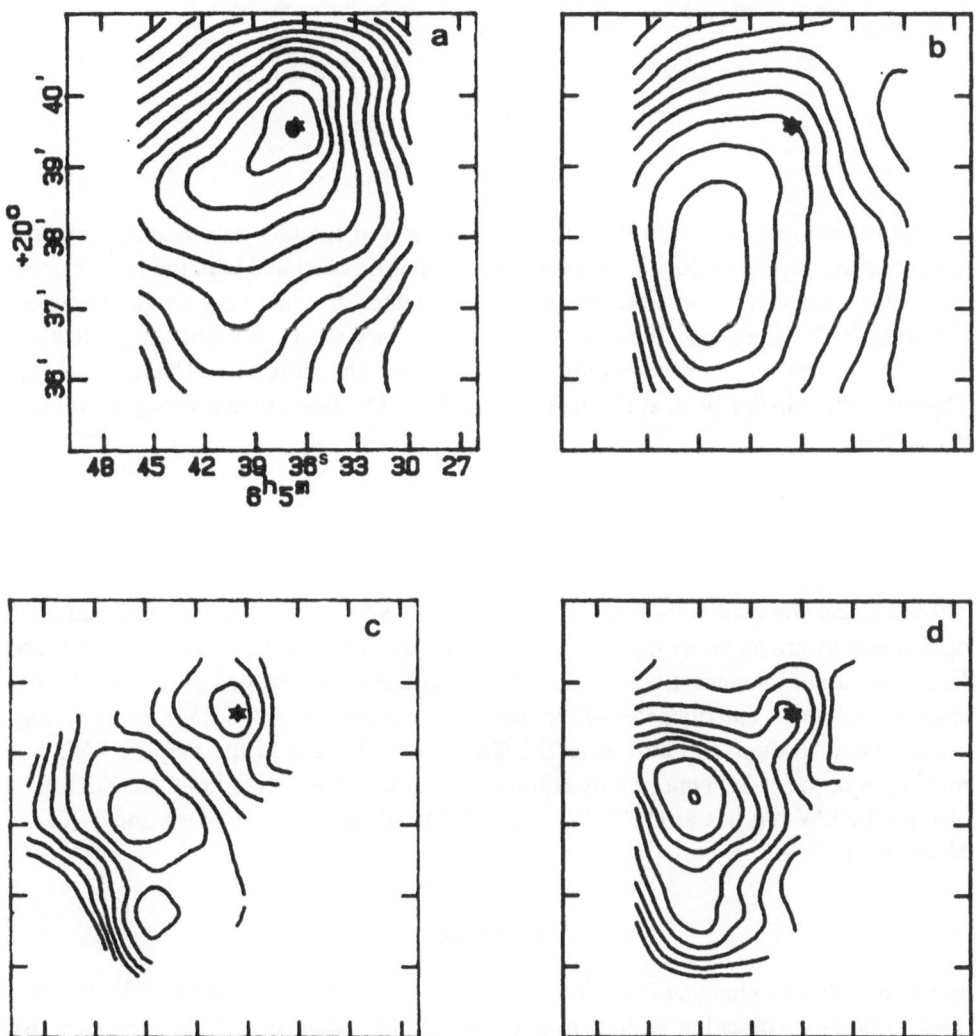

Fig. 4a–c. Integrated line emission of ^{12}CO (a), ^{13}CO (b), $C^{18}O$ (c), NH_3 (d), and HCO $^+$ (e) in arbitrary units. Asterisk shows the position of the H_2O maser. 3′ corresponds to approximately 2 pc.

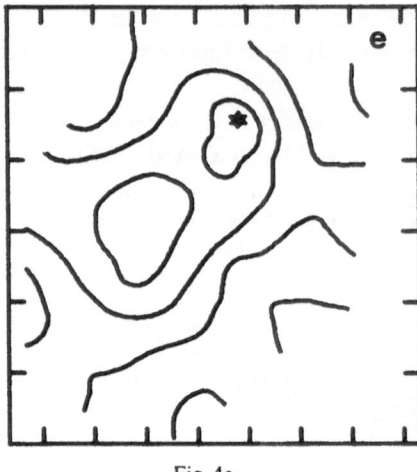

Fig. 4e.

maximum positions, respectively. The interpretation of the CO data is unreliable due to the two overlapping velocity components, which are associated with the two different excitation conditions. Therefore, the calculated values are only very rough estimates.

The $H^{12}CO^+$ lines near the dense core show a very strong self-absorption feature. This self-absorption has been confirmed by observing the more rare $H^{13}CO^+$ isotope. This optically thin line peak at the centre of the $H^{12}CO^+$ line where a strong depression is observed.

5. S 252F

The lines near the ionization front S 252F are more symmetric and the ^{13}CO and NH_3 optical depths are higher ($\tau_{^{13}CO} = \tau_{NH_3} = 0.9$) than at the cloud core (Figure 5(c)). The ^{12}CO excitation temperature is 27 K. The region has been mapped in ^{13}CO with 40″ spatial resolution. The diameter of the density enhancement is 4′ (2.5 pc) with a ridge connecting it to the cloud core near S 252A as already seen in the map by LW. The maximum of the ^{12}CO emission does not coincide with the ionization front S 252F as claimed by LW but lies at $6^h5^m53^s$, $+20°29'40''$ about 4′ to the West and 1′ to the North of it.

6. S 252C

In CO S 252C is a elongated (3′ by 5′) object which has a sharp boundary in the East. This boundary coincides with a bright (reflection) nebulosity. ^{12}CO line has two components (Figure 5(d)). The fainter component does not vary over the face of the cloudlet so it is possibly not associated with S 252C. ^{12}CO excitation temperature in the centre of S 252C 27 K and τ_{13CO} is 0.3. S 252C contains a group of early-type (B0 and later) stars which are the probable excitation sources. The matter density near S 252C is lower than in S 252A and F because no NH_3 nor $C^{18}O$ emission could be detected.

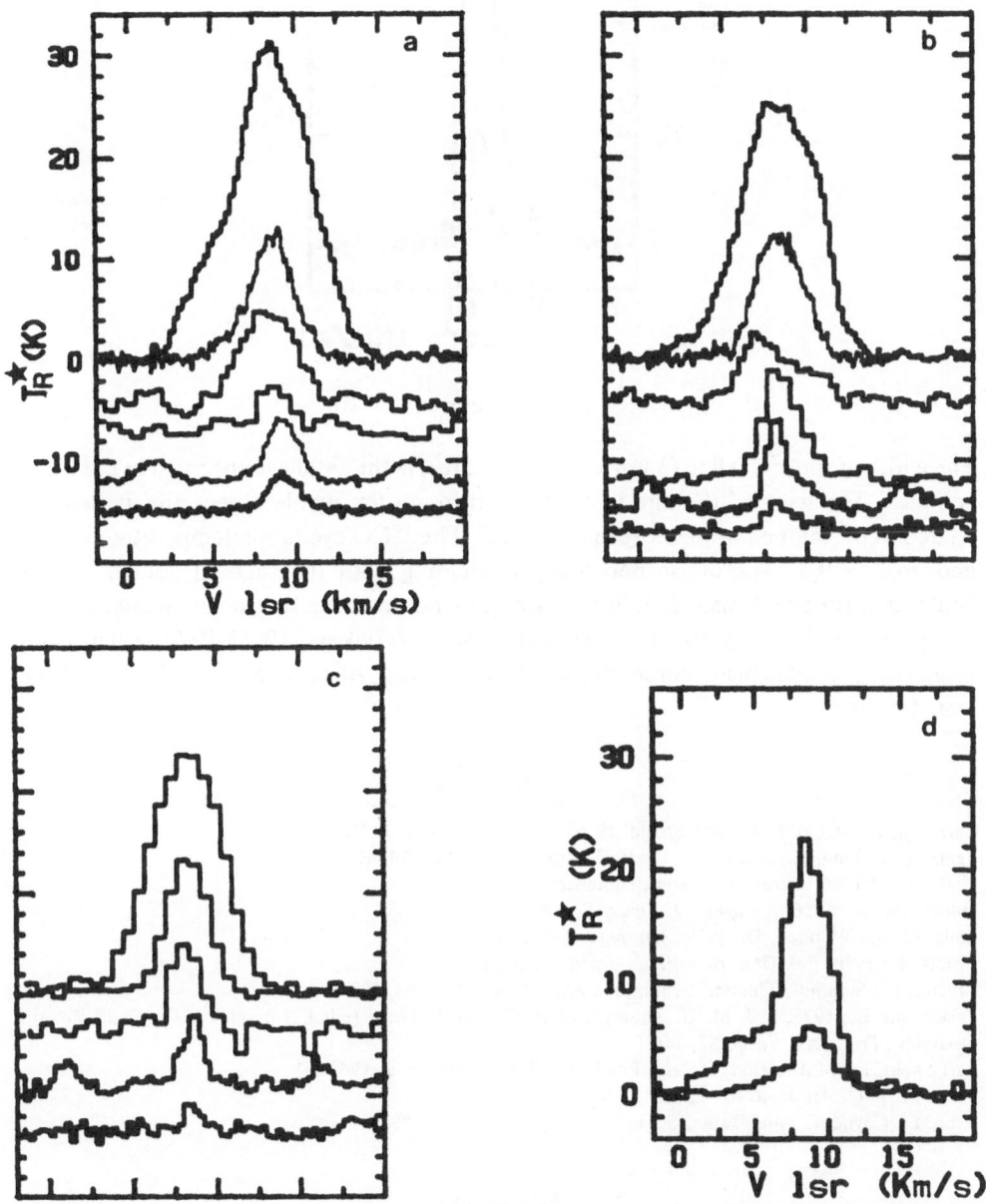

Fig. 5a–e. Individual spectra observed in S 252 (lines from top to bottom are ^{12}CO, ^{13}CO, HCO^+, $C^{18}O$, NH_3 (1, 1), NH_3 (2, 2)) (a) maser position, (b) ^{13}CO peak, (c) S 252F (no HCO^+), (d) and (e) S 252C and E (only ^{12}CO and ^{13}CO).

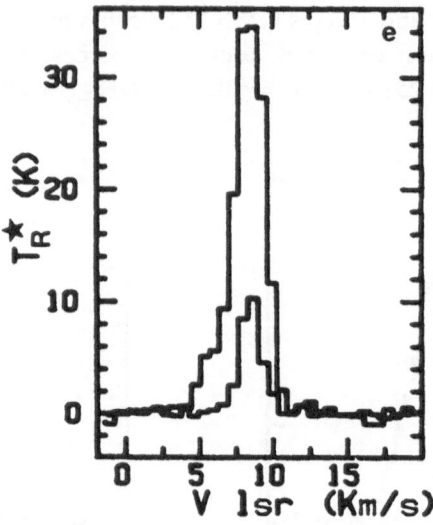

7. S 252E

The width of the ^{12}CO line (3 km s^{-1}, Figure 5(e)) is the smallest observed in the S 252 complex. The wing of the line is observed through the whole object and is possibly related to the faint component seen in S 252C. The ^{12}CO excitation temperature is 35 K and $\tau_{13\text{CO}} = 0.3$. The dense nebulosity coinciding with the thermal source S 252E contains a B0 star (called S 252a). A group of fainter stars lies in the nebulosity and it is probable that they are also associated with it (Haikala, 1986). This stellar group is also the probable heat source for the ^{12}CO emission. As in S 252C no C^{18}O nor NH$_3$ was detected.

References

Carrasco, L., Strom, S. E., and Strom, K. M.: 1973, *Astrophys. J.* **182**, 95.
Felli, M., Habing, H., and Israel, F.: 1977, *Astron. Astrophys.* **59**, 43.
Haikala, L.: 1986, *Astron. Astrophys.*, submitted.
Hiltnes, W. A.: 1956, *Astrophys. J. Suppl. Ser.* **2**, 389.
Lada, C. and Wooden, D.: 1979, *Astrophys. J.* **232**, 158.
Pismis, P.: 1970, *Bol. Obs. Tonanzintla Tacubaya* **5**, 219.
Proctel, K., Schmidt, Th., and Schulz, A.: *Mitt. Astron. Ges.* **43**, 293.
Serkowski, K.: 1973, in J. M. Greenberg and H. C. van de Hulst (eds.), 'Interstellar Dust and Related Topics', *Proc. IAU Symp.* **52**, 144.
Serkowski, K., Mathewson, D., and Ford, V.: 1975, *Astrophys. J.* **196**, 261.
Vrba, F.: 1977, *Astron. J.* **82**, 198.
Vrba, F., Coyne, G., and Tapia, S.: 1981, *Astrophys. J.* **243**, 489.

Discussion

P. G. Mezger: Is there a difference in the structure between the different molecular lines you looked at? In some cases it is always the opacity in other cases, like NH$_3$, it is probably the excitation condition which makes a difference. But if you wanted to use molecules to estimate for example column densities and masses which would you consider the most reliable ones?

L. K. Haikala: For dense molecular cloud cores like in S 252 ammonia is very good, but you can also use CO.

E. Krügel: I noticed that for example the ammonia (1, 1) line was asymmetric and you spoke of self-absorption. I wondered whether you considered the possibility that it is not self-absorption but different clouds which produce the profiles.

L. K. Haikala: In case of HCO^+ we found self-absorption. This is confirmed by the observations of the optically thin $HC^{13}O^+$ line. In NH_3 and $C^{18}O$ it is clear that there are two separate velocity components.

W. Pfau: Don't you think it is possible that your polarization observations refer to foreground dust only? That would explain that λ_{max} and also the polarization efficiency are quite normal and not typical for such a region.

L. K. Haikala: The foreground polarization is between 2 and 3%. In S 252 the polarizations range between 2 and 12%. It is improbable that these high polarizations were due to foreground dust because this would require an extremely variable dust layer in front of the cluster. If you have different λ_{max} inside the molecular cloud you may detect this difference in the λ_{max} also. But if it stays the same then the λ_{max} must be similar both in the foreground and in the molecular cloud and that was my conclusion.

THE RHO OPHIUCHI CLOUD – AN OVERVIEW*

S. KLOSE

Universitäts-Sternwarte Jena, G.D.R.

(Received 27 June, 1986)

Abstract. The relative short distance (~ 165 pc) and considerable mass ($\gtrsim 2000 M_\odot$) of the Rho Ophiuchi cloud enable the study of the physical conditions within the cloud with a fine detail as it is reached in few complexes only. On the basis of the observations between 10^{-14} and 10^{-1} m wavelength, the most remarkable features of the cloud detected are: (1) The cloud represents one of the strongest detectable γ-ray sources for energies greater than 100 MeV. (2) X-ray observations gave no evidence that the cloud is interacting with a supernovae remnant or a neutron star what was supposed to account for the observed γ-ray flux. (3) Ultraviolet observations indicate that a considerable number of small particles is present toward the embedded star HD 147889 being placed at the edge of the cloud. (4) The ratio of total to selective extinction is $R > 4$ toward the denser regions of the cloud. (5) An embedded stellar cluster of about 50 stars was detected by four IR-surveys toward the central region of the cloud. (6) Far-IR observations indicate that there are only 3 B-stars within the cloud. The dust-temperature reaches locally 50 K. (7) More than 10 different molecular species were detected by radio observations. A considerable depletion of heavy elements is observed. There are two dense regions within the cloud which are likely contracting.

1. Introduction

The dark cloud south of the star ρ Oph at the galactic coordinates $l = 354°$, $b = 16°$ has been the subject of detailed examinations since about 15 years. The Rho Ophiuchi cloud (Lynds 1681; Barnard 42, Heiles Cloud 4) is a member of the Sco OB2 association. The general implied distance of the cloud is (165 ± 5) pc, and its estimated age 2–8 million years. The cloud radius implied is ranging between 3.5 pc ($\sim 1°2$; Encrenaz *et al.*, 1975) and 5.8 pc ($\sim 2°$; Myers *et al.*, 1978). On the other hand, most infrared and radio observations concern a small area of only about 0.5 square degrees at the region of highest visual extinction.

2. The Embedded Stellar Cluster

Information about an embedded stellar cluster was obtained from four IR surveys of the central region of the cloud (definition see Figure 2) at 2.2 μm wavelength (Grasdalen *et al.*, 1973; Vrba *et al.*, 1975; Elias, 1978; Wilking and Lada, 1983, hereafter referred to as WL), from recombination line observations of C II and S II (Brown and Knapp, 1974; Brown *et al.*, 1974; Chaisson, 1975; Knapp *et al.*, 1976; Cesarsky *et al.*, 1976a, b; Falgarone *et al.*, 1978), and radio continuum observations between 3 and 21 cm wavelength toward the region of highest visual extinction. By means of IR-photometry

---

* Paper presented at a Workshop on 'The Role of Dust in Dense Regions of Interstellar Matter', held at Georgenthal, G.D.R., in March 1986.

Astrophysics and Space Science **128** (1986) 135–149.
© 1986 by D. Reidel Publishing Company

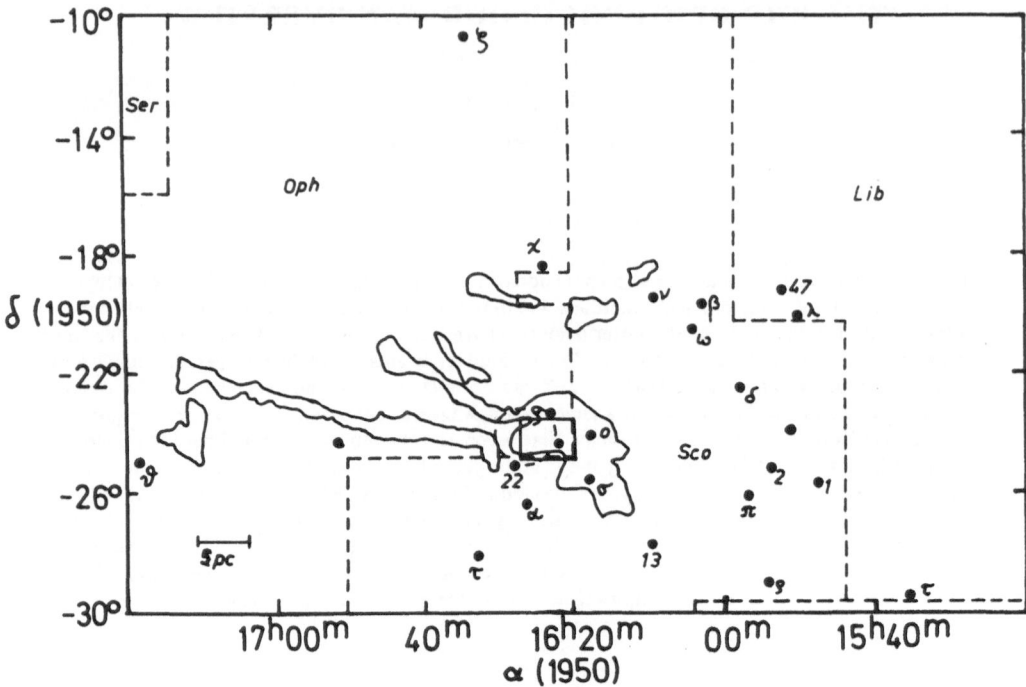

Fig. 1.   Displayed in this figure are the Rho Ophiuchi cloud and some stars of the Sco OB2 association. A remarkable feature of the cloud are two about 10 pc extended streamers in its eastern region. The Rho Ophiuchi cloud is the only obvious feature within the 1.5 deg γ-ray circle (dashed circle; Morfill *et al.*, 1981). The small box at the centre of the cloud encloses the area represented in Figure 2. Map according to Bečvář (1962).

TABLE I

Estimated limits to the age of the cloud

Basic idea	Age in 10^6 yr	Reference
The cloud is a member of the Sco OB2 association; age of this association	6 ± 2	Wouterloot (1984)
CO-column densities observed; necessary production time	>2	Myers *et al.* (1978)
Distribution of T Tauri stars observed; assumed that they have a typical velocity of 1 km s⁻¹ and were formed at the cloud centre	~3	Carasco *et al.* (1973)
Assumed that the two streamers are matter outflows of the cloud	5–7	Vrba (1977)
Embedded star GS 23; its placement on the HR-diagram	~1.5	Lada and Wilking (1984)

up to 1984 ~ 50 stars were determined to be embedded into the cloud (Elias, 1978; Cohen and Kuhi, 1979; Wilking *et al.*, 1979; Chini, 1981; WL). Only five compact H II regions were detected the positions of which coincide with the positions of IR sources (Brown and Zuckerman, 1975; Falgarone and Gilmore, 1981). These compact regions are excited by the embedded stars HD 147889 (double star with a high-luminosity component), S 1, GS 39, GS 23, and likely WL 5 and are indicated as BZ 3 (OPH 3), BZ 4 (OPH 4), BZ 5 (OPH 5), FG 10 (OPH 10), and FG 12 (OPH 12), respectively. The three latter H II regions are likely compact regions collisionally excited by stellar winds from young stars. Completely samples far-IR observations between 40 and 250 μm wavelength by Fazio *et al.* (1976) indicated that there are only 3 high-luminosity stars within the cloud. These observations were only sensitive to stars of spectral type late B and earlier which deliver all their energy to the surrounding dust. Extended far-IR emission was observed only around the stars HD 147889, S 1, and SR 3. The most highly-peaked source was this one excited by the star S 1. Radio observations indicated that the visual extinction toward this star is about 100 magnitudes (WL) and that this star is surrounded by a ~ 0.5 pc extended C II region (Pankonin and Walmsley, 1978). The classification suggested for S 1 is B3V–B5V (Elias, 1978; WL), for SR 3 B9–A0 (WL, Wilking *et al.*, 1985), and for the high-luminosity component of HD 147889 B2V (Garrison, 1967). The star HD 147889 is placed at the edge of the cloud and associated with a reflection nebulae (IC 4603).

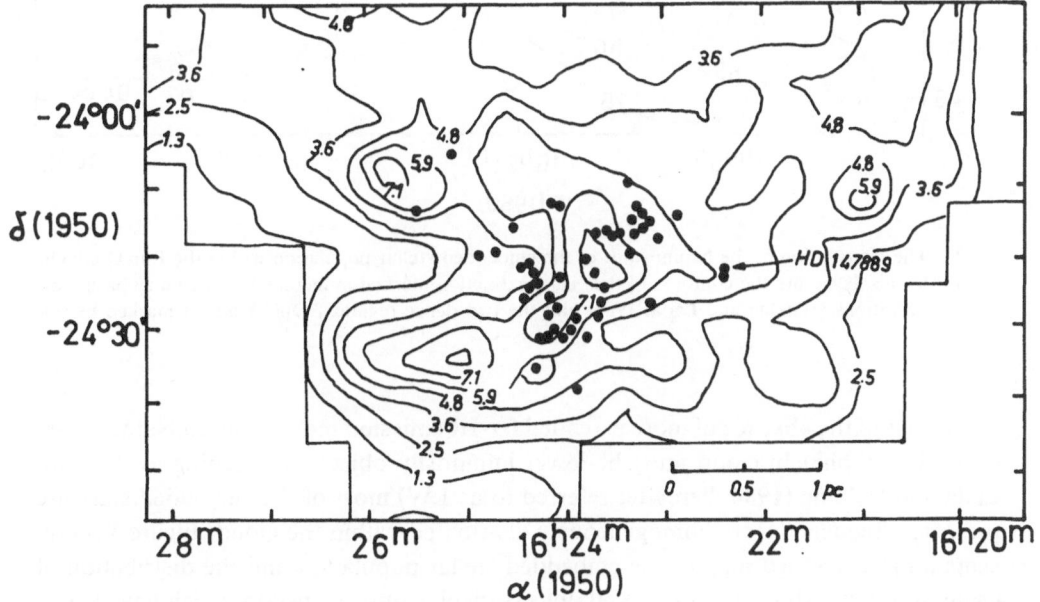

Fig. 2.   Contours of the visual extinction derived from star counts by Encrenaz *et al.* (1975). Also displayed is the embedded stellar population detected by four IR-surveys at 2.2 μm wavelength (Grasdalen *et al.*, 1973; Vrba *et al.*, 1975; Elias, 1978; Wilking and Lada, 1983; limiting magnitudes K = 9 mag, 10 mag, 7.5 mag, and 12 mag, respectively). The term 'central region of the cloud' refers to the area which encloses the highest visual extinction.

In Figure 4 it is shown the temperature distribution of the gas within the cloud obtained from CO observations by Loren *et al.* (1980). Far-IR observations indicate that the dust temperature around S 1 ($>$ 30 K, Cudlip *et al.*, 1984; $\sim$ 50 K, Harvey *et al.*, 1979) as well as around HD 147889 (40–60 K; Cudlip *et al.*, 1984) are higher than the gas temperatures near these sources.

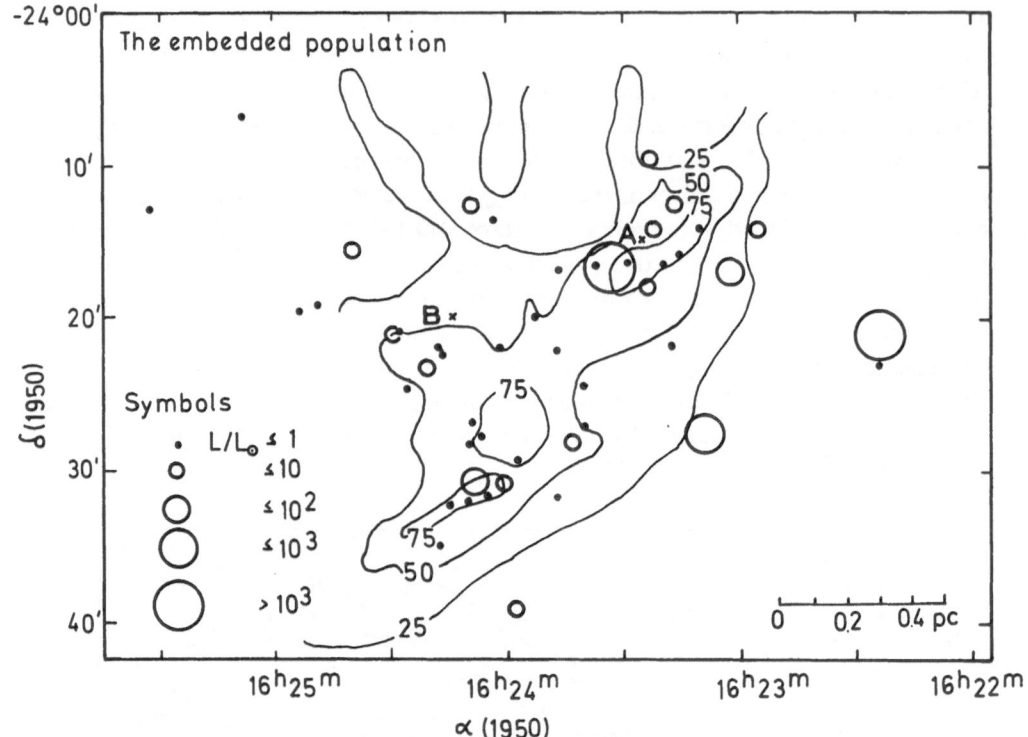

Fig. 3. The distribution of the luminosities of the embedded stellar population within the Rho Ophiuchi cloud. Also displayed are the contours of the visual extinction marked in increments of 25 mag basing on radio observations (Wilking and Lada, 1983) and the two dense regions ρ Oph A and B marked by the crosses.

Because of the absence of more extended far-IR emission most of the embedded stars of the Rho Ophiuchi cloud must be lower luminosity objects. According to WL and Lada and Wilking (1984, hereafter referred to as LW) most of the embedded stars are pre-Main-Sequence stars homogeneously distributed within the cloud. Figure 3 represents a more detailed map of the embedded stellar population and the distribution of its luminosities. About 15 of these 50 objects display optical spectra which have led to their classification as T Tauri stars. No Herbig–Haro objects within the cloud are known (Loren *et al.*, 1979). More than 10 stars are likely X-ray sources (Montmerle *et al.*, 1983). The star density of the embedded cluster is about 15 pc^{-3}. LW estimated the density of stellar matter in the central region of the cloud to be $\sim$ 150 $M_\odot$ pc^{-3} and the star forming efficiency there to be $\sim 25\%$. The total luminosity is about 7000 $L_\odot$. WL and LW pointed out that the observed luminosity function of this embedded stellar population displays a remarkable underabundance of the number of intermediate-luminosity stars compared with the luminosity function for field stars.

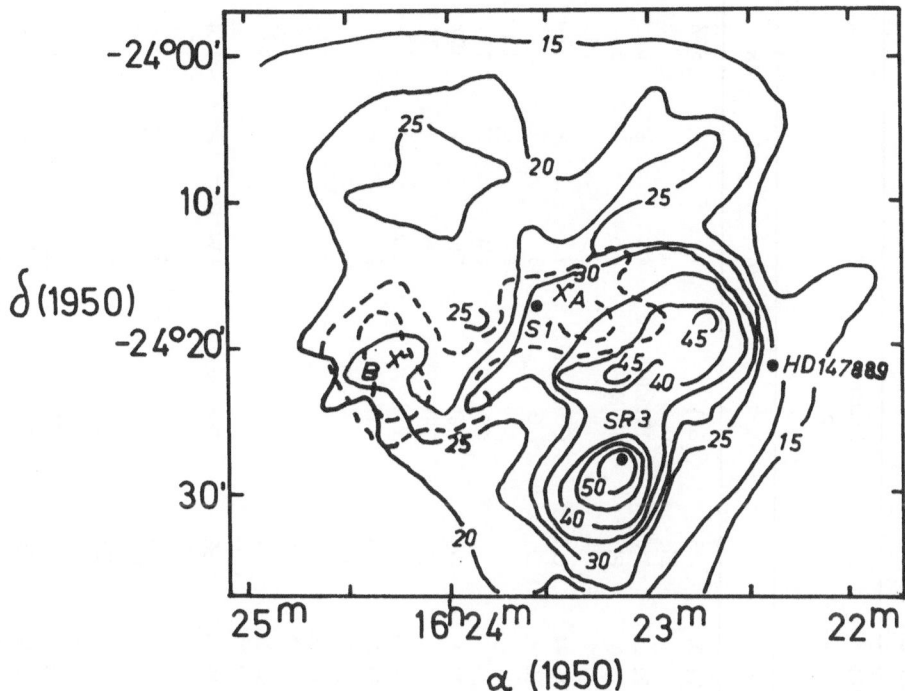

Fig. 4. Contours of peak $T_A^*(CO)$-temperature according to Loren *et al.* (1980). The dashed lines correspond to a density of 5×10^4 cm^{-3} and 2×10^5 cm^{-3}, respectively (Loren *et al.*, 1983). Also displayed are the 3 embedded B-stars and ρ Oph A and B (crosses).

3. Explanations for the Onset of Star Formation

The Rho Ophiuchi cloud represents a rich source of molecule formation; OH, CO, SO, H_2CO, HNO, CN, CS, CH, NH_3, CH_3OH, HCO, H_2O (3.1 μm absorption feature), and some isotopic species were detected. Self-absorption of the $J = 1 \rightarrow 0$ ^{12}CO and ^{13}CO line were discovered (Encrenaz, 1974; Lada and Wilking, 1980). According to Snell and Loren (1977) and Leung and Brown (1977) the CO self-absorption as observed in some molecular clouds can be explained by a contraction at least of the outer regions of a cloud. Myers *et al.* (1978) found some indications that this could be the case for the Rho Ophiuchi cloud. From their radio observations at the transitions of different molecular species they concluded that the Rho Ophiuchi cloud is collapsing with a velocity function $v(r) = -1.1(r/pc)^{0.27}$ km s^{-1}, $0.1 \leq r \leq 5.6$ pc. On the other hand, Javanaud (1980) pointed out that the observed different velocity dispersions of

TABLE II

Likely members of the embedded stellar population

No.	Identification					α(1950)	δ(1950)	Ref.	Spectral type	Ref.		L/L⊙	Ref.
	SR	GS	VS	El	WL	h m s	° ′ ″						
1	2					3		4	5	6	7	8	9
1*	HD 147889	9	–	9	–	16 22 22.8	–24 21 07	GS, El	B2V	1	+	5700	2
2	22	–	–	–	–	16 22 22.8	–24 22 55	3	TT/M0	4	+	0.6	4
3	4	–	–	13	–	16 22 54.8	–24 14 01	El	TT/K7	4,5	+	5.1	4
												2.1	LW
4*	–	23	–	14	–	16 23 02.1	–24 16 44	GS	G5III–K0III	7	+	43–60	2
						23 01.7	16 50	El	K0	3			
									B3	6		1700	6
									G0	LW		25	LW
5*	3	25	–	16	–	16 23 07.7	–24 27 26	GS, El	B6V,	El	+	500	2
									B9V–A0V	LW, 8		125	LW
6	–	26	–	–	–	16 23 08.9	–24 14 13	GS	TT?	LW	–	0.7	LW
7	–	29	–	18	–	16 23 15.7	–24 15 43	GS	TT	3	+	0.8	LW
						23 15.5	15 38	El					
8	–	28	–	19	–	16 23 15.7	–24 13 42	GS	TT	3,5	+	1.1	LW
						23 15.8	13 37	El					
9*	–	–	1	20	–	16 23 16.7	–24 21 29	VS	TT	3	–	0.7	LW
						23 17.5	12 33	El					
10*	–	30	–	21	–	16 23 19.7	–24 16 14	GS	Protostar?	LW, El	–	11.4	LW
						23 19.9	16 18	El				~12	8
11	–	31	–	22	–	16 23 21.4	–24 14 13	GS	TT	3	+	3.7	LW
						23 22.0	14 15	El					
12	S–2	32	–	23	–	16 23 22.5	–24 18 13	GS	TT	LW	+	2.5	LW
						23 22.6	18 04	El					
13	–	–	–	24	–	16 23 22.9	–24 09 29	El	TT?	LW	+	2.7	LW
14	–	–	27	–	–	16 23 28.7	–24 16 14	VS	TT?	LW	–	0.4	LW

Table II (continued)

No.	Identification					α(1950)	δ(1950)	Ref.	Spectral type	Ref.		L/L_O	Ref.
	SR	GS	VS	EI	WL	h m s	° ' "						
1	2					3		4	5	6	7	8	9
15*	S-1	35	-	25	-	16 23 32.7	-24 16 44	GS	B3	WL, 6	+	1700	6
						23 32.8	16 44	EI	B3V–B5V	LW	-	1500	WL, LW
16	-	-	4	-	-	16 23 36.7	-24 16 22	VS	PMS	WL	-	0.5?	WL
17	-	-	-	-	7	16 23 39.8	-24 24 14	WL	PMS	WL	-	0.5	WL
18	-	-	-	-	8	16 23 40.3	-24 26 41	WL	TT?	LW	-	0.2	LW
19	-	-	-	-	12	16 23 42.5	-24 28 04	WL	Protostar?	LW	-	1.4	LW
20*	-	39	-	27	-	16 23 46.3	-24 16 44	GS	TT?	LW	-	0.4	LW
						23 43.3	16 24	EI					
21	-	-	-	-	2	16 23 46.8	-24 21 53	WL	TT?	LW	-	0.1	LW
22	-	-	-	-	18	16 23 47.4	-24 31 34	WL	PMS	WL	-	0.5	WL
23	-	-	5	-	-	16 23 52.0	-24 19 39	VS	PMS	WL	-	0.5	WL
24*	24S	-	-	28	-	16 23 56.5	-24 38 53	EI	TT/K2	4	+	4	4
25	24N	-	-	-	-	16 23 56.5	-24 38 53	9	TT/M0.5	4		3.1	LW
26	-	-	-	-	14	16 23 57.2	-24 29 08	WL	PMS	WL	-	1.2	4
27*	-	-	-	-	16	16 24 00.3	-24 30 44	WL	Protostar?	LW	-	0.5	WL
28	-	-	-	-	1	16 24 01.9	-24 21 48	WL	PMS	WL	-	3.1	LW
29	-	21	-	-	-	16 24 02.8	-24 13 24	VS	F2V	3	+	5	8
									B9V–A2V	7		2.9	2
												95–26	2
30	-	-	-	-	17	16 24 04.8	-24 31 33	WL	PMS	WL	-	0.8	WL
31	-	-	-	-	10	16 24 07.3	-24 27 35	WL	TT?	LW	-	0.3	LW
32*	-	-	-	29	15	16 24 07.7	-24 30 40	EI	TT?	LW, EI	-	0.3	LW
						24 07.8	30 33	VS	Protostar?			20.5	LW
												≤17	8
33	21	-	23	30	-	16 24 08.8	-24 12 24	EI	TT/K0	3	+	3.9	LW
						24 08.9	12 31	VS					
34	-	-	-	-	9	16 24 09.3	-24 26 41	WL	PMS	WL	-	0.5	WL

Table II (continued)

No.	Identification					α(1950)	δ(1950)	Ref.	Spectral type	Ref.		L/L☉	Ref.
	SR	GS	VS	El	WL	h m s	° ′ ″						
1	2					3		4	5	6	7	8	9
35	–	–	–	–	11	16 24 09.5	–24 28 07	WL	PMS	WL	–	0.5	WL
36	–	–	–	–	19	16 24 09.7	–24 31 49	WL	TT?	LW	–	0.1	LW
37	–	–	–	–	20	16 24 13.6	–24 31 59	WL	TT?	LW	–	0.2	LW
38*	–	–	26	–	5	16 24 16.4 16 24 17.0	–24 22 11 22 01	WL VS	Peculiar	LW	–	0.1	LW
39	–	–	–	–	4	16 24 16.8	–24 22 23	WL	TT?	LW	–	0.2	LW
40	–	–	–	–	3	16 24 17.6	–24 22 00	WL	TT?	LW	–	0.1	LW
41	12	–	–	–	–	16 24 17.6	–24 34 59	9	TT/M1	4, 5	+	1.0	4
42	–	–	–	–	6	16 24 19.8	–24 23 08	WL	Protostar?	LW	–	1.1	LW
43	–	–	25	31	13	16 24 25.7 25.4	–24 24 36 24 34	VS El, WL	F2V PMS	3 WL	+	2.9 0.6	2 WL
44*	–	–	18	32	–	16 24 26.8 26.9	–24 20 34 20 37	VS El	PMS	WL	–	0.7	WL
45*	–	–	17	33	–	16 24 28.8 28.6	–24 20 54 21 00	VS El	PMS	WL	–	1.4	LW
46	9	–	–	34	–	16 24 38.8	–24 15 24	El	TT/K7 TT/K5	4 5	+	3.9 2.8	4 LW
47	–	–	14	36	–	16 24 48.8 24 48.3	–24 18 54 19 02	VS El	B9V B7V Peculiar	3 7 LW	+	320 95 1.0	2 2 LW
48	10	–	–	–	–	16 24 53.6	–24 19 39	3	TT/M1.5	4	+	0.6	4
49	15	–	–	–	–	16 25 08.3	–24 06 37	3	A2V	3	+	26	2
50	20	–	–	–	–	16 25 32.4	–24 12 47	3	TT/M	3	+		

Explanations of columns: (2) Numbers refer to the following lists: SR = Struve and Rudkjøbing, 1949; GS = Grasdalen *et al.*, 1973; VS = Vrba *et al.*, 1975; EI = Elias, 1978; WL = Wilking and Lada, 1983. (5) Abbreviations: TT = T Tauri star; PMS = Pre-Main-Sequence star. (7) Visible star? yes (+)/no (–), according to Wilking and Lada, 1983. (8) Bolometric luminosity: EI, WL, and LW basing on IR-photometry (≤ 20 μm wavelength); further information see references to this table.

References to Table II
(1) Garrison (1967); from optical spectra.
(2) Schmidt-Kaler (1982); luminosity refers to the spectral type.
(3) Chini (1981); from IR-photometry, but not in the case of Nos. 29, 43, and 49.
(4) Cohen and Kundi (1979); from optical spectra.
(5) Rydgren *et al.* (1976); from optical spectra.
(6) Cudlip *et al.* (1984); from far-IR photometry.
(7) Chini *et al.* (1977); from IR photometry.
(8) Wilking *et al.* (1985); from far-IR photometry.
(9) Gezari *et al.* (1984).

Remarks to Table II
(1) Far-IR emission (Fazio *et al.*, 1976); H II region BZ 3; double star.
(2) Far-IR emission (Cudlip *et al.*, 1984); H II region FG 10.
(5) Far-IR emission (Fazio *et al.*, 1976).
(9) According to Elias (1978) VS 1 corresponds to El 20. Obviously there is a mistake regarding the δ-coordinate in one of both papers.
(10) Reflection nebulae detected (Castelaz *et al.*, 1985); most luminous object at 20 μm within the cloud (LW).
(15) Far-IR emission (Fazio *et al.*, 1976); H II region BZ 4.
(20) H II region BZ 5.
(24) Double star.
(27) Deep 10 μm absorption feature (LW); far-IR emission (Cudlip *et al.*, 1984).
(32) Variability at 2.2 μm observed (Elias, 1978); possible existence of a reflection nebulae; steeply rising spectra toward 20 μm (Wilking *et al.*, 1985).
(38) WL resolved VS 26 into 3 components; WL 5: deepest 3.1 μm ice band feature for any source within the cloud; possible background star (LW); H II region FG 12.
(44) 3.1 μm ice band feature (Vrba *et al.*, 1975; Harris *et al.*, 1978).
(45) See 44.

---

the line profiles can be caused by contraction as well as by turbulence within the cloud, but one cannot distinguish one effect from the other. Comparing their derived velocity function with the expected one in the case of free fall, Myers *et al.* calculated a retardation factor of about 2, which according to the authors could be explained by the existence of a general magnetic field component of $\sim 3 \times 10^{-5}$ G. Using the virial theorem Vrba (1977) estimated $B \lesssim 2.2 \times 10^{-4}$ G within the Rho Ophiuchi cloud. On the basis of OH Zeeman observations, Chaisson (1975) and Chaisson and Vrba (1978) obtained $B \lesssim 5 \times 10^{-4}$ G toward some regions of the cloud.

A more simple model to explain the observed CO self-absorption in the cloud was proposed by Encrenaz *et al.* (1975). According to this model the observed line profiles could be described by the assumption of two clouds having different radial velocity, temperature, and optical depths and lying one after another with respect to the line-of-sight. The suggestion to distinguish between two clouds is also implied from the study of the position angles of the polarization vectors at optical wavelengths and at 2.2 μm toward some stars in the Rho Ophiuchi cloud. One observes a striking bimodal distribution of position angles at 0 and 50° (Vrba *et al.*, 1976; Wilking *et al.*, 1979). A collision between two clouds was, therefore, suggested to be responsible for the onset of star formation within the Rho Ophiuchi cloud (Wilking *et al.*, 1979). Vrba (1977) proposed that the cloud collaps was initiated by a shock front striking the south–western edge of the cloud resulting in an embedded stellar cluster at the interface, the prevered existence of high-luminosity stars at the south–western edge of the cloud (Figure 3), and the origin of the two streamers in its western region (Figure 1). On the basis of this assumption Vrba (1977) estimated the age of the about 10 pc extended streamers from the observed CO-line velocities to be about 5–7 million years.

The idea to explain the onset of star formation in the Rho Ophiuchi cloud by an infalling shock wave is supported by results obtained from the γ-ray astronomy. The Rho Ophiuchi cloud is one of only few COS-B γ-ray sources which ave been tentatively identified. The cloud is associated with the 100 MeV γ-ray source 2CG 353 + 16. It is the only obvious feature within the $1°5$ γ-ray circle (Figure 1, Morfill *et al.*, 1981; Figure 5 in Bignami, 1985). The γ-ray flux for energies greater than 100 MeV as it is observed from a molecular cloud can be explained by the decay of $\pi°$-mesons which have been produced during the interaction of protons of the cosmic gas cloud with high-energetic ($E > 279.7$ MeV) protons of the galactic background radiation. In this scenario the γ-ray flux emitted from a cloud is an indicator for its mass (Black and Fazio, 1973). But the mass of the Rho Ophiuchi cloud as implies by these observations is larger than its mass as implied from radio observations by a factor of about 5 (Montmerle *et al.*, 1983). Several authors, therefore, suggested that particle acceleration within the cloud or its vicinity is responsible for the observed γ-ray excess. Morfill *et al.* (1981) have argued that the supernova remnant Loop I (North Polar Spur) could provide the necessary acceleration of the cosmic background radiation. Their calculations indeed showed that in this scenario the γ-ray flux would be enhanced by a factor of about 5. However, X-ray observations gave no evidence that Loop I interacts with the Rho Ophiuchi cloud (Apparao *et al.*, 1979). Nevertheless the existence of an expanding shell of neutral hydrogen in the Sco OB2 association was reported by Sancisi (1973). Several authors reported on the existence of a gradient of the H_2CO and H I LSR-velocity of about 0.4 km s^{-1} pc^{-1} crossing the Rho Ophiuchi cloud from north-western to southeastern direction (Myers and Ho, 1975; Myers *et al.*, 1978; Minn, 1981). Minn (1981) proposed to interpret this as an effect which is caused by the rotation of the cloud with a period of $\sim 1.3 \times 10^7$ yr.

In a recent study Meyers *et al.* (1985) reported on the observation of interstellar absorption lines toward some stars in the Sco OB2 association. From their observed line profiles of different molecular species they concluded that the Rho Ophiuchi cloud consists of two distinct different regions delineated by a shock front travelling with a LSR-velocity of $- 10$ km s^{-1}. According to their interpretation a post-shock gas cloud lies in front of a pre-shock gas cloud. Previous observations of several stars in the Sco OB2 association, including HD 147889, also indicated that there are two different regions (e.g., Cohen and Wallerstein, 1974; Danks and Lambert, 1983). All these observations seem to give support for the idea that the onset of star formation in the Rho Ophiuchi cloud was caused by the infall of a shock wave into the cloud. On the other hand, Paul *et al.* (1981) pointed out that at the very least the γ-ray excess can also be explained by the number of T Tauri stars within the cloud. Their stellar winds may be driven by Alfvén waves and could provide the necessary acceleration of the cosmic rays. The possible existence of a neutron star or a black hole within or in the vicinity of the Rho Ophiuchi cloud producing the γ-ray excess was ruled out by the X-ray observations by Montmerle *et al.* (1983) and the radio observations by Andrews and Basart (1984).

4. Visual Extinction, Mass and Density Distribution

Star counts by Bok (1956) and by Encrenaz *et al.* (1975) have led to a visual extinction toward the Rho Ophiuchi cloud up to 7 mag (Figure 2). The values for the visual extinction derived from IR-photometry (Vrba *et al.*, 1975, 1976; Chini *et al.*, 1977: Chini, 1981; WL) as well as from radio observations (e.g., WL) are considerable larger (up to > 100 mag). Frerking *et al.* (1982) have presented a detailed analysis of this problem and treated the uncertainties of these methods. They have shown that star counts lead to low-excitation values for $A_v > 4$ mag. Already Bernes and Sandqvist (1977) pointed out that only the extinction values as derived from IR-photometry or radio observations can explain the relative large H_2CO column densities measured toward the Rho Ophiuchi cloud.

Mass estimates of the Rho Ophiuchi cloud – mainly basing on CO observations – were presented by several authors. WL estimated the mass of gas within an area of about $15' \times 35'$ (0.7×1.7 pc at a distance of 165 pc) enclosed by their $A_v = 25$ mag contour (Figure 3) to be 550 $M_\odot$ ($\pm 50\%$). This is in agreement with the result obtained from Myers *et al.* (1978): $M \sim 450 M_\odot$ within an area with the radius $r = 1$ pc centred at the region of highest visual extinction. Loren *et al.* (1983) have presented a detailed map of the density distribution of almost this whole region. According to their H_2CO observations the density is decreasing toward the 25 mag-contour to about 10^4 cm^{-3}, in agreement with the results obtained from the CO observations by Encrenaz *et al.* (1975) toward the star HD 147889, which is placed near the western edge of the 25 mag-contour. These authors estimated the cloud mass within $r = 3.5$ pc to be about 2000 $M_\odot$ basing on a derived density function $n(r) \sim r^{-1}$, a density function which was also obtained by Bok (1956) from star counts. Myers *et al.* (1978) derived $n(r) \sim r^{-1.3}$ and a mass of 1800 $M_\odot$ within $r = 2.1$ pc. Basing on OH observations, Wouterloot (1984) derived a total mass of the Rho Ophiuchi cloud (including its both streamers) of $(12 \pm 10) \times 10^3 M_\odot$.

SO observations (Gottlieb *et al.*, 1978) and H_2CO observations (Loren *et al.*, 1983) indicated that the spatial density peaks in the Rho Ophiuchi cloud at two relative small regions labelled ρ Oph A (RA(1950) = $16^h23^m25^s$, Dec(1950) = $-24°15'49''$) and ρ Oph B (RA(1950) = $16^h24^m13^s$, Dec(1950) = $-24°19'49''$; according to Loren *et al.*). According to the H_2CO observations the gas density reaches 3×10^5 cm^{-3} toward ρ Oph A. It is decreasing in north–eastern direction over 0.12 pc to 1×10^4 cm^{-3}. On the basis of NH_3 observations, Zeng *et al.* (1984) derived a gas temperature of 45 K toward ρ Oph A. About 0.6 pc (projected distance) away from ρ Oph A is ρ Oph B. Because it displays H_2CO emission at 2 cm wavelength the spatial density there must exceed 10^6 cm^{-3}. The gas density is decreasing to 10^4 cm^{-3} over

0.2–0.3 pc (Loren *et al.*, 1983). Martin-Pintado *et al.* (1983) and Zeng *et al.* (1984) have presented a high-resolution map of ρ Oph B basing on H_2CO and NH_3 observations, respectively. In both observations the mapping of the integrated intensity displays an elliptical structure. On the other hand it does not peak at the same coordinates. Martin-Pintado *et al.* estimated the mass enclosed by an area of about 0.03×0.15 pc, centred at RA(1950) = $16^h24^m10^s$, Dec(1950) = $-24°22'.7$, to be about 6 $M_\odot$. Zeng *et al.* stated similar values. They derived a gas temperature of about 18 K. From the CO observations Loren *et al.* (1980) also concluded that at ρ Oph B the gas is relatively cold. Given the mass implied ρ Oph B would exceed the Jeans-mass by a factor of about 10 (Zeng *et al.*, 1983). Indeed radio observations seem to indicate that ρ Oph A and B are collapsing (Loren *et al.*, 1983). Obviously the star formation in the Rho Ophiuchi cloud is continuing.

5. Indications for Grain Growth

Three effects are known to be indicators for grain growth within a molecular cloud. First the depletion of elements detectable by radio observations, second the occurrence of special absorption features at IR-wavelength caused by molecules which are sticked to the grains and third 'anormal' extinction and polarization of the scattered light from embedded or background stars.

Observational evidence consistent with the depletion of elements have been reported by several authors. Basing on the assumption that sulfur is nearly undepleted within a molecular cloud, Chaisson (1975) concluded from radio observations toward ρ Oph A that carbon is depleted relative to sulfur by a factor of ~ 6, magnesium by $\gtrsim 4$, silicon by $\gtrsim 4$, and iron by $\gtrsim 3$. This was essentially confirmed by Knapp *et al.* (1976). Toward ρ Oph A the line of-sight crosses the C II region excited by the star S 1. On the basis of an assumption that the spectral type of S 1 is B3V, these authors concluded that the detected electron density of $n_e \sim 1.5$ cm^{-3} by Brown *et al.* (1974) as well as their observed [S]/[C] relative abundances indicate that carbon is depleted relative to sulfur by a factor of 5–10. They also estimated the depletion of Mg, Si, and Fe to be greater than a factor of 2, 15, and 10, respectively. Cesarsky *et al.* (1976a) reported on the failure to detect ionized silicon toward the S II region of S 1 about 4' southeast of ρ Oph A. This would indicate that Si is depleted relative to sulfur by a factor of $\gtrsim 5$, because the cosmic abundance of Si is greater than this one of S, as well as its ionization potential is lower. Snow *et al.* (1983) reported on the detection of a considerable depletion of Ca toward HD 147889. The logarithmic depletion observed is $\lesssim -5.1$. Comparing this result with this one by Snow and Jenkins (1980), one finds some evidence that within the Sco OB2 association the depletion of Ca is increasing toward the Rho Ophiuchi cloud. Such an observation was also reported by Crutcher (1976) regarding the elements Na and K. Loren *et al.* (1983) established a decreasing relative abundance of H_2CO toward denser regions of the Rho Ophiuchi cloud. Previous H_2CO-surveys led to the same conclusion if the extinction values derived from radio observations were taken into account (cf. a detailed study by Javanaud, 1979). On the other hand, the H_2CO

molecules can easily be dissociated by ultraviolet-radiation. Thus one expected the denser the region the larger the relative H_2CO column densities, because the denser the region the better these molecules are shielded by grains against the UV-radiation. Therefore, the just contrary observations could be an evidence for a sticking of H_2CO molecules to the grains (where chemical reactions take place) in the denser regions of the Rho Ophiuchi cloud.

Indication for mantle growth of the grains within some regions of the cloud comes from the observation of the 3.1 μm ice band feature toward the embedded stars El 32, El 33, and VS 26 which are placed near ρ Oph B (see remark in Table II; Vrba et al., 1975; Harris et al., 1978).

The third indicator for grain growth is likely the most uncertainty one, because a number of parameters determine the characteristics of the scattered light from the grains. Chini and Krügel (1983) have treated this problem and found for instance no evidence for a unique correlation between the mean grain radius and the ratio of total and selective extinction. But in general it is assumed that the measured ratio of total to selective extinction up to $R = 4.4$ (Carasco et al., 1973; Whittet, 1974; van Breda et al., 1974; Vrba et al., 1975; Whittet and van Breda, 1975, 1980; Rydgren, 1980; Chini, 1981) together with the measured relatively large wavelength of maximum polarization toward some embedded stars of the Rho Ophiuchi cloud (up to 0.81 μm; cf. Carasco et al., 1973; Coyne et al., 1974; Whittet and van Breda, 1975; Whittet, 1981) indicate that the grains have grown within the Rho Ophiuchi cloud. This assumption is supported by observations with the OAO-2 satellite at UV-wavelengths which show a decreasing UV-extinction toward increasing optical depths in the Sco OB2 association (Snow et al., 1974). On the other hand a relatively large UV-extinction was observed toward HD 147889 (Bohlin and Savage, 1981). The presence of a very small particle component in the vicinity of this double star was, therefore, suggested (Crutcher and Chu, 1985).

Acknowledgements

It is a pleasure for me to thank Prof. Zimmermann and Dr Pfau for helpful discussions. I am grateful to Dr Stecklum for pointing out the important references.

References

Andrews, M. D. and Basart, J. P.: 1984, *Astron. J.* **89**, 417.
Apparao, K. M. V., Hayakawa, S., and Hearn, D. R.: 1979, *Astrophys. Space Sci.* **65**, 419.
Bečvář, A.: 1962, *Atlas Coeli 1950.0*, Academic Press, Prague.
Bernes, C. and Sandqvist, Aa.: 1977, *Astrophys. J.* **217**, 71.
Bignami. G. F.: 1985, *Sky Telesc.* **70**, 301.
Black, J. H. and Fazio, G. G.: 1973, *Astrophys. J.* **185**, L7.
Bohlin, R. C. and Savage, B. D.: 1981, *Astrophys. J.* **249**, 109.
Bok, B. J.: 1956, *Astron. J.* **61**, 309.
Brown, R. L. and Knapp, G. R.: 1974, *Astrophys. J.* **189**, 253.
Brown, R. L. and Zuckerman, B.: 1975, *Astrophys. J.* **202**, L125.
Brown, R. L., Gammon, R. H., Knapp, G. R., and Balick, B.: 1974, *Astrophys. J.* **192**, 607.

Carasco, L., Strom, S. E., and Strom, K. M.: 1973, *Astrophys. J.* **182**, 95.

Castelaz, M. W., Hackwell, J. A., Grasdalen, G. L., Gehrz, R. D., and Gullixson, C.: 1985, *Astrophys. J.* **290**, 261.

Cesarsky, D. A., Encrenaz, P. J., Falgarone, E. G., Lazareff, B., Lauqué, R., Lucas, R., and Weliachew, L.: 1976a, *Astron. Astrohys.* **48**, 167.

Cesarsky, D. A., Encrenaz, P. J., Falgarone, E. G., and Lucas, R.: 1976b, *Astron. Astrophys.* **52**, 299.

Chaisson, E. J.: 1975, *Astrophys. J.* **197**, L65.

Chaisson, E. J. and Vrba, F. J.: 1978, in T. Gehrels (ed.), *Protostars and Planets*, University of Arizona Press, Tucson.

Chini, R.: 1981, *Astron. Astrophys.* **99**, 346.

Chini, R. and Krügel, E.: 1983, *Astron. Astrophys.* **117**, 289.

Chini, R., Elsässer, H., Hefele, H., and Weinberger, R.: 1977, *Astron. Astrophys.* **56**, 323.

Cohen, M. and Kuhi, L. V.: 1979, *Astrophys. J. Suppl.* **41**, 743.

Cohen, J. G. and Wallerstein, G.: 1974, *Astrophys. J.* **189**, 259.

Coyne, G. V., Gehrels, S. J. T., and Serkowski, K.: 1974, *Astron. J.* **79**, 581.

Crutcher, R. M.: 1976, *Astrophys. J.* **208**, 382.

Crutcher, R. M. and Chu, Y.-H.: 1985, *Astrophys. J.* **290**, 251.

Cudlip, W., Emerson, J. P., Furniss, I., Glencross, W. M., Jennings, R. E., King, K. J., Lightfoot, J. F., and Towlson, W. A.: 1984, *Monthly Notices Roy. Astron. Soc.* **211**, 563.

Danks, A. C. and Lambert, D. L.: 1983, *Astron. Astrophys.* **124**, 188.

Elias, J. H.: 1978, *Astrophys. J.* **224**, 453.

Encrenaz, P. J.: 1974, *Astrophys. J.* **189**, L135.

Encrenaz, P. J., Falgarone, E., and Lucas, R.: 1975, *Astron. Astrophys.* **44**, 73.

Falgarone, E. and Gilmore, W.: 1981, *Astron. Astrophys.* **95**, 32.

Falgarone, E., Cesarsky, D. A., Encrenaz, P. J., and Lucas, R.: 1978, *Astrophys. J.* **65**, L13.

Fazio, G. G., Wright, E. L., Zeilik, M., II, and Low, F. J.: 1976, *Astrophys. J.* **206**, L165.

Frerking, M. A., Langer, W. D., and Wilson, R. W.: 1982, *Astrophys. J.* **262**, 590.

Garrison, R. F.: 1967, *Astrophys. J.* **147**, 1003.

Gezari, D. Y., Schmitz, M., and Mead, J. M.: 1984, *Catalog of Infrared Observations*, NASA Reference Publ., 1118.

Gottlieb, C. A., Gottlieb, E. M., Litvak, M. M., Ball, J. A., and Penfield, H.: 1978, *Astrophys. J.* **219**, 77.

Grasdalen, G. L., Strom, K. M., and Strom, S. E.: 1973, *Astrophys. J.* **184**, L53.

Harris, D. H., Woolf, N. J., and Rieke, G. H.: 1978, *Astrophys. J.* **226**, 829.

Harvey, P. M., Campbell, M. F., and Hoffmann, W. F.: 1979, *Astrophys. J.* **228**, 445.

Javanaud, C.: 1979, *Monthly Notices Roy. Astron. Soc.* **188**, 203.

Javanaud, C.: 1980, *Monthly Notices Roy. Astron. Soc.* **190**, 487.

Knapp, G. R., Kuiper, T. B. H., and Brown, R. L.: 1976, *Astrophys. J.* **206**, 109.

Lada, C. J. and Wilking, B. A.: 1980, *Astrophys. J.* **238**, 620.

Lada, C. J. and Wilking, B. A.: 1984, *Astrophys. J.* **287**, 610 (LW).

Leung, C. M. and Brown, R. L.: 1977, *Astrophys. J.* **214**, L73.

Loren, R. B., Evans, N. J., II, and Knapp, G. R.: 1979, *Astrophys. J.* **234**, 932.

Loren, R. B., Sandqvist, Aa., and Wootten, A.: 1983, *Astrophys. J.* **270**, 620.

Loren, R. B., Wootten, A., Sandqvist, Aa., and Bernes, C.: 1980, *Astrophys. J.* **240**, L165.

Martin-Pintado, J., Wilson, T. L., Gardner, F. F., and Henkel, C.: 1983, *Astron. Astrophys.* **117**, 145.

Meyers, K. A., Snow, T. P., Federman, S. R., and Breger, M.: 1985, *Astrophys. J.* **288**, 148.

Minn, Y. K.: 1981, *Astron. Astrophys.* **103**, 269.

Montmerle, T., Koch-Miramond, L., Falgarone, E., and Grindlay, J. E.: 1983, *Astrophys. J.* **269**, 182.

Morfill, G. E., Völk, M. J., Drury, L., Forman, M., Bignami, G. F., and Caraveo, P. A.: 1981, *Astrophys. J.* **246**, 810.

Myers, P. C. and Ho, P. T. P.: 1975, *Astrophys. J.* **202**, L25.

Myers, P. C., Ho, P. T. ., Schneps, M. H., Chin, G., Pankonin, V., and Winnberg, A.: 1978, *Astrophys. J.* **220**, 864.

Pankonin, V. and Walmsley, C. M.: 1978, *Astron. Astrophys.* **64**, 333.

Paul, J. A., Cassé, M., and Montmerle, T.: 1981, in G. Setti, G. Spada, and A. W. Wolfendale (eds.), 'Origin of Cosmic Rays', *IAU Symp.* **94**, 325.

Rydgren, A. E.: 1980, *Astron. J.* **85**, 438.

Rydgren, A. E., Strom, S. E., and Strom, K. M.: 1976, *Astrophys. J. Suppl.* **30**, 307.

Sancisi, R.: 1973, in F. J. Kerr and S. C. Simonson, III (eds.), 'Galactic Radio Astronomy', *IAU Symp.* **60**, 115.

Schmidt-Kaler, Th.: 1982, in K. Schaifers and H. H. Voigt (eds.), *Landolt-Börnstein*, New Series, Group VI, Vol. 2, Subvol. b, Springer-Verlag, Berlin, Heidelberg, New York, p. 453.

Snell, R. L. and Loren, R. B.: 1977, *Astrophys. J.* **211**, 122.

Snow, T. P., Jr. and Jenkins, E. B.: 1980, *Astrophys. J.* **241**, 161.

Snow, T. P., Cohen, J. G., and Cohen, J. G., Jr.: 1974, *Astrophys. J.* **194**, 313.

Snow, T. P., Timothy, J. G., and Seab, C. G.: 1983, *Astrophys. J.* **265**, L67.

Struve, O. and Rudkjøbing, M.: 1949, *Astrophys. J.* **109**, 92.

van Breda, I. G., Glass, I. S., and Whittet, D. C. B.: 1974, *Monthly Notices Roy. Astron. Soc.* **168**, 551.

Vrba, F. J.: 1977, *Astron. J.* **82**, 198.

Vrba, F. J., Strom, K. M., Strom, S. E., and Grasdalen, G. L.: 1975, *Astrophys. J.* **197**, 77.

Vrba, F. J., Strom, S. E., and Strom, K. M.: 1976, *Astron. J.* **81**, 958.

Whittet, D. C. B.: 1974, *Monthly Notices Roy. Astron. Soc.* **168**, 371.

Whittet, D. C. B.: 1981, *Monthly Notices Roy. Astron. Soc.* **196**, 469.

Whittet, D. C. B. and van Breda, I. G.: 1975, *Astrophys. Space Sci.* **38**, L3.

Whittet, D. C. B. and van Breda, I. G.: 1980, *Monthly Notices Roy. Astron. Soc.* **192**, 467.

Wilking, B. A. and Lada, C. J.: 1983, *Astrophys. J.* **274**, 698 (WL).

Wilking, B. A., Harvey, P. M., Joy, M., Hyland, A. R., and Jones, T. J.: 1985, *Astrophys. J.* **293**, 165.

Wilking, B. A., Lebofsky, M. J., Rieke, G. H., and Kemp, J. C.: 1979, *Astron. J.* **84**, 199.

Wouterloot, J. G. A.: 1984, *Astron. Astrophys.* **135**, 32.

Zeng, Q., Batrla, W., and Wilson, T. L.: 1984, *Astron. Astrophys.* **141**, 127.

THE RELATION BETWEEN MOLECULAR CLOUDS AND
STELLAR KINEMATICS*

JAN PALOUŠ

Astronomical Institute of the Czechoslovak Academy of Sciences, Prague, Czechoslovakia

(Received 27 June, 1986)

Abstract. The relation between molecular clouds, star clusters, and the stellar component of the galactic disk is investigated. According to Elmegreen (1985) bound stellar systems, e.g., open star clusters, can be formed from molecular cloud of mass $\sim 10^4 \, M_\odot$. A close encounter with a giant molecular cloud or massive black hole disrupts such stellar systems and forms superclusters. This explains why some open star clusters are so mass-deficient. Unbound stellar systems, e.g., expanding OB associations, are formed from molecular clouds of mass $\gtrsim 10^5 \, M_\odot$. When disruptive O-type stars appear the star formation is halted and the cloud is destroyed. An example of the relict of GMC disruption in the solar vicinity is Gould's belt. The velocity dispersion-versus-age relation is also investigated and explained as a consequence of gravitational scattering of stars on GMC, or massive black holes, or as due to recurrent transient spirals.

1. Introduction

The model of our Galaxy created in the 1920's by B. Lindblad and J. H. Oort assumed the differential rotation around the far centre in Sagittarius occurring in the steady-state. The elliptical shape of Lindblad's epicycles in the galactic symmetry plane corresponds to the ellipsoidal hypothesis proposed in 1907 by K. Schwarzschild, which gives an equilibrium statistical representation of the individual stellar velocities.

The basic assumption of this concept, the steady-state of the Galaxy, needs to be modified when we take into account the structures and processes in the ISM, star formation in dense molecular clouds and when we realise the relation between the molecular clouds and stellar kinematics. Stellar motions of young stars record both the motion of their parent cloud and motions of its fragments, the superclusters form from open star clusters due to close encounters with massive molecular clouds or black holes, and the stellar velocity-versus-age relation is produced due to the scattering of stars on GMC or on massive black holes.

Stars are likely to be born from ISM out of the dynamical equilibrium and their approach to some steady-state seems to be much more complicated than believed up to now. The present Galaxy is a dissipative structure which is far from steady-state and the attempts to describe this system in terms of equilibrium statistical physics are incorrect if at all reasonable.

* Paper presented at a Workshop on 'The Role of Dust in Dense Regions of Interstellar Matter', held at Georgenthal, G.D.R., in March 1986.

Astrophysics and Space Science **128** (1986) 151–156.
© 1986 *by D. Reidel Publishing Company*

2. Molecular Clouds and Star Formation

The radio and infrared observations of the interstellar molecules, and mainly the CO observations, discovered molecular clouds, which are the most massive constituents of the Galaxy. More than a half of the ISM is in molecular form and the molecular cloud cores with densities higher than 10^4 cm^{-3} are possible sites of star formation.

The star formation probability depends on the intrinsic stellar mass, but it is also a function of the total mass of the parent cloud. According to Elmegreen (1985) the massive O-type stars seldom form in small or dwarf molecular clouds (DMC) with masses $\sim 10^4 M_\odot$. The DMC produce only less massive stars and the absence of disruptive O stars causes a large fraction of the parent cloud to be transformed into stars, which can form a bound stellar cluster.

But the majority of the molecular mass is in giant molecular clouds (GMC) with masses $\gtrsim 10^5 M_\odot$ (according to Sanders et al., 1984, 1985, about 85%). There the formation of massive O-type stars leads to core disruption which halts the formation of less massive stars. The stellar relicts of GMC evolution are unbound groups of stars, e.g., expanding OB associations. In our opinion the majority of disk stars come from GMC via unbound stellar groups.

The star formation process in GMC is very inefficient. According to Evans II (1985) only 0.1–1.0% of the mass of a parent GMC form stars and the rest is dispersed in aspherical outflows. The outflowing mass, which is expelled from the star forming cloud cores, carry large energies of 10^{44}–10^{46} ergs. These outflows may be driven by gravitational contraction and infall in direction perpendicular to the direction of outflow.

What happens to the non-stellar relict of GMC? The dispersed molecular gas needs to be reorganised into the next generation of molecular clouds. The role played by spiral arms is still unclear. We do not know if GMC are preferably formed in spiral arms because of the transition through the shock-wave connected with spiral arms and subsequent gas contraction, or if arms are formed from chains of GMC, which acquire a spiral shape due to galactic differential rotation. A formative mechanism for molecular clouds has been proposed by Tenorio-Tagle and Palouš (1986).

3. The Young Stars

Stars younger than 10^8 yr – here we consider the evolutionary age of stars after they settle down to ZAMS – spend less than one half of the epicycle in orbit after their formation and, consequently, their orbits have not succeeded in phasemixing completely. Their space and velocity distribution still contains some information on protostellar molecular clouds, their motions, and on the star formation process.

The local young B-type stars are concentrated in Gould's belt, which is inclined to the galactic plane, and which represents the local 'Schindel' in stellar distribution. The uneven distribution within the plane of Gould's belt, when projected onto the galactic plane of symmetry, can be seen in Figure 1. The distinct appearance of the

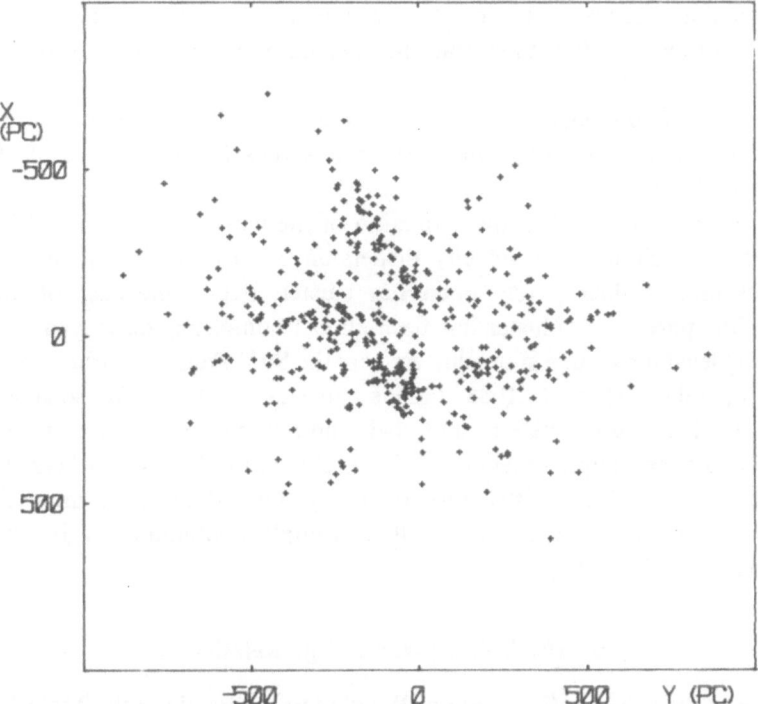

Fig. 1.   The space distribution of B stars younger than 50 Myr in the galactic symmetry plane. The X-, Y-axes are centred on the Sun, the X-axis points toward the galactic centre and the Y-axis in the direction of galactic rotation.

Scorpio–Centaurus and Orion associations shape this space distribution into a 'dragonfly' pattern.

It is obvious that such a remarkably deviating space distribution must be connected with strong deviations from the usual galactic rotation. This has been shown by Palouš (1985), Westin (1985), Lindblad and Westin (1985) and others.

What does this space and velocity distribution represent? In our opinion we observe the stellar relict after a GMC destruction, which led to the creation of our asymmetric 'dragonfly', but we do not know what was the source of the enormous energy ($\sim 10^{52}$ erg) needed for creation of this expanding pattern about 500 pc across.

The motion of its left-wing containing the Orion association differs from the motion of the right-wing where the Scorpio–Centaurus association resides (Palouš, 1986b). This lead us to a picture of the expansion after the GMC destruction.

4. The Superclusters

The star formation in dwarf molecular clouds can produce bound stellar systems, as pointed out in Section 2. Their masses should be comparable to that of the parent clouds, i.e., $\sim 10^4 \, M_\odot$. But masses of some open clusters are lower: Pleiades $\sim 400 \, M_\odot$, Hyades $\sim 200 \, M_\odot$, etc. The missing mass is in these open star clusters.

We assume that these open star clusters were more massive immediately after their formation than now, and that their mass has decreased in the course of their life due to:

(1) internal evolution, e.g., evaporation, formation of binaries, etc.,
(2) the close encounter with a massive objects (GMCs or massive black holes, Wielen, 1985).

Both processes above lead to the formation of the widely separated stellar groups moving with a small internal velocity dispersion almost parallel in space. These superclusters involve their parent open star cluster, but in the case of the Sirius supercluster the parent star cluster has been almost completely destroyed.

There are at least three large superclusters near the Sun: Pleiades, Sirius, and Hyades (Eggen, 1970; Bubeníček *et al.*, 1985; Palouš and Hauck, 1986). The solar vicinity is penetrated by these superclusters, and their motion has to be considered in the assessment of stellar sample kinematics. For example, the Sirius and Hyades super-clusters influence the velocity distribution of stars younger than 10^9 yr, and we assume that the deviation of vertex is the result of sample contamination by these two superclusters (Palouš, 1986a).

5. The Velocity-versus-Age Relation

The increase of the velocity dispersion of Population I stars along the Main Sequence has been interpreted as an indication of an increase of the velocity dispersion with age. Wielen (1974) investigated the motions of stars from Gliese's catalogue of nearby stars within 20 pc of the Sun. Their ages were also classified according to the position in the colour-magnitude diagram, or according to the Ca II emission. The total velocity dispersion of McCormick's K + M dwarfs ranges from 22 km s^{-1} for the youngest stars, about 3×10^8 yr old, to 75 km s^{-1} for the oldest stars of about 9×10^9 yr. The increase of the velocity dispersion is roughly the same in all components and it is approximately proportional to the cube root of their age.

A possible explanation of this heating of the galactic disk is the subsequent accelera-tion of the disk stars after their formation from the interstellar cloud with small velocity dispersion. This heating has been described by Wielen and Fuchs (1983) as a diffusion in velocity space regardless the source of acceleration. The acceleration process is the gravitational scattering of stars by GMC (Lacey, 1984; Villumsen, 1985) or by massive ($\sim 2 \times 10^6 M_\odot$) black holes coming from the dark galactic halo (Lacey and Ostriker, 1985).

An interesting by-product of this is the creation of the power-law tail of the velocity distribution, which is confirmed in the existence of high-velocity A-stars. Another effect is the disruption of bound stellar systems, e.g., open star clusters, due to tidal forces in close encounters between these systems and GMC, or massive black holes, which results in the formation of superclusters.

Carlberg *et al.* (1985) and Palouš and Piskunov (1985) have also investigated the velocity-versus-age relation by an independent method using the individual ages for B-,

A-, and F-type stars from fitting the star position in the HR diagram to the evolutionary tracks from theoretical models of stellar evolution. In both papers the absolute magnitudes, colour excesses and effective temperatures were based on the Strömgren photometry indices.

Their results, however, do not agree with Wielen's. The velocity dispersion-versus-age relation is nearly constant with age for:

(1) A-type stars older than 2×10^8 yr (Paloš and Piskunov, 1985).

(2) F-type stars older than 6×10^9 yr (Carlberg et al., 1985).

These data are in better agreement with models of secular heating of the disk caused by recurrent transient spiral waves as modelled by Carlberg and Sellwood (1985).

In these models a disk builds up as the gas settles down in a dark halo. The continuous accretion of the gas in the course of disk formation could produce cooling which would ensure that the disk never stabilizes completely. Without continuous dynamical cooling, the disk is rapidly heated and becomes completely stable to the formation of spirals. This accretion provokes the transient spirals and the dynamical equilibrium between cooling and heating of the disk is established.

6. Conclusions

Star formation occurs in dense cores of molecular clouds. But the parent clouds are very much more massive that the individual stars, which implies that stars are born in groups. Giant molecular clouds of mass $\sim 10^5 M_\odot$, which also form disruptive O stars, create expanding stellar systems (e.g., OB associations) and the rest of the molecular gas is dispersed in asymmetric expansion. In our opinion the majority of stars were formed from GMC and local example of the relict after a GMC disruption is Gould's belt.

Dwarf molecular clouds of mass $\sim 10^4 M_\odot$ form disruptive O-type stars only seldom. They can produce bound stellar groups, e.g., open star clusters. If a bound group is disturbed in the course of its life in a close encounter with GMC, or with a massive black hole, the tidal forces form the supercluster. This can explain why some open star clusters are so mass-deficient.

The increase of the velocity dispersion with age is another consequence of the gravitational interaction between stars and GMC or massive black holes. The decision between this and an alternative scenario of disk heating via recurrent transient spirals depends on better kinematical data yielding a more definite velocity dispersion-versus-age relation.

The molecular clouds are related to stellar kinematics through star formation and gravitational interactions between stars and GMC. To explain the state and processes in our Galaxy and the above relations we should use methods of non-equilibrium statistical physics.

References

Bubeníček, J., Palouš, J., and Piskunov, A. E.: 1985, Soviet AJ 62, 1073.
Carlberg, R. G. and Sellwood, J. A.: 1985, Astrophys. J. 292, 79.

Carlberg, R. G., Dawson, P. C., Hsu, T., and van den Berg, D. A.: 1985, *Astrophys. J.* **294**, 674.

Eggen, O. J.: 1970, *Vistas Astron.* **12**, 367.

Elmegreen, B. G.: 1985, in D. C. Black and M. S. Mathews (eds.), *Protostars and Planets II*, Univ. of Arizona Press, Tucson, p. 33.

Evans II, N. J.: 1985, in D. C. Black and M. S. Mathews (eds.), *Protostars and Planets II*, Univ. of Arizona Press, Tucson, p. 175.

Lacey, C. G.: 1984, *Monthly Notices Roy. Astron. Soc.* **208**, 687.

Lacey, C. G. and Ostriker, J. P.: 1985, *Astrophys. J.* **299**, 633.

Lindblad, P. O. and Westin, T. N. G.: 1985, in W. Boland and H. van Woerden (eds.), *Birth and Evolution of Massive Stars and Stellar Groups*, D. Reidel Publ. Co., Dordrecht, Holland.

Palouš, J.: 1985, *Bull. Astron. Inst. Czechosl.* **36**, 261.

Palouš, J.: 1986a, in M. S. Mathews, R. Smoluchowski, and J. Bahcall (eds.), *The Galaxy and the Solar System*, Univ. of Arizona Press, Tucson, p. 45.

Palouš, J.: 1986b, in preparation.

Palouš, J. and Hauck, B.: 1986, *Astron. Astrophys.*, **162**, 54.

Palouš, J. and Piskunov, A. E.: 1985, *Astron. Astrophys.* **143**, 102.

Sanders, D. B., Scoville, N. Z., and Solomon, P. M.: 1985, *Astrophys. J.* **289**, 373.

Sanders, D. B., Solomon, P. M., and Scoville, N. Z.: 1984, *Astrophys. J.* **276**, 182.

Tenorio-Tagle, G. and Palouš, J.: 1986, in preparation.

Villumsen, J. V.: 1985, *Astrophys. J.* **290**, 75.

Westin, T. N. G.: 1985, *Astron. Astrophys. Suppl. Ser.* **60**, 99.

Wielen, R.: 1974, in G. Contopoulos (ed.), *Highlights in Astronomy*, D. Reidel Publ. Co., Dordrecht, Holland, p. 395.

Wielen, R.: 1985, in J. Goodman and P. Hut (eds.), 'Dynamics of Star Clusters', *IAU Symp.* **113**, 449.

Wielen, R. and Fuchs, B.: 1983, *Kinematics, Dynamics, and Structure of the Milky Way*, D. Reidel Publ. Co., Dordrecht, Holland, p. 81.

Discussion

B. Stecklum: I have two questions:

(a) You mentioned the scenario of star formation proposed by Elmegreen. In this picture the massive stars form in groups, and so the question arises if the OB associations involve also low-mass stars. You proposed certain processes which can lead to evaporation of low-mass stars. Do you think that these processes work in such a manner that they can drive the low-mass stars out of the associations so that after the removal of the gas we see only the high-mass stars?

(b) What about the total radius of these superclusters? You mentioned a dimension of about 100 pc, whereas their mass should amount to only a few 100 solar masses. Are these clusters stable against tidal disruption?

J. Palouš: Unfortunately, I cannot answer your first question in full. I do not know if there is an evaporation of low-mass stars from OB associations in the Elmegreen picture. But I assume that there must be also some low-mass stars in the associations. We do not see them because of selection effects. To the second question: It is supposed that the superclusters are much younger from the dynamical point of view than from the point of view of internal stellar evolution. They are certainly dynamically unstable. The tidal forces from the gradients of the mean gravitational field of the Galaxy and the irregular forces from massive clumps make the superclusters quite unstable. But in Wielen's picture the superclusters have not been formed so far in the past as their parent open star clusters. The members of a supercluster spend a lot of their life inside an open star cluster and only later a perturbation from a GMC created the supercluster.

THE IRAS SATELLITE – SOME REMARKS ON THE DATA, THEIR AVAILABILITY AND THEIR USE IN LEIDEN*

H. J. HABING

Sterrewacht Leiden, The Netherlands

(Received 27 June, 1986)

1. About IRAS

The Infrared Astronomical Satellite (IRAS) was jointly developed and operated by the US National Aeronautics and Space Administration (NASA), the Netherlands Agency for Aerospace Programs (NIVR), and the UK Science and Engineering Research Council (SERC). It functioned only in 1983 but it completed its main task: the making of a survey of the sky in four infrared wavelength bands. The satellite, its mission and its first results have been described elsewhere (see, e.g., the special issue of *Astrophys. J. Letters*, 1 March, 1984, or an article by Habing and Neugebauer in *Scientific American*, November 1984). In June 1985 a conference was held in Noordwijk, the Netherlands, where scientific results from the IRAS mission were discussed; the proceedings will appear in May 1986 (Israel, 1986); a second 'IRAS conference' will be held in June 1986 in Pasadena (California) and also its proceedings will be published.

The most important part of the mission was to carry out a survey of the whole sky in four broad wavelength bands, centred at 12, 25, 60, and 100 µm, with limited angular resolution (from 0.8×4.5 (arc min)^2 at 12 µm to 3.0×5.0 (arc min)^2 at 100 µm). The survey was repeated 2 times; the three surveys are called HCON1, HCON2, and HCON3 – HCON stands for 'hours confirmation': each scan was reobserved within 36 hours, to confirm detections made the first time. HCON1 observations and HCON2 observations were made usually two weeks apart; HCON3 observations were made 6 months later. Whereas HCON1 and HCON2 observations covered 95% of the sky, HCON3 was terminated at 72% completion when the last drop of coolant (superfluid He) left the dewar on 22 November, 1983. Low-resolution spectra between 8 and 23 µm were obtained of bright, well-isolated point sources by a separate spectrograph that operated simultaneously with the survey instrument. Since the survey operations required only 60% of the available time, the remaining 40% was used for so-called 'additional observations', in which either the survey instrument was used to do longer integrations on small parts of the sky, thus improving the signal to noise ratio, or where a special instrument, the Chopped Photometric Channel (CPC) was used to obtain

* Review presented at a Workshop on 'The Role of Dust in Dense Regions of Interstellar Matter', held at Georgenthal, G.D.R., in March 1986.

Astrophysics and Space Science **128** (1986) 157–162.
© 1986 *by D. Reidel Publishing Company*

maps, with a better angular resolution at the longer wavelengths than the survey instrument provided for.

The analysis of the survey data had two different goals: (1) to provide a catalogue of infrared point sources; (2) to provide sky maps of the infrared emission. Before launch item (2) was considered a very desirable goal but one that would be probably too high; in fact the satellite performed so well in flight, that the goal could be reached quite satisfactorily. The first edition of the *Point Source Catalogue* appeared in November 1984, together with that of the spectra, and with the first edition of the sky maps from the HCON3 data. Successively through 1985 and 1986 further 'data products' have become available. In November 1985 all the 'additional observations' have been made public. Further processing of the original data is continuing both in the U.S. and in The Netherlands. In the U.S. a major goal is to provide a *Point Source Catalogue* that extends to lower sensitivity levels and to provide sky maps of better quality, both with respect to calibration and to the removal of instrumental background effects. In The Netherlands the emphasis is on improved sky maps: this is the so-called 'GEISHA' project that is caried out under supervision of Dr P. Wesselius at the University of Groningen.

Access via computer to practically all the original IRAS data and to the derived products is possible through an IRAS data centre at the Rutherford/Appleton Laboratories in the U.K. and through one at the California Institute of Technology at Pasadena (California) or through university computers at Leiden and at Groningen in The Netherlands. Several derived products, though, have been freely distributed – in continental Europa by the Centre de Données Stellaires in Strasbourg. In all cases these products are in the form of magnetic tapes; in a few cases also in hard copy. To use these distributed data with profit one needs two documents, the first describing the mission, the data analysis, and the *Point Source Catalogue* and the *Sky Flux Maps*; the second the *Small-Scale Structure Catalogue*. These two documents are called (1) *IRAS, Catalogs and Atlases, the Expanatory Supplement* and (2) *IRAS Small-Scale Structures Catalog*. Both documents have appeared in preprint form but they will be printed soon by the U.S. Government Printing Office. For a summary of the more current IRAS data products, see Table I.

Finally, let me give you an illustration of the IRAS results, notably a breakdown of the contents of the *Point Source Catalogue*; I owe this analysis to T. Chester's contribution to the 1985 IRAS conference. The *Point Source Catalogue* contains 158 000 stars (65% of the total number of entries in the catalogue); 35 000 'galactic objects' (14%): compact H II regions, hot cores of molecular clouds, planetary nebulae, etc.; 33 000 (13%) condensations in the thin 'cirrus clouds'; and 22 000 (9%) galaxies. Of the 158 000 stars 18 000 objects have colours indicating a Rayleigh–Jeans tail of a Planck curve; 81 000 objects have colours indicating some circumstellar emission and 59 000 stars are too faint to characterize.

TABLE I

The following sets of IRAS data are present and accessible in the Netherlands. Copies of some of the sets can be obtained through the Centre de Donneées Stellaires in Strasbourg, usually at marginal costs.

The first three items are only in book form:

1. The Explanatory Supplement to the IRAS catalogs and atlases. Freely available in preprint form; only a very few copies have been left.
2. The *IRAS Small-Scale Structure Catalogue – Explanatory Supplement*. Freely available in preprint form.
3. Cataloged galaxies and quasars observed in the IRAS Survey. Freely available in preprint form.

The following data sets are all available on magnetic tape; sometimes a representation on microfiche or a photographic representation exists.

4. *Point Source Catalogue* (2 tapes; 1600 BPI). Also in microfiche. Microfiche and tape are freely available.
5. *Small-Scale Structure Catalogue* (1 tape; 1600 BPI). Also in microfiche. Microfiche and tape are freely available.
6. *Working Survey Data Base and Ancillary File* (2 tapes; 6250 BPI). This data base contains more detailed information on individual point sources from item 4, the *Point Source Catalogue* – e.g., individual detections. Freely available.
7. *Atlas of Low Resolution Spectra* (2 tapes; 6250 BPI). Contains ~ 5000 spectra between 6 and 23 μm of the brighter, well-isolated point sources. Tapes are freely available. The Atlas is in press in the *Supplements to Astronomy and Astrophysics*.
8. *Skyflux HCON1* (27 tapes; 6250 BPI). The brightness distribution of the sky in the four wavelength bands as derived from HCON1 data, organized in plates of 16° × 16°. Freely available. A photographic copy is also available at cost price (~ 1500 Dfl.) from the Laboratory for Space Research, University of Groningen (c.o. Dr P. Wesselius).
9. *Skyflux HCON3* (20 tapes; 6250 BPI). Freely available.
10. *Zody History File* 13 tapes; 1600 BPI). Brightness distribution of the sky as in the skyflux maps, but averaged over bins of 0.5°. Freely available.
11. Maps constructed from 'Additional Observations' taken with the CPC (3 tapes; 1600 BPI). Freely available directly from the Laboratory for Space Research, University of Groningen (c.o. Dr P. Wesselius). A special explanatory supplement is also provided.
12. Maps constructed from 'Additional Observations' taken with the 'Survey Instrument' (66 tapes; 6250 BPI). Freely available in FITS format directly from the Laboratory for Space Research, University of Groningen (c.o. Dr P. Wesselius). A special explanatory supplement is provided.
13. The full data base of the low-resolution spectrograph contains approximately 100 000 spectra, most of poor quality, from which the *LRS Atlas* (item 7) has been constructed. This data base will not be distributed, but can be consulted in the Netherlands.
14. The CRDD data base (CRDD = Calibrated Raw Detector Data) (450 tapes; 250 BPI). Almost raw data. Not available for distribution, access is possible via Leiden University (the Observatory). Data reduction software is also available.
15. Raw detector data base (approximately 450 tapes; 6250 BPI). The raw original data. Not available for distribution, but access possible via Groningen University (Laboratory for Space Research). In a project called 'GEISHA' the data are being rearranged into an easily accessed data base and new calibration data are provided. In the future (> 1987.5?) these data sets can be quickly accessed; maps can then be derived of significantly higher quality than the present 'sky flux' products.

2. Some Studies Based on IRAS Data and Carried out in Leiden

IRAS data are often valuable by themselves; however, they still gain in importance if they are combined with those obtained by other astronomical instruments. Studies based on IRAS data are being carried out at each of the major astronomical research centres in The Netherlands (Amsterdam, Dwingeloo, Groningen, Utrecht, and Leiden);

the objects range from nearby stars to 'starburst' galaxies. Here I will describe briefly a few IRAS related research topics at the Observatory in Leiden. First is the work by C. de Vries and R. S. le Poole, who discovered that the faint, extended 100 μm emission, named 'Infrared Cirrus' corresponds very closely to faint emission on high-quality sky survey plates (notably the plates of the ESO/SRC southern survey). De Vries and le Poole showed that the optical emission is in fact general galactic light that is scattered back by high-latitude clouds of low extinction ($A_v \sim 0.5$ to 1.5). Analysis of all available data suggests that the grains have an albedo of ~ 0.6 – actually a confirmation of earlier results. Another conclusion is that one of the cirrus clouds is unexpectedly strong at 12 μm, but another is not – the difference is very significant. Excess emission at 12 μm is observed in quite a few IRAS sources – however, its cause is unknown, but it may be due to very small dust particles (< 10 nm). De Vries's thesis of May 1986 gives a full report. Second, I mention the work by E. Deul, H. Walker, and B. Burton on the zodiacal light. The zodiacal emission is a very prominent feature in all IRAS maps, but especially so in the short wavelength maps: for example galactic structure studies are seriously affected by this solar system dust. The derivation of the zodiacal light model from the data is in first instance a gigantic data processing job; it is proceeding quite well. The interpretation in terms of an ecliptic dust distribution will be done in collaboration with R. Wolstencroft from Edinburgh (and in competition with other groups in Europe and in the U.S.).

Third, surprisingly nice maps were obtained by R. Braun (Thesis, Leiden, 1985) who for the first time described the emission from several supernova remnants. If one assumes that the infrared emission is from interstellar dust, swept up by the supernova remnant (and that in the ejected gases dust has not yet formed), then the infrared radiation gives a clue to the total amount of swept up gas – a quantity that is otherwise difficult to measure. A major open question is, of course, how much of the dust has been destroyed by the supernova remnant.

Still other work involves external galaxies: P. Schwering and F. Israel have derived maps of the two Magellanic Clouds. The maps are very carefully calibrated and optimized in angular resolution. A comparison with optical and radiomaps is being made. A similar comparison is made for M31 by R. Walterbos. Schwering and Walterbos will each present their results in their theses (1987, and September 1986). More distant are the Seyfert galaxies, and other active galaxies, detected first by, but now studied by R. de Grijp, W. Keel, and G. Miley. Selecting sources from the *Point Source Catalogue* according to certain criteria and, next, taking optical spectra, these astronomers discovered several hundreds of previously unknown Seyfert galaxies. The work is now extended to the additional observations; galaxies are detected out to $z = 0.5$.

3. Highly Evolved Stars in the Bulge of the Galaxy

If one selects sources from the point source catalogue with the criteria that at 12 μm the flux density is below 6 Jy and that at 25 μm the flux density is the same as at 12 μm

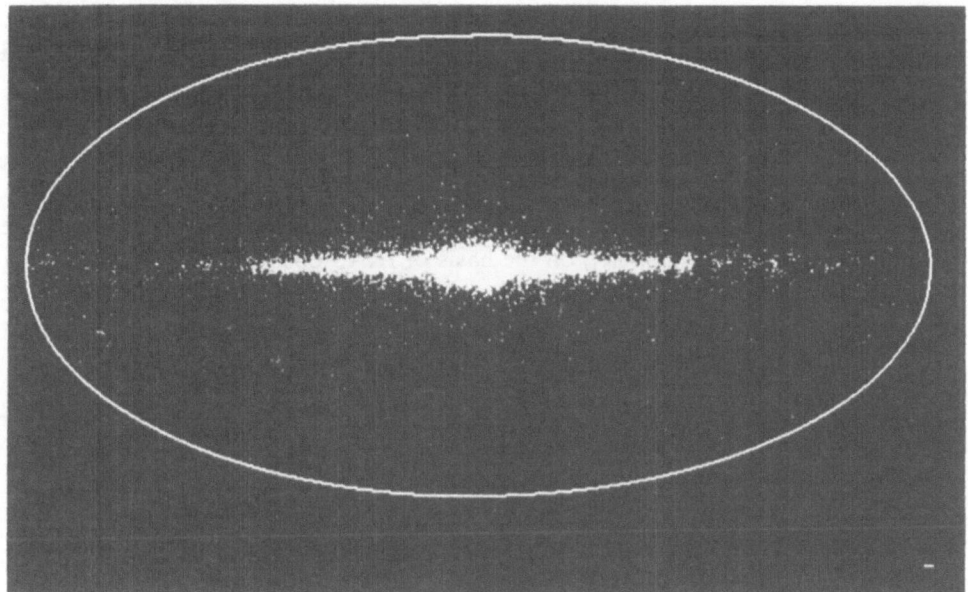

Fig. 1.   Distribution on the sky of sources from the *Point Source Catalogue*, selected according to the criterium that the 12 and 25 μm flux densities are equal within a factor of two and that the 12 μm flux density is below 6 Jy.

within a factor of 2, then one derives a total sample of more than 7000 stars, distributed as in Figure 1; it shows very distinctly a disk and a bulge – clearly a spiral galaxy seen edge on! The selection criteria have to be chosen with some care, if one wants to obtain a clear picture. We found them in Leiden in the spring of 1984 by using the colours of the IRAS counterparts of OH maser stars (now called OH/IR stars). Recent 18 cm observations in Parkes (Australia) and in Effelsberg (BRD) show that more than $\frac{1}{3}$ of the IR sources are indeed OH masers with the maser line profile characteristic of an expanding shell. Therefore, I am convinced that the fast majority of sources in Figure 1 are similar to OH/IR stars: stellar objects in their very last stage of evolution, on the so-called Asymptotic Giant Branch. For the stars near the galactic centre (assumed to be at 8.7 kpc distance) we can estimate the luminosities and we find these to be between 2000 and 6000 $L_\odot$. The existence of such stars in the disk of the Galaxy is to be expected, but not their existence in the bulge: the point is that the brighter stars (say $L > 4000 \, L_\odot$) are estimated to originate from Main-Sequence masses $> 1.1 \, M_\odot$. Unless these stars are metal rich (say $Z \gtrsim 0.10$), they cannot be as old as stars in the bulge are supposed to be, i.e., 15 Gyr. So the question appears to be: is the bulge old, but metal rich; or young? Clearly this question has been asked before and has been tentatively answered – for example by Arp in 1965 (*Astrophys. J.* **141**, 43). But I doubt whether any earlier picture has shown the problem better defined than Figure 1. An analysis of Figure 1 is at present being made in Leiden by Wil van der Veen, who will write a thesis about it (ready in 1988?). A proper analysis not only covers the question: 'what stage of evolution is this of what type of star', but also the question: 'what does their presence in the Bulge tell us about the formation of the Bulge'. To us it is an exciting task to find answers to these questions.

Discussion

P. G. Mezger: Did you have to correct any of your earlier statements about the distribution, number of stars, etc., on the basis of IRAS observations? We think that there are more those stars in the Galaxy with a distribution different for their luminosities.

H. J. Habing: I do not expect that there will be more than 10 000 or 15 000 in the Galaxy. Their total luminosity will be relatively small, about 10^8 solar luminosities.

CIRCUMSTELLAR DUST SHELLS AROUND VERY YOUNG AND MASSIVE STARS*

J. GÜRTLER and TH. HENNING

Universitäts-Sternwarte Jena, G.D.R.

(Received 23 June, 1986)

Abstract. In this paper we investigate the properties of dust in circumstellar shells around very young massive compact IR sources (Becklin–Neugebauer objects).

We found no correlation between the optical depth in the centre of the 10-μm band and the 3.1-μm ice band. An inverse correlation between the strength of the silicate feature and the colour temperature for the 8–13 μm interval was detected. Our sample of BN objects extends this kind of relation already known for Mira stars and OH/IR stars to higher optical depths.

We present a radiative transfer model for BN objects and discuss its main properties. Using this model, the interpretation of the observations led to the conclusion that the type of silicates present in the dust shells of very young stellar objects is different from that type around oxygen-rich giants and supergiants. These different silicates may be tentatively identified with pyroxenes and olivines, respectively.

We studied the influence of the adopted dust model in deriving source parameters of BN objects. The object W3–IRS5 was discussed in some detail.

1. Introduction

The definition of the class of BN objects given in the paper by Henning and Gürtler (1986) includes the presence of absorption features in their mid-infrared spectra. It is obvious that the analysis of these features should provide valuable information on the dust properties in the shells surrounding these objects and some additional clues on the physical parameters of the embedded very young stellar sources. In what follows we will focus our attention on the properties of the silicate component of the circumstellar dust.

2. Observed Properties of the Silicate Features

The observations are virtually restricted to the silicate feature at 10 μm. The only exception is the Becklin–Neugebauer object in Orion where the 20-μm band was observed, too. The strength of the 10-μm band is primarily a measure for the amount of silicate dust in the envelope. Gürtler *et al.* (1985) have calculated the optical depths for 20 BN objects by means of a simple two-component model originally suggested by Gillett *et al.* (1975). In this model the source is assumed to consist of a warm core with optically thin thermal emission of silicate dust surrounded by a cool shell of absorbing silicate dust. This modeling is only a convenient way for approximating the 'continuum' within the wavelength region of the 10-μm band. It is of great interest for the investigation of BN objects to find relations between various observed source parameters. The optical

---

* Paper presented at a Workshop on 'The Role of Dust in Dense Regions of Interstellar Matter', held at Georgenthal, G.D.R., in March 1986.

Astrophysics and Space Science **128** (1986) 163–177.
© 1986 by *D. Reidel Publishing Company*

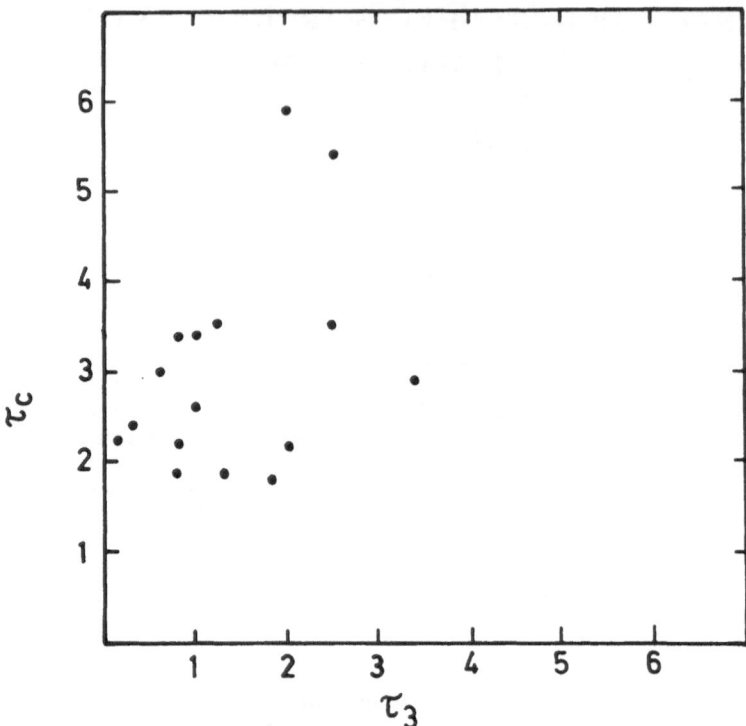

Fig. 1.   Optical depth τ_c of the 10-μm silicate band vs optical depth τ_3 of the 3.08-μm ice band for
BN objects.

depths in the centres of the ice and silicate bands (τ_3 and τ_c) are key parameters concerning physical state and chemical composition of the envelopes. Figure 1 shows τ_c vs τ_3. A linear regression analysis results in a correlation coefficient $r = 0.65$. One reason for that poor correlation is that the inner zone of the shells is ice-free. Ice particles can only exist in the outermost cool regions of these objects. It seems tempting to derive from the depth of the ice band the portion of the 10-μm optical depth due to this region. However, a standard value of the ratio τ_3/τ_c is needed for that, but is unknown up to now. Even worse, it seems doubtful that such a standard value exists at all, because we expect variations of both the density ratio and the size distribution from cloud to cloud.

Our procedure in calculating the optical depth of the silicate band also provides a temperature T characterizing the thermal emission of the object in the band region. It is a kind of colour temperature for the 8–13 μm interval. In Figure 2 τ_c was plotted vs T. Apart from S235–IRS2, for which only poor observations were available, the remaining 19 points clearly indicate the existence of a relationship of the kind

$$\tau_c \propto T^{-p}, \tag{1}$$

with $p = 0.97$.

An inverse correlation between the strength of the silicate feature and the colour temperature in the near IR was found by several investigators for Mira and OH/IR stars (Evans and Beckwith, 1977; Werner *et al.*, 1980a; Engels *et al.*, 1983; Herman *et al.*, 1984). In Figure 2, these stars are located in the hatched area. Our sample extends this relation to higher optical depths.

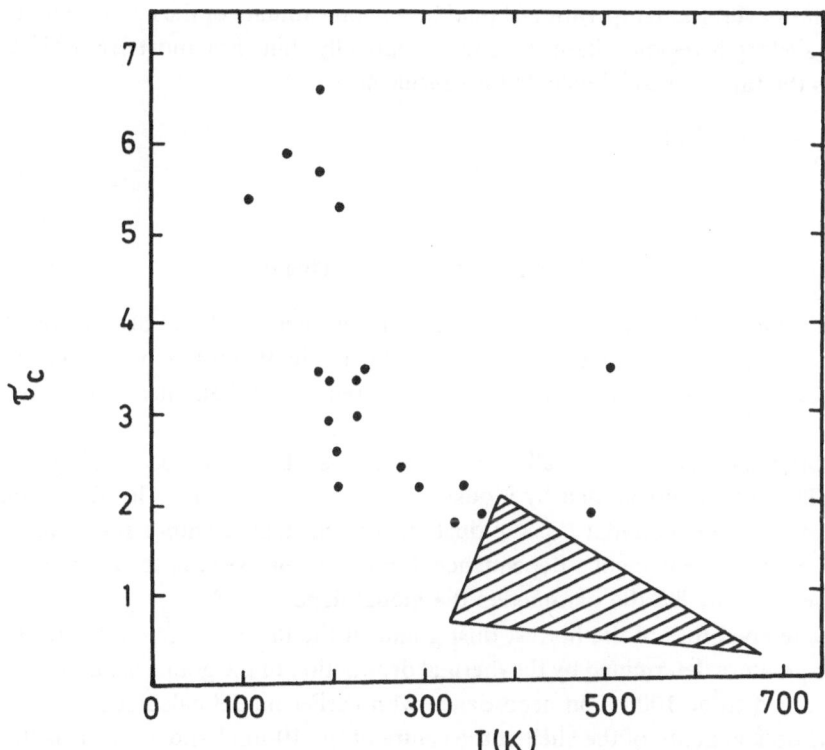

Fig. 2.   Optical depth τ_c of the 10-μm silicate band vs colour temperature T for the 8–13 μm interval. The hatched area is the region where Mira stars and OH/IR stars are situated.

Such a relationship is expected for dust shells having a radial temperature gradient. For smaller optical depths we look deeper into the shell. Therefore, the temperature of the radiating dust will be higher. The exact form of the relationship between τ_c and the colour temperature in the near IR is sensitive to the optical properties of the dust and depends also on the shell structure and the density distribution.

Radiative transfer models enable the derivation of a mass-averaged temperature defined by

$$\langle T_d \rangle = \frac{\int_{r_1}^{r_2} T_d(r) r^{(2-d)} \, dr}{\int_{r_1}^{r_2} r^{(2-d)} \, dr} \, , \tag{2}$$

where $T_d(r)$ is the temperature distribution within the shell and the density is assumed to be given by

$$\rho(r) \propto r^{-d}. \tag{3}$$

The quantities r_1 and r_2 are the inner and outer radii of the dust shell, respectively.

The mass-averaged temperature should be a good estimate of the colour temperature in the wavelength region where the shell is optically thin. For most BN objects that region is the far IR. Model calculations result in

$$\tau_c \propto \langle T_d \rangle^{-5}. \tag{4}$$

3. Parameters of the Model

To extract more information about structure and properties of BN objects, one has to make use of more sophisticated models than the simple two-component model used to derive the optical depths in the silicate feature. Henning (1983a) has developed such a model.

In short, it assumes a spherically-symmetric dust shell illuminated by a hot star at its centre. The star is surrounded by a dust-free cavity of radius r_1. The Eddington flux $H_\lambda^d(r_2)$ at the outer boundary of the dust shell is calculated with a radiative-transfer model without explicit angular dependence. For the details we refer to Henning (1983a). Here, we will only list the parameters the model depends on.

– The temperature of the hottest dust grains at the inner boundary of the shell T_1. This temperature determined by the thermal destruction of the grains is not well known. It is assumed to be 1000 K in accordance with earlier model calculations.

– The optical depth of the shell at the centre of the 10-μm band λ_c, τ_c. It is the most influential parameter.

– The absorption efficiency at the centre of the 10-μm band, $Q_a(\lambda_c)$.

– The surface brightness $F = L_*/r_2^2$, where L_* is the luminosity of the star. In our models are $F = 0.577$ W m^{-2}, 4.453 W m^{-2}, and 71.249 W m^{-2}. These values correspond to a temperature T_2 at the outer boundary of about 30, 50, and 100 K, respectively, in the limit of a black body.

– The density distribution. We adopt a power law (Equation (3)) with exponents $d = 0$, 1, and 2.

The influence of the model parameters on the shape of the 10- and 20-μm bands are extensively discussed in Henning (1983b). Here, we only want to point out four major results.

(1) Depending on the optical depth τ_c, three different regimes can be distinguished: emission, self-absorption, and absorption. Figure 3 shows examples of spectra for these three regimes.

While IR sources in general are observed to belong to each of the three regimes, it is clear from the defining criteria that BN objects are preferably in the absorption regime.

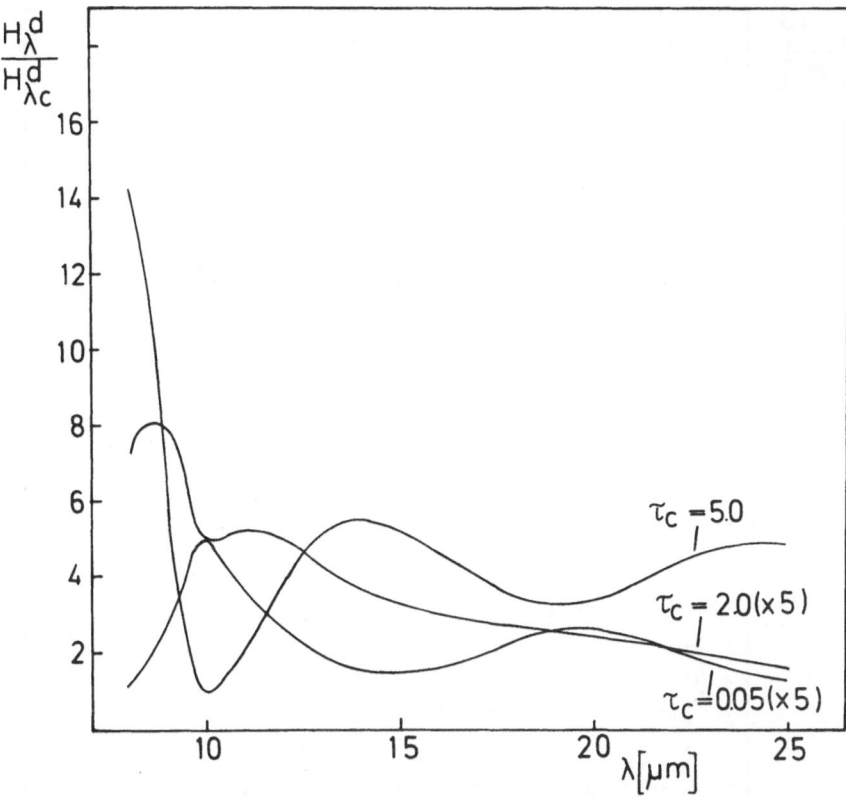

Fig. 3.   Emission, self-absorption, and absorption in the 10-μm band for a model with the parameters
T_1 = 1000 K, F = 0.577 W m^{-2}, $Q_a(\lambda_c)$ = 0.1, and d = 1.

There are, however, a few cases where the self-absorption regime is realized. Typical examples are GL 961 and OMC2–IRS3 with optical depths of about 2.

(2) A shift of the wavelength position of the band centre is never observed with the exception of models with unrealistically high optical depths. To the contrary, when Cohen and Witteborn (1985) fitted the Trapezium profile to their observations of T Tauri stars, the fitted model spectra show shifts in the wavelength position of the emission features. This surprising result is due to inappropriately assuming that the emitting dust slab remains isothermal even for optical depths approaching unity.

(3) The temperature of the star T_* has little effect on the IR spectrum of the shell in the case of the BN objects because the star radiates primarily in the ultraviolet and the shell is optically thick. Inversely, that means that it is not possible to infer the evolutionary stage of the central source from the IR spectrum alone.

(4) We found that the temperature at the inner boundary T_1 has little influence on the shell spectrum. Decrease of the value of T_1 results in a slightly deeper absorption in the 10-μm band.

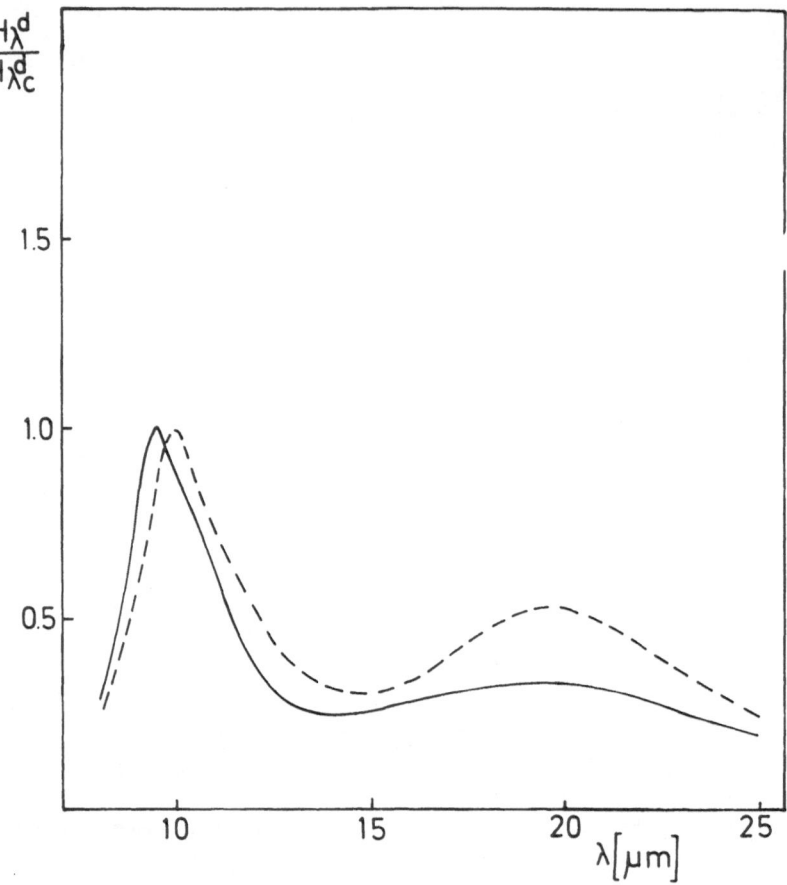

Fig. 4. Emission profiles (τ_c = 0.05) for the pyroxene-type dust (solid line) and the olivine-type dust (dashed line; $Q_a(\lambda_c)$ = 0.1). Model parameters are T_1 = 1000 K, F = 0.577 W m^{-2}, and d = 1.

4. Role of the Optical Properties of the Dust

In addition to the model parameters discussed up to now, the wavelength-dependent absorption efficiency of the dust particles of course determines the exact spectral appearance of the IR sources. At present our knowledge of the absorbing properties of interstellar and circumstellar dust is reaching such a level that the influence of different dust properties has to be more carefully accounted for.

Comparison of laboratory spectra of a large number of silicates with astronomical observations has led to the conclusion that the cosmic silicate particles are very probably amorphous and singled out the groups of olivines and pyroxenes as especially promising candidate materials in many respects. From both the cosmic elemental abundance and modeling of the condensation processes in the circumstellar environment the magnesium silicate seem to be of primary importance. An important difference between the olivines

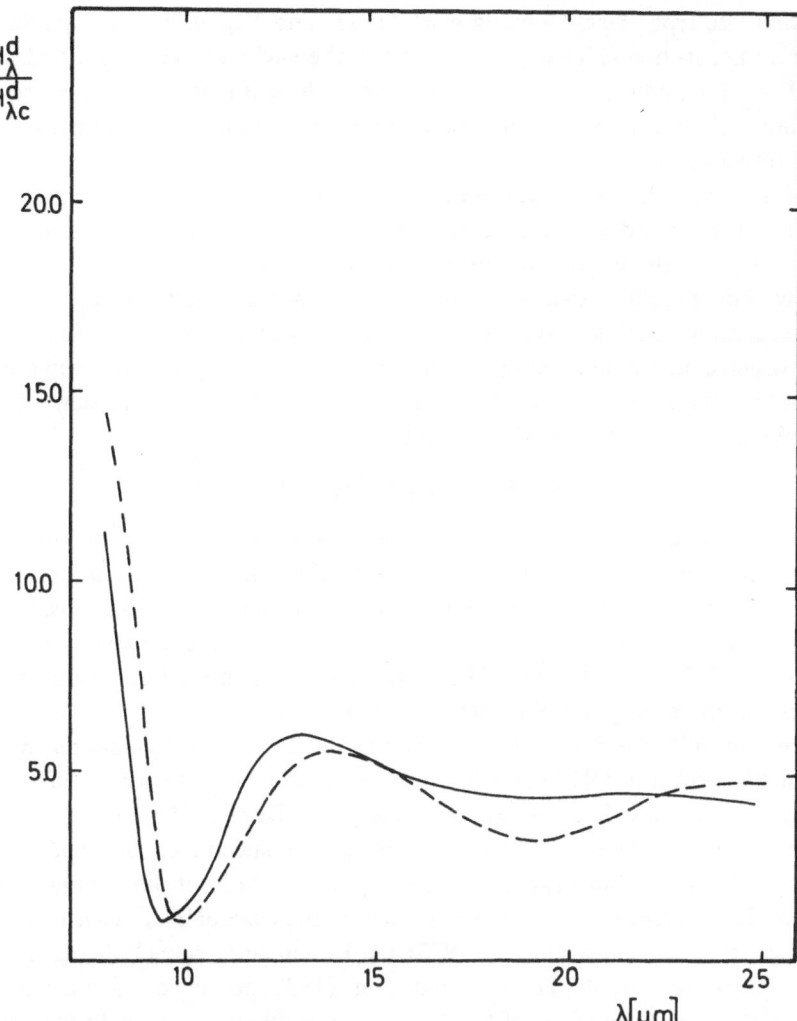

Fig. 5. Absorption profiles (τ_c = 5.0) for the pyroxene-type dust (solid line) and the olivine-type dust (dashed line; $Q_a(\lambda_c)$ = 0.1). Model parameters are T_1 = 1000 K, F = 0.577 W m^{-2}, and d = 1.

and the pyroxenes is the wavelength position of the absorption peak. The 10-μm feature of the latter centres at about 9.5 μm, whereas that of the olivines peaks at about 10 μm.

Such a wavelength difference may be present in the astronomical observations, too. The published spectra seem to indicate that the 10-μm feature in BN objects (Willner *et al.*, 1982) and T Tauri stars (Cohen and Witteborn, 1985) can be well represented by an absorption efficiency peaking at 9.5 μm. Combining experimental measurements on silicates with astronomical observations of the Ney–Allen nebula, Draine and Lee (1984) derived the optical constants of a so-called 'astronomical silicate' appropriate for the above mentioned group of IR objects. On the other hand, from the IR spectra of oxygen-rich late-type stars, Henning *et al.* (1983) and Pégourié and Papoular (1985) deduced an absorption efficiency of the dust in the shells of these stars that peaks at about 10 μm. That difference may imply that two different kinds of silicates are present in astronomical objects, which may be tentatively identified with the pyroxenes and olivines, respectively.

Therefore, we will for convenience refer to the two absorption efficiencies as pyroxene-type dust and olivine-type dust. Surely, we do not want to anticipate a final clarification of the mineralogical composition of circumstellar dust.

In view of the possible existence of the two kinds of silicate dust, it seems reasonable to take secondary chemical reactions into consideration, e.g., hydration and dehydration, which cause, among others, a certain coexistence of pyroxenes and olivines by transmutating the former into the latter and *vice versa*. In this connection we point to the following possible reaction discussed by Larimer (1979)

$$5 \, MgSiO_3 + H_2O \Leftrightarrow Mg_3Si_4O_{10}(OH)_2 + Mg_2SiO_4 \, .$$

Using the radiative transfer model introduced above, we have calculated the emerging spectra for both an optically thin and an optically thick model and using pyroxene-type dust as well as olivine-type dust. The wavelength shift of the band centre in the absorption efficiencies is clearly seen in both the emission and the absorption spectra. The observed shift cannot be caused by radiative transfer effects but must be due to real differences in the absorption efficiencies (Figures 4 and 5).

Evidence for different optical grain properties was also found by Schutte and Tielens (1985) in IR spectra of OH 26.5 + 0.6, IRC + 10°011, and R Cassiopeiae, excluding an interpretation of the differences by different grain sizes. On the other hand, pointing out different halfwidths of the 10 μm feature in the spectra of late-type giants and supergiants, Papoular and Pégourié (1983) argued in favour of different grain sizes as a cause of the differences. Larger particles than in the general interstellar medium were also suggested by Rowan-Robinson (1975) and Rouan and Léger (1984) as cause of the very wide Trapezium profile. Tielens and de Jong (1979) pointed out that a considerable amount of the flux in the wings of the Trapezium profile may be due to emission from carbon grains which are a widely accepted second component of interstellar dust in addition to the silicates. This emission causes the silicate band to appear broader. We have found that in fitting synthetic spectra to the observed spectra of BN objects a good match of the width is generally achieved if pyroxene-type dust and a grain size of $a = 0.1$ μm are assumed.

If the varying bandwidth is a size effect it can be expected that the ratio of the absorption efficiencies in the wings is constant whereas the absorption efficiency in the near IR should vary relative to the absorption efficiency in either wing with changing the grain size as long as submicrometre-sized particles are concerned.

Amorphous instead of crystalline silicates are a natural explanation of the rather great width and lack of internal structure of the observed silicate features. Very broad features can be accounted for by grains with larger sizes than the averages. Of course, it is also possible to get broad profiles peaking between the limits 9.5 and 10.0 μm if mixtures of pyroxene- and olivine-type silicates are permitted. Such a mixture can provide a good fit to the notoriously broad silicate feature observed with Comet Kohoutek. This possibility seems very promising if it is realized that both groups of silicates are constituents of carbonaceous chondrites. Recently, Sandford and Walker (1985) tentatively identified both minerals as major components of interplanetary particles collected in the upper atmosphere. This finding supports the view that such mixtures may indeed exist in comets and elsewhere. However, it must be kept in mind that larger grains than in interstellar space are common for comets and will have their effect on the appearance of the silicate bands.

5. Derivation of Source Parameters from Observed Spectra

Fitting synthetic spectra to observations allows to derive a number of physical parameters of the IR sources. Here, we aim at the influence the adopted dust model exerts on the source parameters found. This is especially important since we have to restrict ourselves to the wavelength region between 8 and 13 μm. Good spectrophotometric data for a larger sample of sources is available only in that wavelength region. Given the observed fluxes at the Earth $\pi S_\lambda(d_e)$, the values of the model parameters are determined by minimizing the quantity

$$\chi^2 = \sum_{i=1}^{N} \left(\frac{EH_{\lambda i}(r_2) - \pi S_{\lambda i}(d_e)}{SB_{\lambda i}} \right)^2 , \tag{5}$$

where d_e denotes the distance to the object, $SB_{\lambda i}$ is the observational error of the flux, mainly assumed to be 5%, N and i are the total number of considered wavelengths and the ith wavelength in the spectral profile, respectively. E accounts for the geometrical dilution of the radiation from the object. In some models we tested whether the introduction of an additional absorption by cold dust in the intervening molecular cloud ('foreground'), expressed by $\exp(-\tau_f(\lambda))$, results in a better representation of the observed spectra.

Table I summarizes the parameters derived for four sources using the two different dust models; O stands for olivine-type dust and P for pyroxene-type dust. The following parameters are shown:

– the foreground absorption $\tau_f(\tau_c)$ giving the best fit,
– the optical depths at the band centre τ_c and in the ultraviolet τ (UV). Here ultraviolet means that wavelength region where the absorption efficiency becomes nearly independent of the wavelength. The region begins at 0.4 μm for our dust model and at 0.18 μm for the dust model of Draine and Lee (1984).
– the absorption efficiency at the band centre $Q_a(\lambda_c)$ and the mass absorption coefficient $\kappa_{m,a}(\lambda_c)$. The mass absorption coefficient was calculated by assuming an internal density of 3.3 g cm^{-3}. The value in parentheses is calculated with a density 2.5 g cm^{-3} given by Perry et al. (1972) for lunar rocks and adopted sometimes in model calculations. The radius of the grains was assumed to be 0.1 μm,

TABLE I

Model parameters for BN objects

No.	Object/Ref.	d_e (kpc)	Dust model	χ^2/N	$\tau_f(\lambda_c)$	τ_c/τ(UV)	$Q_a(\lambda_c)/\kappa_{m,a}(\lambda_c)$ (m² kg⁻¹)	d	F (W m⁻²)	T_2 (K)	V
1	2	3	4	5	6	7	8	9	10	11	12
1	W3–IRS5/(1)	2.3	O	519.8/11	–	50/1667	0.03/68(90)	2	4.453	59.4	50
			O	326.8/11	3	36/1200	0.03/68(90)	1	71.249	98.2	12
			P	138.5/11	–	7.6/51.2	0.13/300(390)	0	0.577	39.0	352
10	BN/(2)	0.5	O	159.7/11	0	4.2/140	0.03/68(90)	2	71.249	80.6	12
			O	161.7/11	0	3.2/32	0.1/230(300)	0	0.577	38.0	252
			P	14.7/11	–	3.4/22.9	0.13/300(390)	0	71.249	91.7	32
28	W33A/(3)	3.7	O	502.2/10	–	19.6/653	0.03/68(90)	1	71.249	96.5	12
			O	96.0/10	2	16.6/166	0.1/230(300)	0	71.249	102.4	23
			P	209.1/10	–	8.4/56.6	0.13/300(390)	1	0.577	41.3	352
32	GL 2591/(1)	1.5	O	278.7/11	0	6.8/227	0.03/68(90)	2	71.249	83.7	12
			P	27.3/11	–	4.6/31.0	0.13/300(390)	1	71.249	94.1	32

References:
(1) Willner et al. (1982).
(2) Aitken et al. (1981).
(3) Capps et al. (1978).

– the exponent of the density distribution d,
– the surface brightness F,
– the temperature at the outer boundary T_2,
– the ratio between outer and inner radii V,
– the outer radius r_2,
– the angular dimensions, θ_2 is the angular radius corresponding to r_2, whereas θ(400 K) denotes the angular radius of the shell having a temperature of 400 K,
– the dust density at the outer boundary $\rho(r_2)$ for internal density 2.5 g cm⁻³,
– the total mass of dust M_d,
– the luminosity of the embedded star L_* and its temperature T_*, the spectral type, and radius r_*, assuming luminosity class V.

We wish to make a note of caution. As always in model fitting a single criterion such as minimizing the χ^2-value may not enable a clear decision between concurring models. Then, the reliability of the derived parameters must be checked against additional observational constraints. We shall discuss this procedure in some detail later in connection with W3–IRS5. To demonstrate the ambivalence involved we list in Table I two models for the Becklin–Neugebauer object with similar χ^2-values.

The data summarized in Table I show the effects of the adopted dust properties very distinctly. Besides the differences in the values calculated for the various parameters a major result is that the synthetic spectra based on Draine and Lee's dust model give a significant better fit to the observed spectra. We consider this point as strong support for the existence of the two types of dust discussed earlier in this paper.

$\theta_2/\theta(400\ K)$ (")	$\rho(r_2)$ (kg m⁻³)	M_d ($M_\odot$)	L_* ($L_\odot$)	T_*/Sp. (K)	r_* (m)	Remarks
14	15	16	17	18	19	20
$\times 10^{15}$ 27.7/0.62	1.19×10^{-18}	6.39	1.06×10^6	48 900/O4	9.95×10^9	Probably
$\times 10^{15}$ 9.8/0.95	4.70×10^{-17}	5.66	2.12×10^6	57 800/O3	1.0×10^{10}	binary
$\times 10^{15}$ 29.0/0.43	1.91×10^{-18}	3.98	1.50×10^5	35 200/O8	7.24×10^9	source
$\times 10^{13}$ 1.3/0.11	4.22×10^{-17}	2.22×10^{-4}	1.74×10^3	18 300/B3	2.89×10^9	(I) Observed flux at
$\times 10^{15}$ 23.3/0.40	6.15×10^{-18}	6.83×10^{-2}	4.58×10^3	21 300/B2–B3	3.45×10^9	(II) 9.5 μm
$\times 10^{14}$ 1.4/0.14	8.25×10^{-17}	2.08×10^{-4}	2.10×10^3	19 000/B2 – B3	2.94×10^9	from (1)
$\times 10^{14}$ 0.9/0.01	1.71×10^{-16}	6.88×10^{-2}	4.74×10^4	29 600/B0–B1	5.75×10^9	
$\times 10^{14}$ 1.6/0.15	6.39×10^{-17}	0.10	1.53×10^5	35 300/O8–O9	7.27×10^9	
$\times 10^{15}$ 6.1/0.04	1.06×10^{-18}	1.29×10^{-1}	1.72×10^4	25 700/B0–B1	4.60×10^9	
$\times 10^{14}$ 1.5/0.13	1.97×10^{-17}	4.30×10^{-3}	2.09×10^4	26 400/B0–B1	4.80×10^9	
$\times 10^{14}$ 1.4/0.10	1.09×10^{-17}	9.69×10^{-4}	1.72×10^4	25 700/B0–B1	4.60×10^9	

The case of the objects W3–IRS5 and W33A is particularly noteworthy. Using olivine-type dust properties, a fit of the observed spectra could not be reached or with rather large an optical depth only. Models based on the pyroxene-type dust lead to very realistic source parameters. Interestingly, the assumption of a foreground absorption results for W33A in a good fit of the observations with olivine-type dust, but at the expense of a rather high luminosity.

We will now discuss W3–IRS5 in more detail. This object is very likely a double source. The diameter of the two sources together is 1".26. The diameter of the individual sources is ≤ 0".25 with an angular separation of 0".9 (Howell *et al.*, 1981; Dyck and Howell, 1981; Neugebauer *et al.*, 1982). Both sources seem to be very similar in respect of their spectra and their luminosities.

Willner (1977) derived a total luminosity of $4 \times 10^5\ L_\odot$, while Werner *et al.* (1980b) give $2 \times 10^5\ L_\odot$. Using pyroxene-type dust we find $1.5 \times 10^5\ L_\odot$. If we assume a single Main-Sequence star, this luminosity corresponds to a spectral type O8. Adopting olivine-type dust, however, we obtain an unreasonably high luminosity of $1 \times 10^6\ L_\odot$. Introducing a foreground extinction leads to a slightly better fit, but the luminosity increases to an even higher value. The optical depth τ_c was found to be 7.6 ± 0.2 with pyroxene-type dust. This is in accordance with the results of Willner (1977), Willner *et al.* (1982), and Gürtler *et al.* (1979) who found 7.6 ± 0.2, 7.64, and 6.8, respectively. With olivine-type dust a minimum χ^2-value could not be found up to an unrealistically high optical depth of 50. The total mass of dust derived by adopting pyroxene-type dust properties is $4.0\ M_\odot$. This is comparable to the $1\ M_\odot$ Willner (1977) estimated by means of a very preliminary model.

Summarizing this discussion we can state that the synthetic spectrum based on pyroxene-type dust gives not only a lower χ^2-value if fitted to the observed spectrum but leads to model parameters which are in general agreement with the current ideas of this object.

Figure 6 shows that the more reasonable source parameters derived by means of the more plausible dust properties go along with the better representation of the observed profile. In this figure triangles mark the observations, the circles are the model with the pyroxene-type dust, whereas crosses represent the model with olivine-type dust.

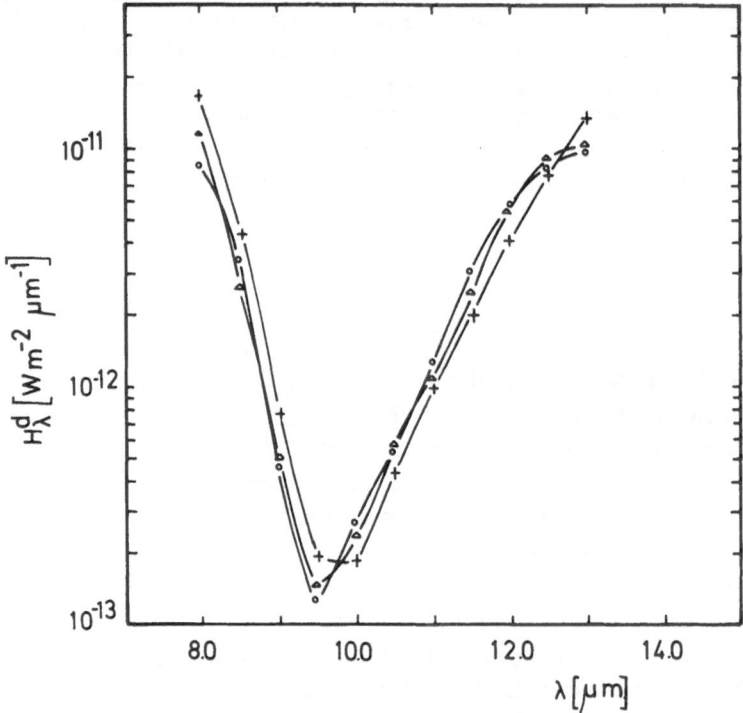

Fig. 6. Representation of the observed profile of W3–IRS5 (triangles). Circles are the best-fitting model with pyroxene-type dust, whereas crosses represent the model with olivine-type dust.

6. Conclusions

Analyzing the observed properties of the IR spectra of BN objects we found an inverse correlation between the optical depth of the silicate feature and the colour temperature of the objects in the band region. No correlation was found between the silicate and the ice features.

Using a radiative transfer model for spherically-symmetric dust shells, we have calculated synthetic spectra for the mid-IR and investigated the influence of the various model parameters on them. The optical properties of the grains play a critical role in fitting the synthetic spectra to the observed profiles. Clear evidence is found that the dust responsible for circumstellar emission in late-type supergiants and giants is different from the dust in BN objects. While the dust around evolved stars has optical properties similar to amorphous olivine-type silicates, pyroxene-type silicates are good candidates for the dust in BN objects.

Acknowledgement

This work constitutes a part of Thesis B (University of Jena) of Th. H.

References

Aitken, D. K., Roche, P. F., Spencer, D. M., and Jones, B.: 1981, *Monthly Notices Roy. Astron. Soc.* **195**, 921.
Capps, R. W., Gillett, F. C., and Knacke, R. J.: 1978, *Astrophys. J.* **226**, 863.
Cohen, M. and Witteborn, F. C.: 1985, *Astrophys. J.* **294**, 345.
Draine, B. T. and Lee, H. M.: 1984, *Astrophys. J.* **285**, 89.
Dyck, H. M. and Howell, R. R.: 1981, *Astron. J.* **87**, 400.
Engels, D., Kreysa, E., Schultz, G. V., and Sherwood, W. A.: 1983, *Astron. Astrophys.* **124**, 123.
Evans II, N. J. and Beckwith, S.: 1977, *Astrophys. J.* **217**, 729.
Gillett, F. C., Forrest, W. J., Merrill, K. M., Capps, R. W., and Soifer, B. T.: 1975, *Astrophys. J.* **200**, 609.
Gürtler, J., Dorschner, J., and Friedemann, C.: 1979, *Astron. Nachr.* **300**, 17.
Gürtler, J., Henning, Th., Dorschner, J., and Friedemann, C.: 1985, *Astron. Nachr.* **306**, 311.
Henning, Th.: 1983a, Thesis, University of Jena.
Henning, Th.: 1983b, *Astrophys. Space Sci.* **97**, 405.
Henning, Th. and Gürtler, J.: 1986, *Astrophys. Space Sci.* **128**, 199 (this issue).
Henning, Th., Gürtler, J., and Dorschner, J.: 1983, *Astrophys. Space Sci.* **94**, 133.
Herman, J., Isaacman, R., Sargent, A., and Habing, H. J.: 1984, *Astron. Astrophys.* **139**, 171.
Howell, R. R., Low, F. J., and McCarthy, D. W.: 1981, *Astrophys. J.* **251**, L21.
Larimer, J. W.: 1979, *Astrophys. Space Sci.* **65**, 351.
Neugebauer, G., Becklin, E. E., and Matthews, K.: 1982, *Astron. J.* **87**, 395.
Papoular, R. and Pégourié, B.: 1983, *Astron. Astrophys.* **128**, 335.
Pégourié, B. and Papoular, R.: 1985, *Astron. Astrophys.* **142**, 451.
Perry, C. H., Agrawal, D. K., Anastassakis, E., Lowndes, R. P., and Tornberg, N. E.: 1972, *Geochim. Cosmochim. Acta* **3**, 3077.
Rouan, D. and Léger, A.: 1984, *Astron. Astrophys.* **132**, L1.
Rowan-Robinson, M.: 1975, *Monthly Notices Roy. Astron. Soc.* **172**, 109.
Sandford, S. A. and Walker, R. M.: 1985, *Astrophys. J.* **291**, 838.
Schutte, W. and Tielens, A. G. G. M.: 1985, in M. Morris and B. Zuckerman (eds.), *Mass Loss in Red Giants*, D. Reidel Publ. Co., Dordrecht, Holland, p. 87.
Tielens, A. G. G. M. and de Jong, T.: 1979, *Astron. Astrophys.* **75**, 326.
Werner, M. W., Beckwith, S., Gatley, L., Sellgren, K., Berriman, G., and Whiting, D. L.: 1980a, *Astrophys. J.* **239**, 540.
Werner, M. W., Becklin, E. E., Gatley, I., Neugebauer, G., Sellgren, K., and Thronson, H. A., Jr.: 1980b, *Astrophys. J.* **242**, 601.
Willner, S. P.: 1977, *Astrophys. J.* **214**, 706.
Willner, S. P., Gillett, F. C., Herter, T. L., Jones, B., Krassner, J., Merrill, K. M., Pipher, J. L., Puetter, R. C., Rudy, R. J., Russell, R. W., and Soifer, B. T.: 1982, *Astrophys. J.* **253**, 174.

Discussion

H. J. Habing: Would the large luminosity you get for the last object not rule out your dust?

J. Gürtler: Yes, this dust is not suitable for modelling BN objects. It is just these improbable results which point to the difference between the dust around late-type stars and the dust in molecular clouds or around young objects.

J. Staude: What about the wavelength? You have a difference in the position of the feature. If you exclude olivine, you need another material to explain the feature in BN objects.

J. Gürtler: One needs a substance that peaks at the right wavelength. The pyroxenes with a maximum at about 9.5 μm do this.

J. M. Greenberg: What happens to the dust once you get it to the interstellar medium? For the interstellar silicate I expect 9.7 μm as the peak position. Is that not correct?

J. Gürtler: It seems that a value somewhat smaller than 9.7 μm, say 9.5 μm, is better.

H. J. Habing: You still have spherical models. Don't you see any need for disks or things like that? They are so popular now.

J. Gürtler: Yes, but I think that geometry does not have a strong effect on the emergent spectrum if you do not look along the axis, of course, where the outflow is coming out. If you have an angle between the polar axis and your line-of-sight I think it does not matter.

J. Staude: But if you look along the axis you do not see the feature!

J. Gürtler: In this case you would not spot the presence of a BN object.

H. J. Habing: What grain size distribution did you use?

J. Gürtler: We use a mean radius of 0.1 μm.

J. M. Greenberg: At 10 μm it would not make any difference to have twice this size.

J. Staude: But would not the shape of the feature depend on the size of the grains?

J. Gürtler: As long as the radius remains small compared to the wavelength of 10 μm the shape of the feature is not very sensitive.

W. Krätschmer: In general, the form of the band is a product of the radius and the shape of the grains.

J. M. Greenberg: There is an interesting difference if you calculate the extinction efficiencies for spherical particles and for elongated ones, in that there is a shift in the peak position depending on the elongation. Did you use Mie theory for small particles?

J. Gürtler: For the P-type dust we calculated the absorption efficiencies using the optical constants given by Draine and Lee. In the case of O-type dust we used the observationally based absorption efficiencies given by Henning *et al*. Here the absorption efficiency in the band centre is an additional parameter in the model computations.

BIPOLAR NEBULAE AND JETS FROM YOUNG STARS: CONTRIBU-
TIONS FROM CALAR ALTO*

H. J. STAUDE

Max-Planck-Institut für Astronomie, Heidelberg-Königstuhl, Bundesrepublik Deutschland

(Received 23 June, 1986)

Abstract. The topic is reviewed with emphasis on observations in the optical and near infrared spectral range.

1. Introduction

Bipolar nebulae are defined morphologically. They consist of two optically bright, axially-symmetric polar lobes, which are separated by a dark equatorial lane. A central star is embedded in this lane, whose light propagating in the equatorial plane is heavily obscured, while the radiation directed towards the poles suffers much less extinction.

In a number of cases the material in the bright polar lobes is found to be flowing away from the central star at high velocity (up to about 100 km s^{-1}). In some objects, highly collimated gaseous jets are embedded in the polar lobes, which are streaming towards the poles at velocities up to 400 km s^{-1}.

It is now well established, that bipolar nebulae occur both, in the phase of early stellar evolution, as well as at the evolved stages of planetary nebulae, postnovae and symbiotic stars. The phenomenon thus results from quite fundamental, although still basically unknown processes related to stellar rotation, stellar winds and their interactions with circumstellar matter and magnetic field.

At the Calar Alto Observatory a variety of observing programmes has been dedicated to the phenomenon of bipolar nebulae during the past years. This review will be restricted to studies of bipolar nebulae associated with young stars. In addition to the characteristics mentioned above, they are situated in regions of active star formation. They are associated with massive molecular clouds and, in some cases, with bipolar molecular flows on larger scales, which are originating at the same central source and oriented along the same axis as the bipolar nebulae. When highly collimated jets are present, they are frequently found to be associated with Herbig–Haro objects embedded in the polar lobes of the nebulae.

Figure 1 summarizes all typical features of bipolar nebulae around young stars, which, however, in individual objects seldom can be observed in this completeness. For instance, in some cases only one of the two bright polar lobes can be observed; the nebula is then called a cometary, but for a number of such objects it could be shown,

* Paper presented at a Workshop on 'The Role of Dust in Dense Regions of Interstellar Matter', held at Georgenthal, G.D.R., in March 1986.

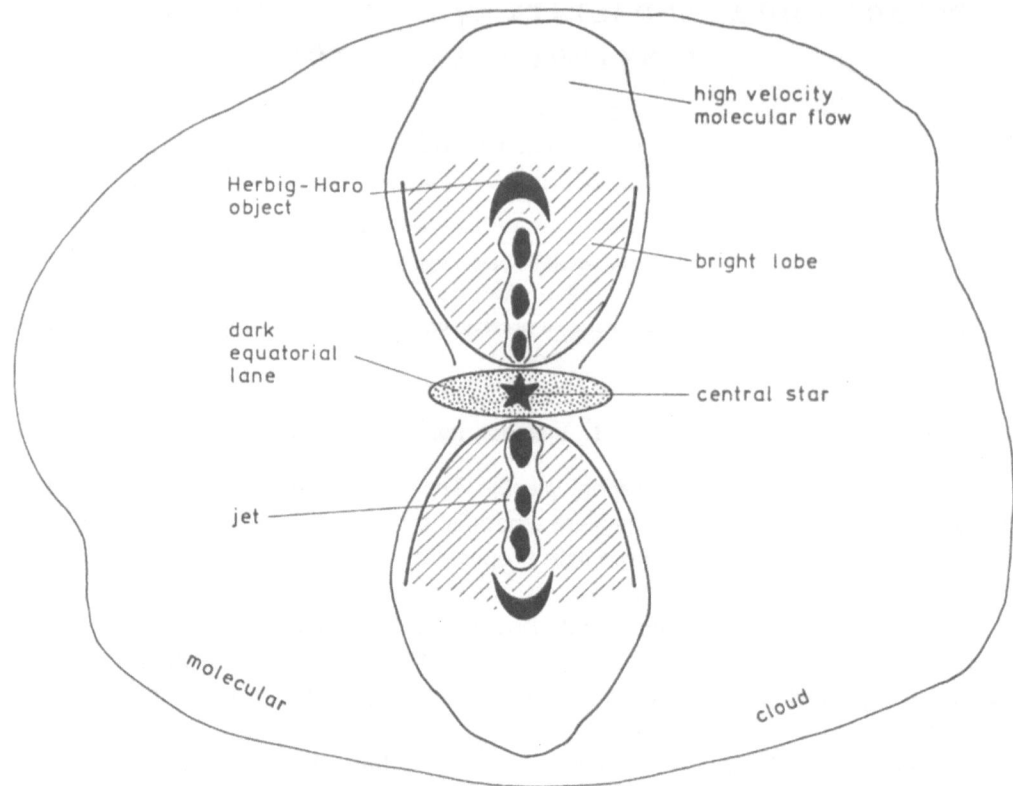

Fig. 1.   Idealized bipolar nebula around a young star.

that a second lobe was hidden by heavy local observation, the nebula thus being intrinsically bipolar. Also, jets emitted by young stars are observed, which apparently are not associated with bright lobes and equatorial dark lanes. But there is now strong evidence for circumstellar dust disks oriented perpendicularly to these jets, and bright polar lobes, which were hidden by local dust, became visible in objects exhibiting jets or high-velocity molecular flows, observed at infrared wavelengths.

Below some well-studied bipolar nebulae associated with young stars will be described in detail, the observations of which have guided the empirical research during the last years. Further, the survey work will be summarized, which was aimed at extending the observational basis and at exploring the relevance of peculiar properties found in individual objects, with respect to the general phenomenon.

2. The Compact H II Region S 106

This bipolar nebula is located at the centre of a molecular cloud (Figure 2) of about $10^4 \, M_{\odot}$ (Lucas *et al.*, 1978). Its bipolar morphology, in particular the stellar nature of the central source, was established by Eiroa *et al.* (1979) from photography in the near

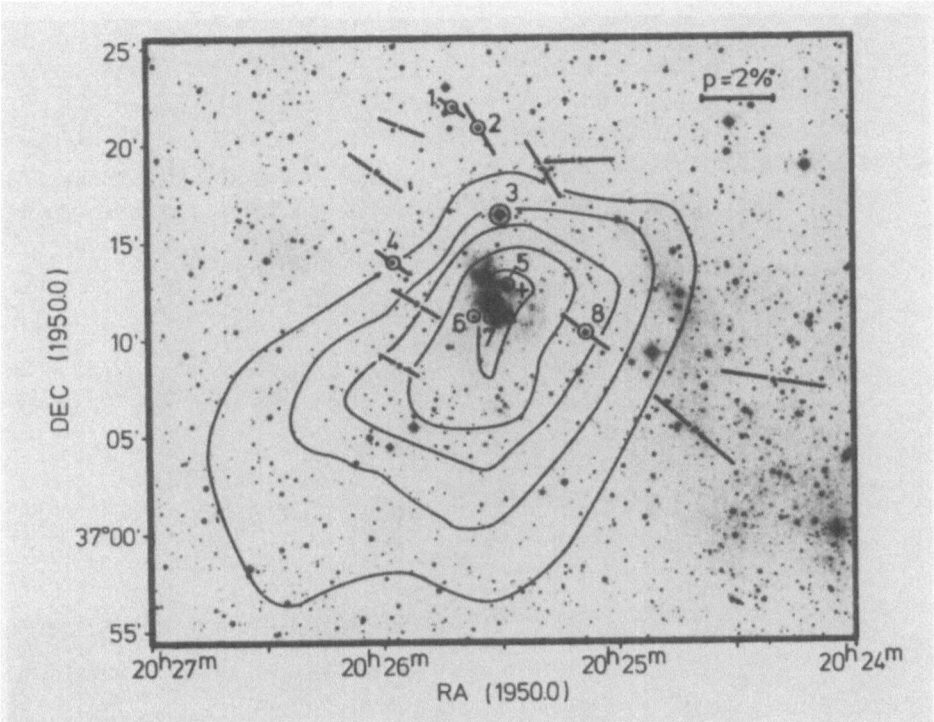

Fig. 2.   The bipolar nebula S 106, a compact H II region embedded in its parent molecular cloud. On the POSS red print the CO map by Lucas *et al.* (1978) is superimposed. Polarimetry of field stars from Staude *et al.* (1982).

IR (see also Figure 3). These authors derived a distance of about 500 pc and a spectral type O9–B0 of the exciting star, which is obscured by a visual extinction of $A_v = 21$ mag. Staude *et al.* (1982) determined the direction of the interstellar magnetic field from polarimetry of neighbouring stars, which is essentially parallel to the axis of symmetry of the bipolar nebula (Figure 2). The map of the molecular cloud in NH_3 (Little *et al.*, 1978; Torrelles *et al.*, 1983) revealed the presence of two broad maxima located approximately in the equatorial plane of S 106 and peaking about 2' east and west of the central star.

Felli *et al.* (1984) presented a VLA map of the bipolar nebula at $\lambda = 1.3$ cm (Figure 4), which reveals that only about 0.1% of the total radio continuum is emitted by the unresolved central source, and that the gap between the two ionized lobes is only about 2 arc sec wide in the central part, much narrower than what is seen in the optical and near infrared, where heavy extinction is effective.

The explanation for the radio gap must be sought in the immediate surroundings of the central star. Ionizing and heating radiation, which is emitted into the equatorial plane, is absorbed within few stellar radii: the diameter of the central source is less than 0.05 arc sec at $\lambda = 2.2$ μm, as derived from speckle interferometry at the 3.5 m telescope

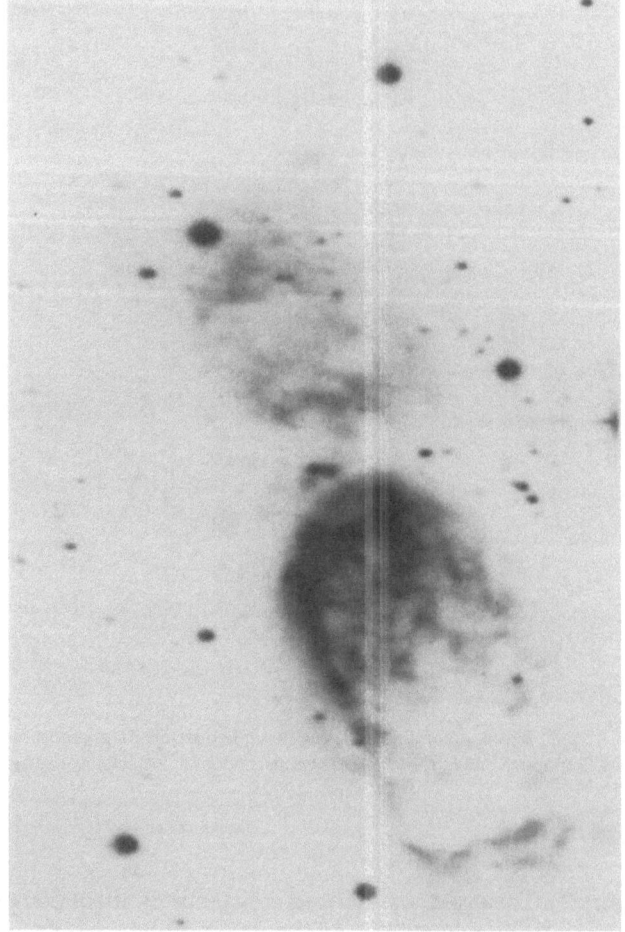

Fig. 3.   A CCD picture of S 106 in the I band, taken by Lenzen with the 2.2 m telescope. The central star embedded in the dark lane (A_v = 21 mag) is completely hidden in the visual domain. The fuzzy nebulosities protruding from the star mainly consist of scattering dust, while in the lobes strong line emission is also present.

on Calar Alto by Leinert (1985), and less than 0.15 arc sec at λ = 1.3 cm, according to Felli *et al.* (1985). But the radiation is free to escape towards the polar lobes. Felli *et al.* (1984) suggest that a mass loss envelope surrounds the star, whose density is lower in the polar regions than around the equator, where it becomes optically thick to most of the stellar radiation.

A detailed map of the dust distribution in the equatorial belt of S 106 was derived by Felli *et al.* (1984) from comparison of the VLA map with CCD pictures taken at the 2.2 m telescope on Calar Alto through a narrow band filter centered on the [S III] λ9532 Å emission line. The extinction is highest near the equator ($A_v \gtrsim$ 18 mag)

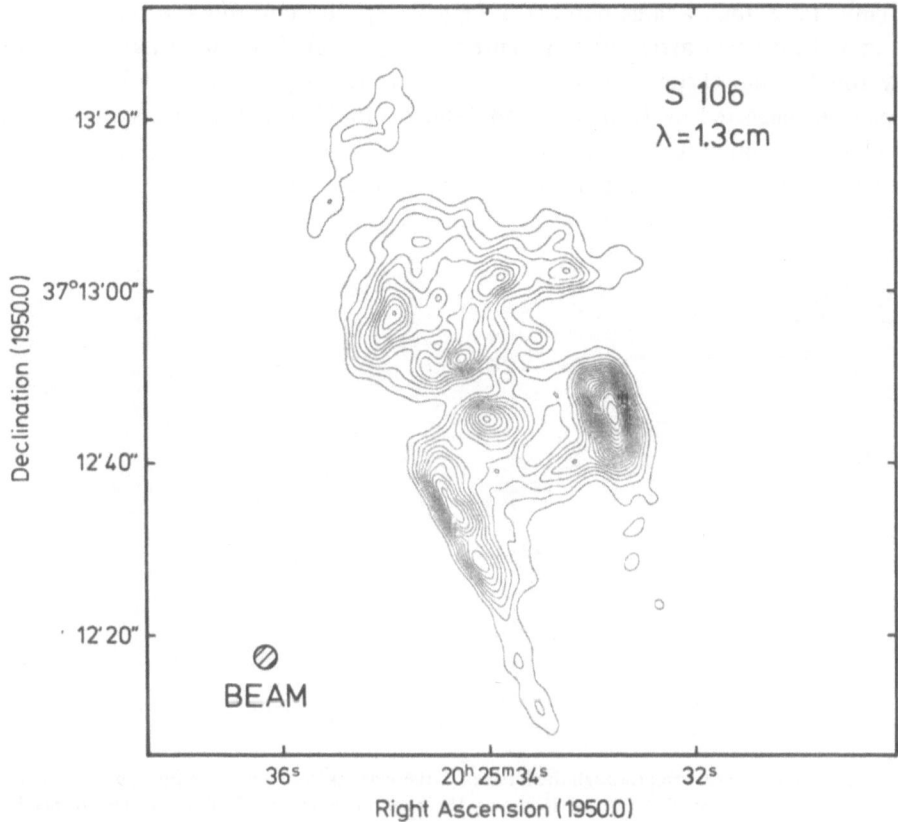

Fig. 4.   VLA map of S 106 at λ = 1.3 cm, from Felli *et al.* (1984). The weak unresolved source centered in
the gap between the two lobes coincides with the O9 star seen in the I picutre (Figure 3).

and falls gradually below A_v = 4 mag in front of the bright lobes. This dust distribution
is consistent with the above mentioned structures of the molecular cloud seen in NH_3
emission east and west of the H II region, as well as with the map of thermal emission
of cool dust observed in the far infrared (Harvey *et al.*, 1982).

Wavelength-dependent polarimetry of the lobes of S 106 was obtained by Staude
et al. (1982), and confirmed that the central star is the sole source illuminating the
nebula. The polarisation of the scattered continuum reaches p = 50% in the K band:
this requires that only a small range of scattering angles near 90 deg is effective, and thus
suggests that the scattering dust is concentrated in a narrow sheet at the surface of the
H II region.

Kinematic studies of S 106 were performed by Hippelein and Münch (1981), who did
Fabry–Pérot interferometry of the [S III] $\lambda 9531$ line at the 1.2 m telecope on Calar Alto.
They suggest that in both lobes the ionized gas streams within a thin layer tangentially
along the surface of an ovoidal cavity produced by a stellar wind of $2 \times 10^{-5} M_\odot$ yr^{-1}.
By use of the Coudé spectrograph with the 2.2 m telescope on Calar Alto, Solf and

Carsenty (1982) took a large number of long-slit spectra covering the H II region with high spatial and spectral resolution (Figure 5). They concluded that the ionized gas flows away from the central star at about 75 km s^{-1} towards the poles, and confirmed in detail the picture suggested by Hippelein and Münch (1981), which also fits well with the spatial distribution of the dust around the H II region, as derived from polarimetry, and with the morphology of the ionized gas as seen in the VLA map.

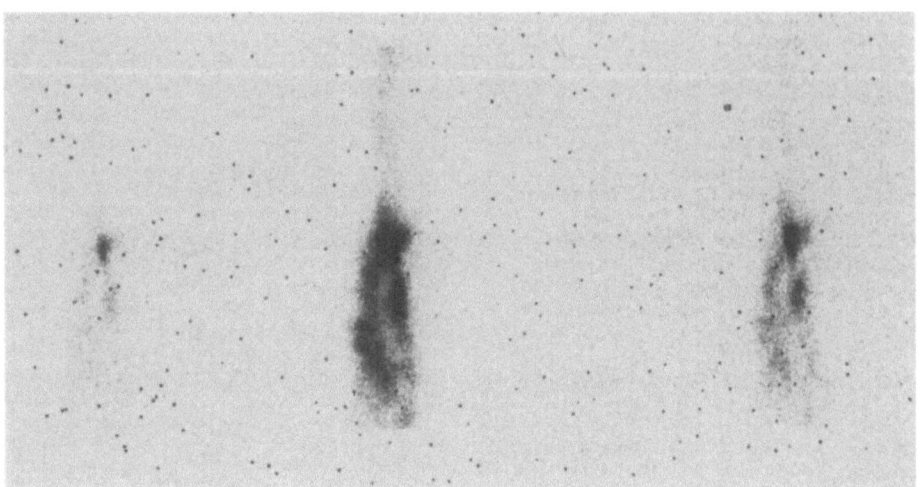

Fig. 5. This long-slit spectrum through the lobes of S 106 was taken by Solf with the coudé spectrograph at the 2.2 m telescope on Calar Alto. It shows Hα and the adjacent [N II] lines. The dispersion is 1.8 Å mm^{-1}, the slit covers 72 arc sec and is oriented parallel to the polar axis of the nebula. The northern lobe is quite faint. South of the bright knot at the centre, which corresponds to the strongest emission in the southern lobe (Figures 3 and 4), the lines are very broad (total width up to 220 km s^{-1}) and show a line splitting of about 100 km s^{-1}.

In summary, in S 106 we observe the alignment and axisymmetrical orientation of a variety of phenomena, from the interstellar magnetic field (on a scale of more than 10 pc) to the distribution of gas and dust in the molecular cloud (1 pc), the bright ionized lobes with their dusty shells (0.5 pc), the equatorial dust belt and the gap in the H II region (0.2–0.01 pc), and the 'circumstellar nozzle' ($< 10^{-4}$ pc), which determines the asymmetry of the radiation field and the bipolar flow of the ionized gas.

3. An Optical Search for New Bipolar Nebulae

While detailed observations of S 106 were under way, we started an optical search for hitherto unknown bipolar or cometary nebulae. The search was performed on the red POSS prints and was therefore restricted to objects larger than about 15 arc sec. It covers nearly the total Milky Way north of δ = −30° and about 100 fields outside the Milky Way. Twenty candidate bipolar and cometary nebulae were found (Neckel and

Staude, 1984), and studied with *UBVRI* and infrared photometry, infrared photography, and intermediate resolution spectroscopy.

From these data the spectral types of the central stars, as well as their extinctions and distances could be estimated. With two exceptions, the derived spectral types are O and B. This, however, reflects a selection effect, since only stars earlier than B3 can produce conspicuous emission nebulae, and only luminous stars can give origin to reflection nebulae sufficiently bright and extended to be detected at greater distances.

In 10 cases the derived visual extinction is substantially larger than the expected interstellar extinction. The extinction A_v = 21 mag, which is associated with the central star in S 106, lies near to the upper limit of the range found in the survey. Such a wide range of extinction values is not unexpected, since the observed extinction depends on the amount of dust associated with the object, as well as on its orientation relative to the observer. The upper limit again is due to the fact, that we were searching in the optical range: more heavily obscured objects could not be found.

All linear dimensions were found to lie between 0.1 and 1.3 pc. In this respect S 106 ($\approx$ 0.6 pc) is just average. This range is almost identical with that found by Bally and Lada (1983) for the size of high-velocity molecular flows exhibiting bipolar structure.

Recently, White and Gee (1986) have studied our sample with the VLA: their results are in good agreement with our estimates. At present some of the new objects are being observed in more detail (Neckel *et al.*, 1986), but under the optically selected bipolar nebulae the prominent position of S 106 remains unchallenged. The reason for this becomes evident, when we consider the sources selected due to their molecular bipolar flows. It turns out that most of them are so deeply buried into their parent cloud, that they would never have been noticed in an optical search.

4. Bipolar Nebulae Associated with Molecular Flows

During systematic searches for high-velocity molecular gas associated with star forming molecular clouds, in the last few years a number of massive bipolar high-velocity CO flows have been discovered, where the gas is flowing away from a deeply buried infrared source in two oppositely directed streams. A comprehensive discussion of the radio observations is given by Bally and Lada (1983).

In the following, optical and near-infrared observations of three such sources are presented. Although the obscuration involved is extreme in most cases, these tools can work out some essential aspects of the phenomenon.

L 1551

Figure 6 shows the dark cloud L 1551, which is located in the Taurus complex at a distance of 160 pc, with its high-velocity bipolar CO flow according to Snell *et al.* (1980). In the southwestern lobe the radial CO velocity averages to v_r = 1.2 km s^{-1}, in the northeastern lobe to v_r = 11.5 km s^{-1}. That the two lobes are moving away from a common origin, is confirmed by the proper motion of the Herbig–Haro objects HH 28 and HH 29: Cudworth and Herbig (1979) found that their transversal velocity is

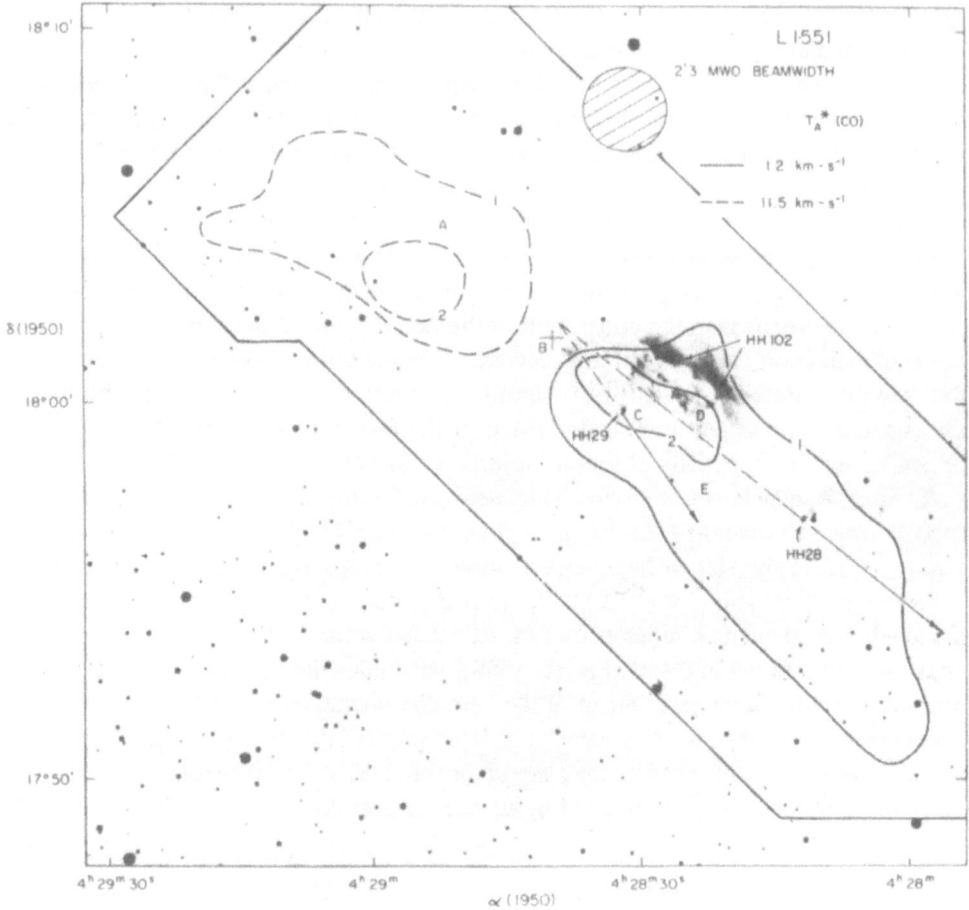

Fig. 6.   The bipolar CO flow in the dark cloud L 1551, from Snell *et al.* (1980). The southern lobe is approaching the observer, relative to the central source L 1551 IRS 5 (marked by a cross), the northern lobe is receding. The HH objects Nos. 28 and 29 are also moving away from the same origin.

v_t = 150 km s^{-1}, and that 3000 years ago they started at a common origin coincident within errors with the cross between the lobes. This cross marks the position of an infrared point source, called L 1551 IRS 5, of about 25 $L_\odot$ (Storm *et al.*, 1976).

Mundt and Fried (1983) took the red CCD picture of the central source and its immediate surroundings, which is shown in Figure 7. Here a sharply confined parabolic lobe appears (see also Figure 8), which is open towards the southwest, and at whose apex the stellar source is located. Its overall shape recalls the southern lobe of S 106 (Figure 3), but in addition along its axis-of-symmetry a bright knotty jet is seen, which emanates immediately from the central star, in the same direction as the southwestern CO lobe and the remote HH objects.

Spectroscopy of the jet and optical lobe was obtained by Sarcander *et al.* (1985). The emission lines in the spectra are showing the features of a shock excited gas in a state

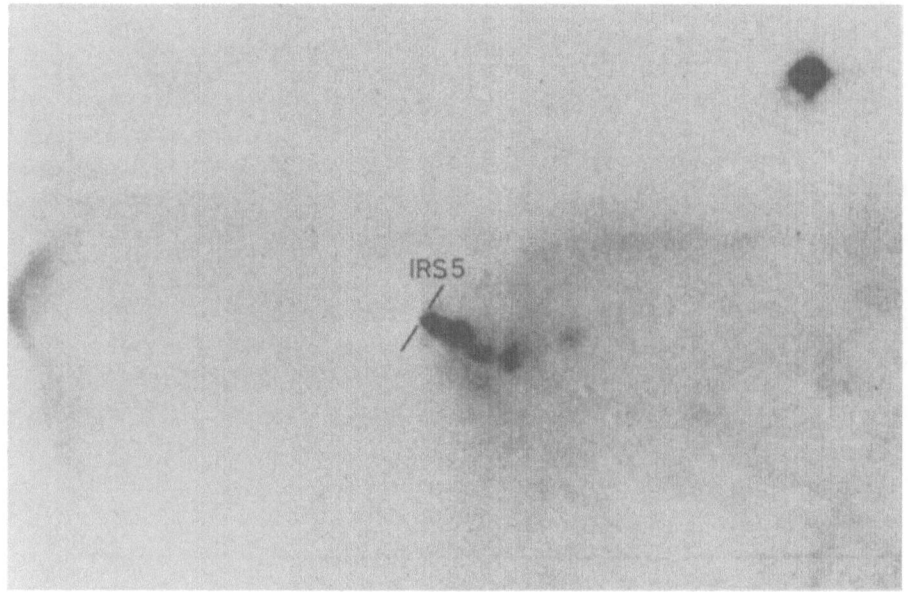

Fig. 7. This red CCD picture was taken at the 2.2 m telescope (Mundt and Fried, 1983), exposure time was 1.5 hr. The central source IRS 5 is located at the apex of a tenuous parabolic lobe and emits a highly collimated, bright and knotty jet.

like that of low-excited HH objects. The derived radial velocities reach $v_r = -210$ km s^{-1} in the jet and lie between -20 and -90 km s^{-1} in the optical lobe. From the Hα/Hβ line ratios the extinction in front of jet and lobe was derived: it is surprisingly low, $A_v \approx 1\text{--}3$ mag, whereas in front of the central star it is estimated to be $A_v \gtrsim 20$ mag. This indicates the presence of a sharp increase of the dust density over a few arc sec, such as would be expected if the central star were located in a dense circumstellar disk seen nearly edge on. The absence of a northwestern optical lobe and jet is naturally explained, if this disk is slightly tilted with respect to the line-of-sight, in accordance with the observed radial and tangential outflow velocities.

The hypothesis of the circumstellar disk is strengthened by the fact that the central star is highly polarized in HKL ($p = 20\%$, Nagata *et al.*, 1983; Hodapp, 1984), with its polarization vector nearly perpendicular to the jet. According to the bipolar model presented by Elsässer and Staude (1978), a spatially unresolved source, consisting of a star embedded in a dense dust disk seen nearly edge on, with low-density polar lobes, is polarized by scattering in the lobes, the polarization being oriented parallel to the disk. A further confirmation was provided by CS observations of Kaifu *et al.* (1984), who reported the direct detection of a gaseous disk around L 1551 IRS 5.

A polarization map at $\lambda = 1$ μm of the southwestern lobe was obtained by Lenzen (1986a) at the prime focus of the 3.5 m telescope (Figure 8). In the whole lobe the polarization is about 50%; the highly-ordered pattern localizes the illuminating source

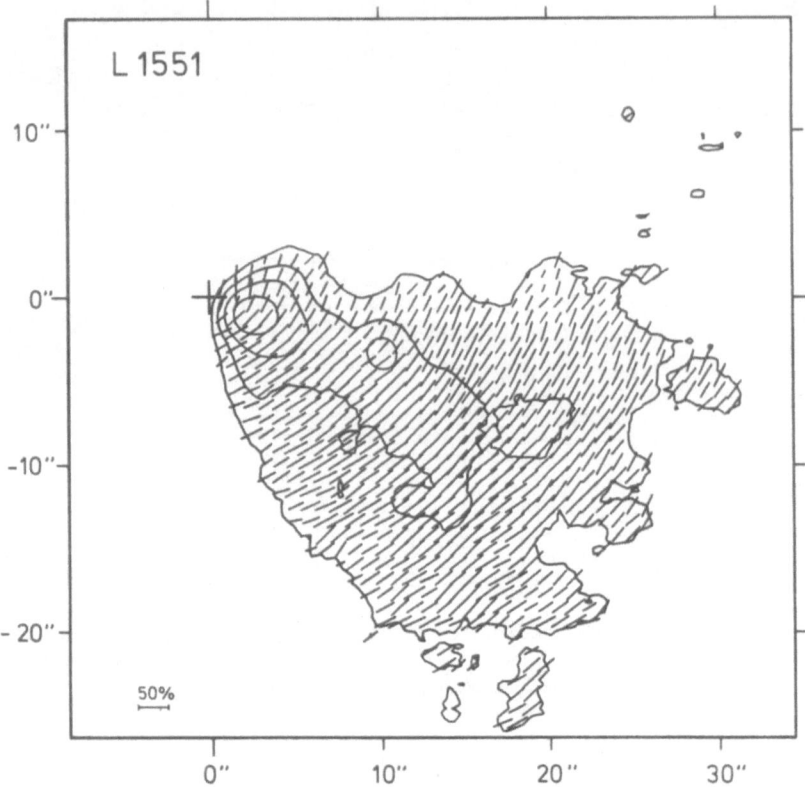

Fig. 8.   This CCD polarimetry of the field including L 1551 IRS 5 was obtained at the prime focus of the 3.5 m telescope by Lenzen (1986a). Jet and lobe show the same strong polarization, which defines the position of the illuminating source. This position coincides within errors with IRS 5.

within about 1 arc sec. This source is clearly distinct from the first bright knot in the jet (which is polarized as all other sections of the lobe) and possibly it is not visible at all at this wavelength. On the other hand, the transversal extinction between illuminating source and scattering dust in the lobe must be very small, since multiple scattering would not allow the extremely high polarization observed. This result again supports the disk hypothesis and the validity of the Elsässer–Staude model.

Since the lobe is such an effective reflection nebula, medium resolution spectroscopy of its brightest parts in the visual (near HH 102, see Figure 6) could be used by Mundt *et al.* (1985) to study the spectrum of the central star. They concluded that its spectral type is G–K, and found its spectral characteristics to be quite similar to those of FU Orionis objects.

Cepheus A

This is a compact molecular cloud in the Cep OB3 association at a distance of 730 pc, in which Beichmann *et al.* (1979) found one of the coldest 8–25 μm sources yet detected by ground based observations. The source is associated with H_2O maser spots (Lada et al., 1981; Wouterloot et al., 1980) and with a high-velocity bipolar CO outflow (Ho et al., 1982), whose projected polar axis is oriented roughly in east–west direction.

On the red POSS plate (Figure 9) the field is almost empty. West of the 8–25 μm source a diffuse nebulosity is seen, which was suspected to be a Herbig–Haro object (GGD 37 of Gyulbudaghian et al., 1978). Lenzen et al. (1984) mapped the field at 2.2 and 3.6 μm and obtained surface polarimetry at 3.6 μm. On the 2.2 μm map (Figure 9) a bright extended source appears in the eastern half of the field, and some extended emission is seen in the west, just below GGD 37. The extremely high 3.6 μm polarization (Figure 10) reveals, that the eastern extended source is an infrared reflection nebula. The typical polarization pattern allows the precise localization of the illumination source, which coincides within errors with the 8–25 μm source.

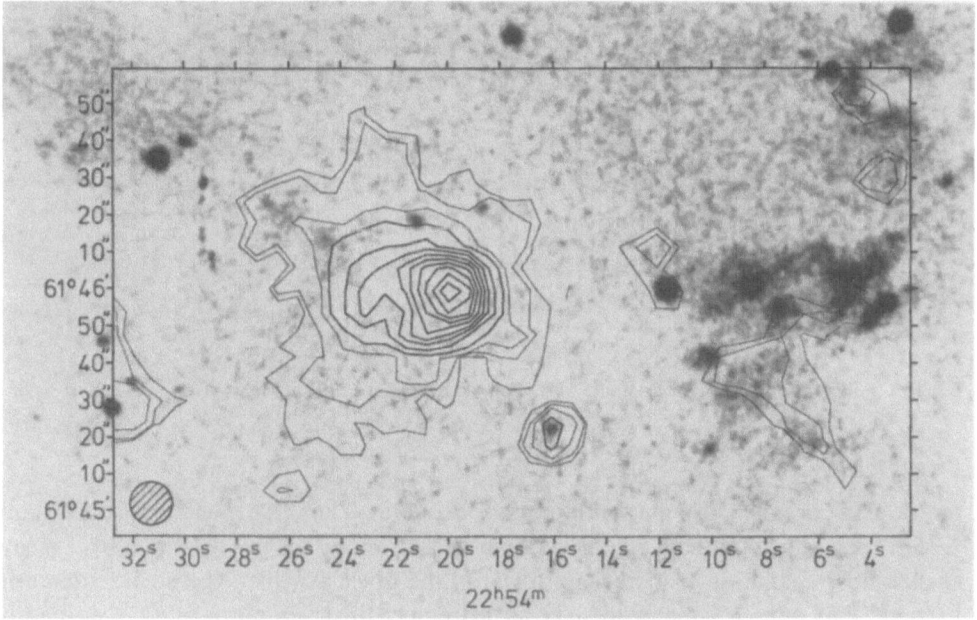

Fig. 9.   The field of Cepheus A on the red POSS print, with the 2.2 μm isophotes obtained by Lenzen et al. (1984) at the 2.2 m telescope on Calar Alto.

From the analysis of the brightness distribution across the reflection nebula, Lenzen et al. conclude that the transversal extinction between source and scattering dust is very low, while the comparison between the 2.2 and 3.6 μm maps indicates, that the extinction along the line-of-sight increases from east to west, reaching $A_v \approx 30$ mag two arc sec east of the illuminating source. Here again a dramatic jump in the extinction is observed: from the total brightness of the reflection nebula and the marginal or non-detection of the illuminating source at 3.6 μm, a lower limit $A_v \geq 75$ mag follows

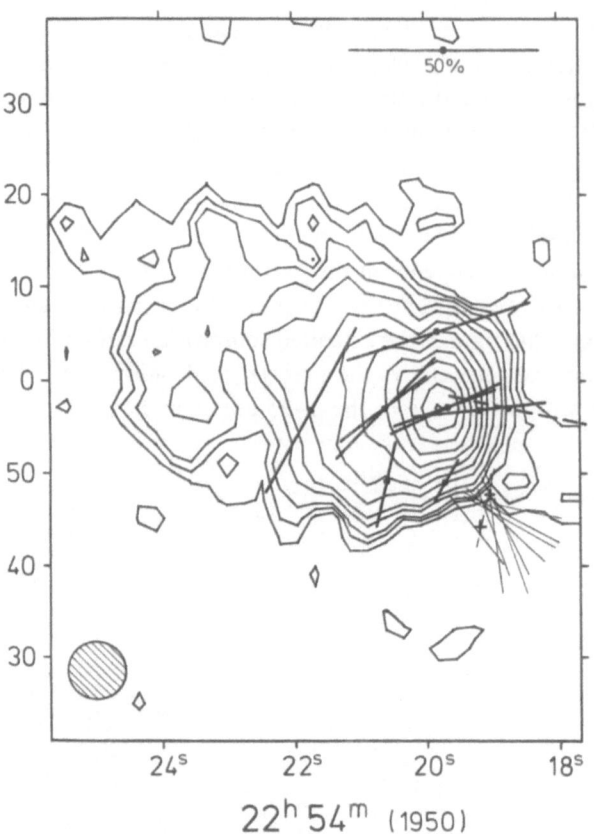

Fig. 10.   Polarization and outer contours at 3.6 μm of the eastern infrared lobe of Cep A (Lenzen *et al.*, 1984).

for the extinction between the source and the observer. Thus, like the lobes of S 106 and L 1551, this reflection nebula consists of an almost empty cavity delimited by a dusty shell, whose thickness is increasing towards the apex at which the stellar source is located. This source is embedded in an extremely thick equatorial disk seen nearly edge-on, but radiation and stellar wind are free to escape towards the poles.

In fact, the Coudé spectroscopy of GGD 37 by Lenzen *et al.* yielded a clearly shock-excited spectrum typical for HH objects, with radial velocities up to 300 km s^{-1}. At the northeastern end of the eastern lobe, a deep CCD image in Hα, taken at the prime focus of the 3.5 m telescope (Lenzen, 1986b) revealed a bright, compact nebulosity showing the morphology typical for 'bow shocks' (similar to HH 34, see Figure 12 below). Hughes and Wouterloot (1984) presented 6 cm VLA observations, which show several compact radio sources aligned along the edges of the infrared reflection nebula. Since no infrared point sources are found at these positions, they probably have to be interpreted as being shock ionized H II regions. All these elements point to the presence of a rapid stellar wind associated with the bipolar morphology of the source.

PV Cephei

This highly reddened F star exhibiting a strong infrared excess is associated with a massive molecular cloud and with a bipolar molecular outflow (Levreault, 1984) with its polar axis at position angle ≈ 130 deg. It is associated also with a highly variable cometary reflection nebula located north of the star, which was described extensively by Cohen *et al.* (1977, 1981). The nebula appeared in our survey as object No. 19 (Neckel and Staude, 1984) and has since then regularly been monitored at Calar Alto. On the first images taken at the prime focus of the 3.5 m telescope (Neckel *et al.*, 1986; see Figure 11), besides the known northern lobe a much weaker southern lobe became visible, showing that the nebula is in fact bipolar. This southern lobe is heavily reddened, indicating that it penetrates deeply into the molecular cloud. A series of long-slit spectrograms in the red, covering the northern lobe, shows the scattered stellar continuum with superimposed chromospheric emission lines, and in addition broad HH emission lines, which are shifted to -225 km s^{-1} with respect to the chromospheric emission. These lines can be observed only at two narrow spots along the northern polar axis: therefore they must be emitted there, since they are absent in other parts of the reflection nebula.

Thus here again, as in L 1551 and Cep A, we find that infrared sources associated with bipolar molecular flows are illuminating a reflection nebula of characteristic morphology, which may appear cometary, due to its peculiar location near the surface of the parent molecular cloud, but which is probably always intrinsically bipolar. Along

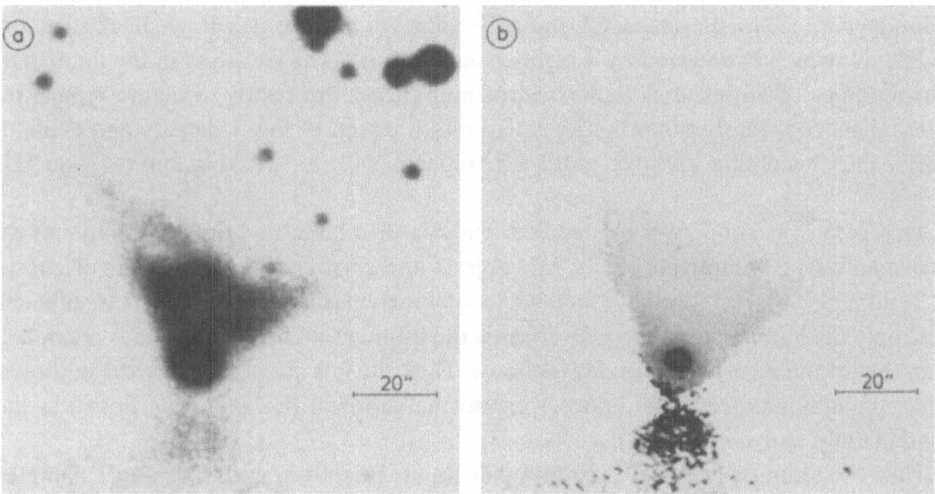

Fig. 11.   CCD images of the PV Cephei nebula. (a) *I* picture, taken with the 3.5 m telescope in December 1985; (b) the colour index *R–I*, derived from (a) and from an *R* frame taken at the same time: the reddening strongly increases from north to south. The star appears inside the northern lobe, because we look at it from above the equatorial plane of the bipolar nebula, as is indicated by the reddening distribution and by kinematic data.

the polar axis of the lobes, and in some cases also along their rims (e.g., Cep A), signposts of shock excitation and of collimated flows with velocities up to several hundred km s^{-1} are typical findings.

5. Jets from Young Stars and Circumstellar Disks

The discovery of the jet in L 1551 by Mundt and Fried (1983) happened at the early stages of a systematic search for optical jets around young stars. Until now it has led to the discovery of more than half of the about 20 known jets of this kind. The topic has been reviewed extensively by Mundt (1985a, b, 1986): here we will briefly sketch the essential properties of these jets.

The search was directed at the surroundings of young T Tauri stars, characterized by strong emission lines, and of known HH objects, as indicators of high-velocity mass flows in star forming regions. Of the selected fields deep CCD images were taken, mainly with the 2.2 m telescope, through narrow band filters including and excluding Hα and the strongest lines characterizing shock excited emission. In both kinds of fields the search was highly successful. Some of the jets are diametrically opposed pairs, some are associated with known bipolar molecular flows. The jets show emission-line spectra essentially similar to those of the HH objects; some of their brightest knots were known as HH objects since years. The essential difference between jets and HH objects consists in the high collimation of the flow in the jets, which is performed in the immediate surroundings of the stellar source (within less than 70 AU in L 1551), while the HH objects can be understood as the working surfaces of the jets boring through the tenuous outskirts of the molecular cloud. This picture is best visualized in Figure 12 (Mundt *et al.*, 1986): the deep CCD image reveals the characteristic bow-shock structure of HH 34, which is powered by a highly collimated jet. The jet arises in the immediate surroundings of an infrared stellar source and shows the knotty structure typical for internal shocks. Farther out, the jet enters into a region of lower density and expands freely, thus becoming invisible, until it reappears with its working surface: the HH object.

This picture is supported by detailed analysis of a large body of observations: the derived physical parameters of jets, HH objects and surrounding material are discussed by Mundt (1986). Here we note that, while the kinetic luminosity of the jets is sufficient to supply the luminosity of the HH objects, the momentum flux of the jets is much less than that of the massive molecular outflows. Thus the jets are not driving the outflows. Also, the momenta of these outflows cannot be supplied by radiatively driven stellar winds (Bally and Lada, 1983).

Thus the ultimate origin of the molecular flows, bright bipolar lobes and collimated jets near to the stellar surface is still unknown. However, the spatial distribution of circumstellar matter in the immediate surroundings of the central star (as derived above for S 106 and L 1551 IRS 5) certainly plays an essential role in determining the morphology of the objects. This has been impressively demonstrated by the work of Hodapp (1984), who performed a near-infrared polarization survey of T Tauri stars

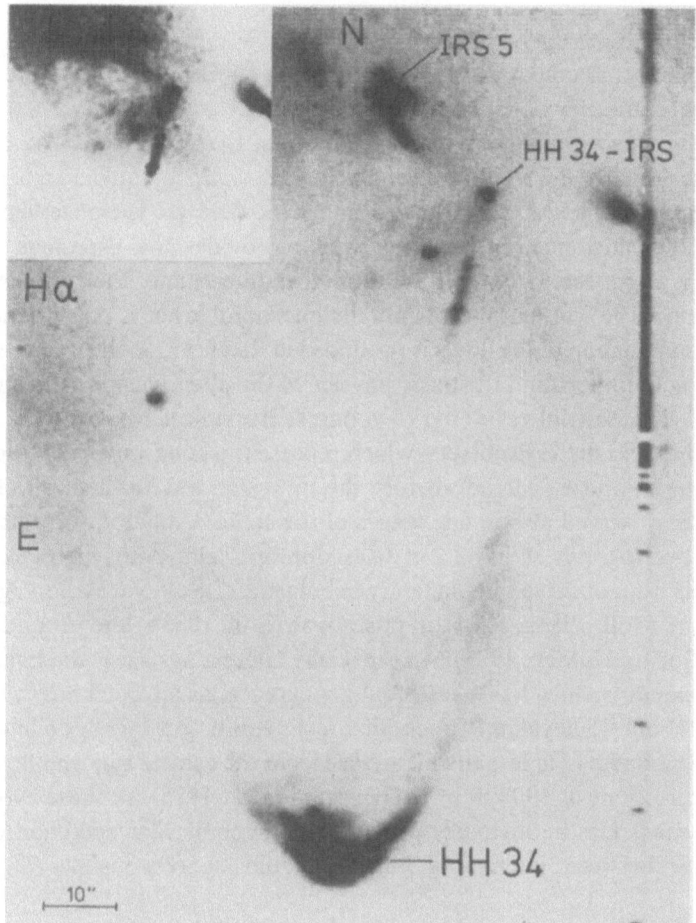

Fig. 12. CCD picture of HH 34 and its environment in the light of the Hα-line. From a faint red star (HH 34–IRS) a jet is pointing towards HH 34. The HH object shows a characteristic parabolic structure, which is interpreted as the bow shock of the jet's working surface (from Mundt *et al.*, 1986). The inset in the upper left corner shows the environment of the stellar source in more detail.

producing optical jets, as well as of infrared sources associated with bipolar molecular outflows. High degrees of polarization are found in all cases. The polarization vectors are preferentially oriented perpendicular to the outflow directions, as is expected for scattering in spatially unresolved bipolar structures (Elsässer and Staude, 1979) with the outflows directed towards the poles. Furthermore, the inferred circumstellar disks associated with both kinds of outflows (jets as well as massive CO flows) are oriented preferentially perpendicular to the magnetic fields in their neighbourhoods, which means that the outflows tend to be in the direction of the local magnetic field: S 106 (Figure 2) is representative also in this respect.

6. Conclusions

Bipolar structures and collimated flows represent a short lived stage in the early stellar evolution, when the newborn stars are still strongly interacting with their parental clouds. Strong anisotropic mass loss by the stars, the topology of the circumstellar material on scales from the whole molecular cloud down to few stellar radii and possibly to the stellar surface, and finally the local magnetic field are the observed ingredients determining the phenomenon. The link which provides the alignment of all these components is supposed to be the total angular momentum of the complex. But many aspects remain to be worked out, mainly the questions, what is driving the mass loss, and how the collimation of the flows is produced in detail within the circumstellar disks.

When trying to understand the basic physics of the phenomenon, one always should keep in mind that it is not restricted to young stellar objects. Apart from the massive molecular flows and the HH objects, which represent a consequence of the interaction with the material present already before the new star was formed, all characteristic aspects can be observed also in late stages of the stellar evolution. For instance, using the Coudé spectrograph at the 2.2 m telescope on Calar Alto, high-velocity bipolar expansion has been observed in planetary nebulae (e.g., Solf and Weinberger, 1984), in symbiotic stars (Solf, 1983a, 1984), in postnovae (Solf, 1983b) and around the peculiar star R Aqr (Solf and Ulrich, 1985). An especially interesting case is the Eskimo Nebula, a classical planetary which has a spherical appearence, as anyone knows, but a bipolar heart: since about 1500 years it is continuously emitting a highly collimated, knotty bipolar jet, which sets in at less than 2 arc sec from the central star and flows outwards in opposite directions at 190 km s^{-1} (Gieseking et al., 1985). In these evolved objects again heavy mass loss is obviously present, and circumstellar accretion disks due to mass transfer between interacting binaries could be responsible for the bipolar morphology.

References

Bally, J. and Lada, C. J.: 1983, *Astrophys. J.* **265**, 824.
Beichmann, C. A., Becklin, E. E., and Wynn-Williams, C. G.: 1979, *Astrophys. J.* **232**, L47.
Cohen, M., Kuhi, L. V., and Harlan, E. A.: 1977, *Astrophys. J.* **215**, L127.
Cohen, M., Kuhi, L. V., Harlan, E. A., and Spinrad, H.: 1981, *Astrophys. J.* **245**, 920.
Cudworth, K. M. and Herbig, J.: 1979, *Astron. J.* **84**, 548.
Eiroa, C., Elsässer, H., and Lahulla, J. F.: 1979, *Astron. Astrophys.* **74**, 89.
Elsässer, H. and Staude, H. J.: 1978, *Astron. Astrophys.* **70**, L3.
Felli, M., Staude, H. J., Reddmann, T., Massi, M., Eiroa, C., Hefele, H., Neckel, T., and Panagia, N.: 1984, *Astron. Astrophys.* **135**, 261.
Felli, M., Simon, M., Fischer, J., and Hamann, F.: 1985, *Astron. Astrophys.* **145**, 305.
Gieseking, F., Becker, I., and Solf, J.: 1985, *Astrophys. J.* **295**, L17.
Gyulbudaghian, A. L., Glushov, Y. I., and Denisyuk, E. K.: 1978, *Astrophys. J.* **224**, L137.
Harvey, P. M., Gatley, J., Thronson, H. A., and Werner, M. W.: 1982, *Astrophys. J.* **258**, 568.
Hippelein, H. H. and Münch, G.: 1981, *Astron. Astrophys.* **99**, 248.
Ho, P. T. P., Moran, J. M., and Rodriguez, L. F.: 1982, *Astrophys. J.* **262**, 619.
Hodapp, K. W.: 1984, *Astron. Astrophys.* **141**, 255.

Hughes, V. A. and Wouterloot, J. G. A.: 1984, *Astrophys. J.* **276**, 204.
Kaifu, N., Suzuki, S., Hasegawa, T., Morimoto, M., Inatani, J., Nagane, K., Miyazawa, K., Chikada, Y., Kanzawa, T., and Akabane, K.: 1984, *Astron. Astrophys.* **134**, 7.
Lada, C. J., Blitz, L., Reid, M. J., and Moran, J. M.: 1981, *Astrophys. J.* **243**, 769.
Leinert, C.: 1985, private communication.
Lenzen, R.: 1986a, *Astron. Astrophys.* (submitted).
Lenzen, R.: 1986b, in preparation.
Lenzen, R., Hodapp, K. W., and Solf, J.: 1984, *Astron. Astrophys.* **137**, 202.
Levreault, R. M.: 1984, *Astrophys. J.* **277**, 634.
Little, L. T., Macdonald, A. M., Riley, P. W., and Matheson, D. M.: 1978, *Monthly Notices Roy. Astron. Soc.* **188**, 429.
Lucas, R., Le Squéren, A. M., Kazés, J., and Encrenaz, P. J.: 1978, *Astron. Astrophys.* **66**, 155.
Mundt, R.: 1985a, in D. C. Black and M. S. Matthews (eds.), *Protostars and Planets II*, Univ. of Arizona Press, Tucson.
Mundt, R.: 1985b, in G. Serra (ed.), 'Nearby Molecular Clouds', *Lecture Notes in Physics* **237**, Springer-Verlag, Berlin.
Mundt, R.: 1986, *Can. J. Phys.* (to be published).
Mundt, R. and Fried, J. W.: 1983, *Astrophys. J.* **274**, L83.
Mundt, R., Brugel, E. W., and Bührke, T.: 1986, in preparation.
Mundt, R., Stocke, J., Strom, S. E., Strom, K. M., and Anderson, E. R.: 1985, *Astrophysics* **297**, L41.
Nagata, T., Sato, S., and Kobayashi, Y.: 1983, *Astron. Astrophys.* **119**, L1.
Neckel, T. and Staude, H. J.: 1984, *Astron. Astrophys.* **131**, 200.
Neckel, T., Staude, H. J., Sarcander, M., and Birkle, K.: 1986, *Astron. Astrophys.* (submitted).
Sarcander, M., Neckel, T., and Elsässer, H.: 1985, *Astrophys. J.* **288**, L51.
Snell, R. L., Loren, R. B., and Plambeck, R. L.: 1980, *Astrophys. J.* **239**, L17.
Solf, J.: 1983a, *Astrophys. J.* **266**, 113.
Solf, J.: 1983b, *Astrophys. J.* **273**, 647.
Solf, J.: 1984, *Astron. Astrophys.* **139**, 296.
Solf, J. and Carsenty, U.: 1982, *Astron. Astrophys.* **113**, 142.
Solf, J. and Ulrich, H.: 1985, *Astron. Astrophys.* **148**, 274.
Solf, J. and Weinberger, R.: 1984, *Astron. Astrophys.* **130**, 269.
Staude, H. J., Lenzen, R., Dyck, H. M., and Schmidt, G. D.: 1982, *Astrophys. J.* **255**, 95.
Strom, K. M., Strom, S. E., and Vrba, F. J.: 1976, *Astron. J.* **81**, 320.
Torrelles, J. M., Rodriguez, L. F., Cantó, J., Carral, P., Marcaide, J., Moran, J. M., and Ho, P. T. P.: 1983, *Astrophys. J.* **274**, 214.
White, G. J. and Gee, G.: 1986, *Astron. Astrophys.* **156**, 301.
Wouterloot, J. G. A., Habing, H. J., and Herman, J.: 1980, *Astron. Astrophys.* **81**, L11.

Discussion

P. G. Mezger: I would like to know, where do you place this type of bipolar nebulae in the evolution of stars?

H. J. Staude: As described above, the morphological characteristics of bipolar nebulae can be found in young as well as in evolved objects. Unless we achieve a deeper theoretical understanding of the evolution of young objects through the stage of bipolar nebulae, we must refer to the usual signposts of star forming activity, such as association with dark clouds, masers, mass loss, etc., to guess their evolutionary stage.

H. J. Habing: The fact that you see all these things (jets, etc.) also in planetary nebulae, suggests that it is not the process of contraction in which the circumstellar disk is formed, that determines the outflow structure.

H. J. Staude: Our present understanding is, that the circumstellar disk is an essential ingredient in both, young as well as evolved bipolar nebulae. In the young objects the disk develops while the parental cloud is contracting. In the evolved objects the suspicion is, that there is a binary system with mass exchange, in which an accretion disk is formed. In the case of postnovae and of symbiotic stars this hypothesis is supported by independent evidence. In other cases it is very difficult to check observationally.

J. Dorschner: I am not very happy that you used the abbreviation 'BN' for bipolar nebula, because this abbreviation is already used for Becklin–Neugebauer objects.

H. J. Staude: I agree with you, I will not do it again!

J. Palouš: May I ask you what are the energies which are liberated by the outflows?

H. J. Staude: The strongest energetic requirements are set by the bipolar high-velocity molecular flows. The kinetic luminosities required to accelerate the flows are in several cases of the order of the stellar luminosities ($L_{kin}/L_{star} \approx 0.01-0.5$). Because of this difficulty it has been suggested that the outflows are driven not by radiation, but rather by the gravitational and magnetic energy of matter accreting onto these objects (Draine, 1983; Pudritz and Norman, 1983; Ushida and Shibata, 1984). The observations described above support the notion that the magnetic field plays an essential role.

P. G. Mezger: I think that Königl has proposed a quite general model.

H. J. Staude: Yes, but this model encounters difficulties with those objects, in which the focusing is observed to occur not further out than a few hundred AU.

P. G. Mezger: Is there beyond what you have demonstrated, namely high extinction and collimation, any other observational evidence for the existence of an accretion disk around some of those young objects?

H. J. Staude: Besides the observed extinction, collimation, and polarization (both, spatially resolved and unresolved) and the radio maps with high-angular resolution (Kaifu *et al.*, 1984), the strongest pieces of direct evidence for circumstellar disks are provided by Speckle interferometry in the near infrared. Beckwith *et al.* (1984) report the detection of solar system sized mass distributions around the stars HL Tau and R Mon, which are both emitting highly collimated jets (Mundt and Fried, 1983; Brugel *et al.*, 1984). See also Leinert (1986) for the Be star MWC 349.

References

Beckwith, S., Zuckermann, B., Skrutskie, M. F., and Dyck, H. M.: 1984, *Astrophys. J.* **287**, 793.
Brugel, E. W., Mundt, R., and Bührke, T.: 1984, *Astrophys. J.* **287**, L73.
Draine, B. T.: 1983, *Astrophys. J.* **270**, 519.
Leinert, C.: 1986, *Astron. Astrophys.* **155**, L6.
Pudritz, R. E. and Norman, C. A.: 1983, *Astrophys. J.* **274**, 677.
Ushida, Y. and Shibata, K.: 1984, *Publ. Astron. Soc. Japan* **36**, 105.

References

BN OBJECTS - A CLASS OF VERY YOUNG AND MASSIVE STARS*

TH. HENNING and J. GÜRTLER

Universitäts-Sternwarte Jena, G.D.R.

(Received 27 June, 1986)

Abstract. The Becklin–Neugebauer objects are identified by means of a wide range of observable features as a separate class of very young and massive stars surrounded by optically thick dust shells. We show that they evolutionarily connect real protostars to compact H II regions. We give criteria which should be appropriate to segregate the BN objects from compact H II regions. Finally, we describe the structure of a typical BN object.

1. Introduction

Infrared astronomy made an important observational contribution to the study of early stages of stellar evolution by the discovery of numerous compact IR sources in dense molecular clouds. A subgroup of about 30 objects may be defined which are generically similar to the Becklin–Neugebauer object in Orion. Their main properties include a spectrum dominated by infrared emission, lacking of optically visible features such as an exciting star and biconical lobes, and only weak radio continuum emission, if present at all, because a developed H II region does not yet exist. This class of objects is the earliest stage of newly formed massive stars observationally identified up to now. Generally spoken these objects we will refer to as BN objects from now are OB stars at or approaching the Main Sequence surrounded by optically thick circumstellar shells or deeply embedded in their parent molecular cloud. They are evolutionarily connected with other young IR sources, such as compact H II regions and OB stars with IR excesses which are later stages on the evolutionary path of massive stars.

The optical properties of the dust grains play a crucial role in determining the spectral appearance of these IR objects (Gürtler and Henning, 1986). Therefore, appropriate analysis of their radiation provides clues on the physical and chemical nature of the dust. This is especially true since the observation of a number of discrete absorption features gives more ample information than the more or less featureless visible part of the interstellar extinction curve.

In the following we will focus on the observational characteristics of BN objects (Section 2), on their evolutionary status (Section 3), and a source model of them.

* Paper presented at a Workshop on 'The Role of Dust in Dense Regions of Interstellar Matter', held at Georgenthal, G.D.R., in March, 1986.

2. Observational Characteristics of BN Objects

The most comprehensive compilations of very young and compact massive IR sources in star-forming regions were provided by Wynn-Williams (1982) and Henning *et al.* (1984). The latter catalogue attempted to select the youngest objects and gives additional information on the properties of the dust shells. The observed properties of these sources are as follows.

BN objects are very compact sources in the near and middle IR. The apparent diameters are $\leq 1''$ even for the nearest ones. Models give characteristic linear diameters for the envelopes which are typically of the order of 10^{14}–10^{15} m. The envelopes are optically thick with the dust in them thermalizing the radiation of the central source. BN objects are found to be located in the densest parts of (mostly giant) molecular clouds. High-resolution IR continuum maps show very often several peaks embedded in extended emission. This is mostly interpreted as evidence for multiple systems. But a study by Wynn-Williams *et al.* (1984) of the Kleinmann–Low nebula shows that such an interpretation should be regarded with caution because some of the peaks may be simply light scattering irregularities.

The continuum generally shows a steep increase in the near IR region and a broad peak between 50 and 200 μm. Some BN objects (e.g., GL 490) investigated by McGregor *et al.* (1984) display extremely red continua in the 0.6–1.0 μm region. The BN object itself is too faint to be observed in this wavelength region with present-day equipment. The fit of black-body curves in the wavelength region 2–20 μm yields colour temperatures of about 300 to 700 K, whereas mid- and far-IR spectra lead to much lower values of 30–90 K. Deduced luminosities range from 10^2 to $10^5 L_\odot$. Objects with luminosities smaller than about $10^2 L_\odot$ are excluded from our considerations because they are definitely not related to the formation of OB stars.

The temperature of the central stellar source must be known, of course, for more precisely establishing the evolutionary state of the source. However, Yorke and Shustov (1981) as well as Henning (1983) showed that the IR spectra do not allow to derive this temperature.

The spectra exhibit strong absorption bands at 9.5 and 3.08 μm generally attributed to silicates and 'ices', respectively. Additional weak absorption features are seen in some objects at 3.3–3.6, 3.9, 4.6, 4.9, 6.0, 6.8, and 18 μm. Suggested identifications are summarized in Table I.

Substantial near-IR polarization is measured for many BN objects. Two mechanisms for producing this polarization have been proposed: scattering and absorption by aligned particles. Heckert and Zeilik (1985) concluded from an analysis of polarization data for the JHKLM bands that scattering produces the observed near-IR polarization in most cases. On the other hand, spectropolarimetric observations of the BNKL region between 8–13 μm by Aitken *et al.* (1985) show evidence for grain alignment by the angular momentum of photons from IRc 2. This mechanism is expected to work in other star-forming regions, too. We suggest that the strong polarization in the near IR is mainly by scattering, whereas the polarization in the absorption bands is due to

TABLE I

Observed absorption features in the spectra of BN objects

Wavelength	Identification	Mechanism	Remarks	Other studies relevant for BN objects
9.5 μm	Amorphous silicate (1), (2), (3)	Si–O stretching mode	Detected in all BN objects	(4), (5), (6), (7)
18 μm	Amorphous silicate (1), (2), (3)	Si–O–Si deformation mode	Detected in the spectrum of the Becklin–Neugebauer object only	
3.08 μm	Amorphous H_2O ice (8), (9), (10)	O–H stretching mode		(4), (6), (7)
2.98 μm (??)	NH_3 ice (8)	N–H stretching mode	Wing of 3.08-μm band; the 2.98-μm band in BN detected by (8) is very probably a misinterpretation of the data (24)	
3.3–3.6 μm	(a) NH_3 + H_2O ice (8)	Vibration of an NH–O bond	Wing of 3.08-μm band	(4)
	(b) Molecular mixture (9)	C–H stretching mode, NH–O bond vibration		
	(c) Hydrocarbons, e.g., CH_3OH ice (??) (11), (12), (13)	Stretching mode of $-CH_2$ and $-CH_3$ groups	If the identification (c) is correct the bending modes at ≈ 6.8 μm should also be present and correlated with 3.3–3.6 μm; no correlation was found (4)	
3.9 μm	H_2S (14)	H–S stretching mode	Detected in W 33A only	
4.6 μm	CO + CN ice (15), (16)	(i) C–N stretching mode (ii), (iii) C–O stretching mode	Consists of three features: (i) a broad absorption feature at 4.62 μm; (ii) a weaker band at 4.68 μm; (iii) a sharper feature at 4.67 μm; clearly detected only in W 33A	
4.9 μm	Unidentified, maybe sulfur-containing molecule in grain mantles (14)	–	Detected in W 33A only	
6.0 μm	H_2O, NH_3, H_2CO, X–CN ice (4), (8), (9), (13), (17)	Bending modes	Mon R2-IRS2, BN, and NGC 2264 show a broad and shallow feature in the 6.0/6.8 μm region possibly due to a mixture of different hydrocarbons (17)	(18)

Table 1 (continued)

Wavelength	Identification	Mechanism	Remarks	Other studies relevant for **BN** objects
6.8 μm	(a) Hydrocarbons, e.g., CH_3OH ice (19), (11), (17), (20); (??)	Scissoring and bending modes of $-CH_2$ and $-CH_3$ groups	See remark 3.3–3.6 μm	(4), (18)
	(b) Carbonate minerals (?) (21), (22)	Stretching mode of the carbonate ion	Bands at 11.4 and 13.6 μm should be present	
	(c) NH_4^+ (8) (?)	Deformation mode	Identification with H_2O ice excluded	
42 μm	H_2O ice (23)	Transverse optical transformation	Detected in the KL spectrum only	

References:

(1) Woolf, N. J. and Ney, E. P.: 1969, *Astrophys. J.* **155**, L181.

(2) Dorschner, J.: 1971, *Astron. Nachr.* **293**, 53.

(3) Henning, Th., Gürtler, J., and Dorschner, J.: 1983, *Astrophys. Space Sci.* **94**, 333.

(4) Willner, S. P., Gillett, F. C., Herter, T. L., Jones, B., Krassner, J., Merrill, K. M., Pipher, J. L., Puetter, R. C., Rudy, R. J., Russell, R. W., and Soifer, B. T.: 1982, *Astrophys. J.* **253**, 174.

(5) Draine, B. T. and Lee, H. M.: 1984, *Astrophys. J.* **285**, 89.

(6) Henning, Th., Friedemann, C., Gürtler, J., and Dorschner, J.: 1984, *Astron. Nachr.* **305**, 67.

(7) Gürtler, J., Henning, Th., Dorschner, J., and Friedemann, C.: 1985, *Astron. Nachr.* **306**, 311.

(8) Knacke, R. F., McCorkle, S., Puetter, R. C., Erickson, E. F., and Krätschmer, W.: 1982, *Astrophys. J.* **260**, 141.

(9) Hagen, W., Tielens, A. G. G. M., and Greenberg, J. M.: 1983, *Astron. Astrophys.* **117**, 132.

(10) Léger, A., Gauthier, S., Défourneau, D., and Rouan, D.: 1983, *Astron. Astrophys.* **117**, 164.

(11) Hagen, W., Allamandola, L. J., and Greenberg, J. M.: 1980, *Astron. Astrophys.* **86**, L3.

(12) Strazzulla, G., Cataliotti, R. S., Calcagno, L., and Foti, G.: 1984, *Astron. Astrophys.* **133**, 77.

(13) Van de Bult, C. E. P. M.: 1984, Thesis, University of Leiden.

(14) Geballe, T. R., Baas, F., Greenberg, J. M., and Schutte, W.: 1985, *Astron. Astrophys.* **146**, L6.

(15) Lacy, J. H., Baas, F., Allamandola, L. J., Persson, S. E., McGregor, P. J., Lonsdale, C. J., Geballe, T. R., and van de Bult, C. E. P. M.: 1984, *Astrophys. J.* **276**, 533.

Table I (continued)

(16) Geballa, T. R.: 1985, in preparation.
(17) Tielens, A. G. G. M., Allamandola, L. J., Bregman, J., Goebel, J., d'Hendecourt, L. B., and Witteborn, F. C.: 1984, *Astrophys. J.* **287**, 697.
(18) Evans, N. J., Carr, J. S., Beckwith, S., Skrutskie, M., and Wyant, J.: 1983, *Publ. Astron. Soc. Pacific* **95**, 648.
(19) Puetter, R. C., Russell, R. W., Soifer, B. T., and Willner, S. P.: 1979, *Astrophys. J.* **228**, 118.
(20) Allamandola, C. J.: 1984, in M. F. Kessler and J. P. Phillips (eds.), *Galactic and Extragalactic Infrared Spectroscopy*, D. Reidel Publ. Co., Dordrecht, Holland, p. 5.
(21) Knacke, R. F. and Krätschmer, W.: 1980, *Astron. Astrophys.* **92**, 281.
(22) Sandford, S. A. and Walker, R. M.: 1985, *Astrophys. J.* **291**, 838.
(23) Erickson, E. F., Knacke, R. F., Tokunaga, A. T., and Haas, M. R.: 1981, *Astrophys. J.* **245**, 148.
(24) Krätschmer, W.: 1986, private communication.

absorption by aligned grains. This explains why only a weak correlation exists between the maximum polarization in the near IR and both the depth of the silicate and ice bands. For instance, Gürtler et al. (1985) found correlation coefficients much smaller than 0.5.

The detection of high-velocity molecular line emission in star-forming regions show that bipolar outflow of matter instead of inflow phenomena dominates in the environments of BN objects (Bally and Lada, 1983). Large extent of these flows combined with high velocities suggest that we do not observe gravitationally caused infall of matter. Henning (1986) showed that these outflows have energies of the order of $10^{38}-10^{40}$ W s, time-scales of 10^3-10^5 yr, and masses of 2–100 $M_\odot$. The mass loss rates inferred are $10^{-4}-10^{-3}\ M_\odot\ \mathrm{yr}^{-1}$.

At present two main lines for explaining the anisotropic mass flows are discussed. In the first group of models the bipolar structure comes from the interaction of a spherically-symmetric wind with an anisotropic confining medium (e.g., Cantó, 1980). Within the second group several mechanisms related to the star and/or its circumstellar disk have been proposed (e.g., Draine, 1983; Torbett, 1984). In these models the conversion of rotational, magnetic, and gravitational energy into expansion may naturally lead to the anisotropic flows. The momentum rate of the molecular outflows greatly exceeds the momentum rate of the stellar radiation. Therefore, these outflows cannot be driven by radiation pressure on dust grains.

Extended IR emission from vibration-rotation lines of molecular hydrogen has been detected in roughly half of the sources. The line emission is generally distributed over a region of 0.1–0.3 pc around the IR source, suggesting that the H_2 emitting region and the IR object are physically associated (Oliva and Moorwood, 1986). In all cases the H_2 emission is associated with bipolar CO high-velocity flows. In the Orion source a close spatial connection between the emission region and the high-velocity gas component could be proved (Nadeau et al., 1982). The observed characteristics of the emissions is best explained by shocks formed where the highly supersonic outflows plunge into the quiescent molecular cloud gas. Observations strongly point to thermal excitation of the lines. The excitation temperature for the $v = 1$ and 2 vibrational states is typically 2000 K.

Several BN objects show lines of the Paschen, Brackett, and Pfund series of atomic hydrogen in emission and/or weak radio continuum emission. The Brackett lines have very extended wings with a FWHM of ≥ 1000 km s^{-1}, indicating that the emitting gas is blown away. The intensities of the lines are best explained by stellar-wind models. Simon et al. (1983) proposed that both the IR lines and the radio continuum originate from an ionized stellar wind. Supporting evidence is given by Persson et al. (1984), Gürtler et al. (1985), and Henning (1985) from the analysis of several line ratios. McGregor et al. (1984) observed Paschen lines in a few BN objects. The line ratios are not consistent with Case B recombination, supporting the conclusions drawn from the ratios of the other IR hydrogen lines. The extent of the ionized wind region is about 10^{10} m. So, we can consider this region as an extended stellar atmosphere. The electron density is $\leq 10^{17}$ m^{-3}. This is greater by 7 orders of magnitude than the density of a normal H II region. The mass loss rates in the ionized winds are estimated at

10^{-6}–10^{-7} $M_\odot$ yr^{-1}. These mass loss rates are orders of magnitude lower than the rates of the molecular outflows. Therefore, the ionized wind near the star cannot drive the molecular outflows into the cloud.

Water and/or hydroxyl main–line maser sources are typically found in the vicinity of BN objects. Our sample contains much more H_2O masers than OH masers. The masers consist of individual spots with diameters of about 10^{12}–10^{13} m forming groups with typical sizes of the order of 10^{14} m.

In the vicinity of OMC 1–IRc2 two types of water masers must be distinguished (Downes *et al.*, 1981). Shell-type masers are obviously physically related to IRc2. Besides these masers, weaker high-velocity water masers are observed at a distance of 4×10^{15} m, which are not directly related to OMC 1–IRc2. For the other BN objects we computed as mean separations between IR and OH sources 4.7×10^{15} m and between IR and H_2O sources 2.6×10^{15} m. Although the significance of the derived mean values is hampered by the positional uncertainties, the H_2O masers seem to be situated closer to the IR sources than the OH masers do. Since the outer boundaries of the envelopes according to model computations (Henning, 1983, 1985) range from 10^{14} to 10^{15} m, the mean separation found here supports the view that the masers are located in the outer regions of the BN envelopes.

McGregor *et al.* (1984) found in some BN objects not embedded too deeply (GL 490, GL 961, M 17–IRS1, M 8E, and GL 437S) the O I λ844.6 nm and the Ca II IR triplet lines at $\lambda\lambda$849.8 nm, 854.2 nm, and 866.2 nm, plus other weaker lines. The O I emission line is excited by Lβ fluorescence. The emission-line region should lie at the boundary of the ionized-hydrogen region since oxygen cannot remain neutral in an ionized hydrogen region and the excitation mechanism requires that O be mostly neutral.

In those cases where they are detected, the Ca II triplet lines have equal strength, which points to high optical depth in these lines. These lines must also arise in a dense, warm ($T \leq 5000$ K), and only partially ionized medium. The neutral material must be either in the form of a disk or the emission-line region may be clumpy, with neutral material with a small filling factor in the ionized region. As McGregor *et al.* (1984) discussed the immediate circumstellar regions of BN objects appear to be quite similar to the circumstellar regions around many other well-known classes of visible emission-line stars, e.g., Be stars and T Tauri stars. This does by no means imply that the embedded stars are simply Be stars.

3. Evolutionary State of the Objects

The evolutionary state of the young objects can be determined only by a confrontation of the observed properties with theoretical evolutionary paths. After the second hydro-static core has been formed during the protostellar collapse, we suggest to distinguish three evolutionary stages.

3.1. PROTOSTARS (PURE ACCRETION PHASE)

The term protostar has quite different meanings in the literature. We will call protostar a stellar-like object in the pure accretion phase. The infalling matter is stopped in an accretion shock. There virtually the whole kinetic energy is transformed into heat yielding a luminosity of

$$L = \varepsilon G M_c \dot{M}/R_c ,$$ (1)

where G is the constant of gravitation, M_c the mass of the hydrostatic core, $\dot{M}$ the accretion rate, and R_c the core radius. The letter ε signifies an efficiency factor which depends on the relative rotation of core and inner disk. It is often assumed to be about 1. Numerical computations by, e.g., Yorke (1984) show that the contribution of the accretion shock to the total luminosity can exceed the Helmholtz–Kelvin luminosity of the core by an order of magnitude. This means that a protostar can appear quite luminous before hydrogen burning begins. But objects with $M \geq 3 \, M_\odot$ are completely obscured by optically thick dust shells during their whole pre-Main-Sequence evolution. The reason for that is that the Helmholtz–Kelvin time-scale, the evolutionary time-scale of the core, is shorter than the accretion time-scale of the envelope. This also explains why no observational evidence for pre-Main-Sequence objects with $M \geq 3 \, M_\odot$ on a Hayashi track could be found. Models of the hydrodynamic and spectral evolution of massive protostars by Yorke (1979, 1980) and others predict that even OB stars spent part of their Main-Sequence lifetime surrounded by optically thick dust-gas shells.

As mentioned already it is not possible to derive the temperature of the central source from the infrared spectra. Therefore, one cannot decide on the exact evolutionary state of the embedded source. Nevertheless, there are strong arguments that the BN objects discussed in the foregoing section are not real protostars but have reached the Main Sequence already. The earliest phases of protostellar evolution are observable in the far IR only because the available energy is not enough to heat the shell sufficiently. The low spatial resolution of present-day observations and the diffuse far IR background in the star-forming regions render the detection of compact sources difficult. The period for which a massive protostar is very luminous and perhaps becomes visible in mid or near IR is short (time-scale 10^3–10^4 yr) if compared with the lifetime of a massive Main-Sequence star with a dust shell (time-scale 10^5 yr). In addition, it seems highly probable that real protostars are deeply embedded in molecular clouds ($A_V > 30$ mag) and cannot be detected at wavelengths shorter than 20 μm. In accordance with this reasoning observations have not revealed real candidates for massive protostars. Especially the detection of recombination lines in a number of sources excludes them from being protostars. Failing in the detection of recombination lines is by no means a conclusive proof that an object is not yet producing ionizing radiation. On the one hand, the extinction in near IR may be still very high and, on the other hand, the sensitivity of the instrument may not be great enough.

3.2. BECKLIN–NEUGEBAUER OBJECTS (YOUNG STELLAR OBJECTS, YSO)

From the discussion up to now it seems highly probable that the BN objects we defined by means of observed characteristics in the second section are OB stars surrounded by optically thick dust shells. The BN phase ends by definition when the OB stars set up a developed H II region observable through an intense radio continuum flux.

Among the observational features, such as deep absorption bands, weak radio continuum emission, etc., the high-velocity outflows are one of the most important characteristics. Found in many BN objects, these flows show that the pure accretion phase has come to an end and the central source has started to react upon the cloud environment. This phenomenon has nothing to do with a simple reversion of the accretion flow by radiation pressure or a developing H II region as predicted by spherically-symmetric models of early stellar evolution. Hydrodynamic calculations, cf. for references, e.g., Henning and Gürtler (1985) suggest that rotation-modified collapse results in a disk-like or toroidal accretion structure. The energy liberated in the accretion shock may cause a mass loss which, as a consequence of a inhomogeneous density distribution, is bipolar. Several mechanisms have been proposed for driving these bipolar outflows (see Henning, 1986, for a critical review).

The inhomogeneously distributed gas observed in the vicinity of a number of BN objects with dimensions of 0.1 pc, e.g., in NGC 2071 and GL 490 (Takano *et al.*, 1984; Kawabe *et al.*, 1984) are not the disk-like accretion structures we spoke about. The latter should be much smaller than the observed structures and have radii of the order of 10^{13} m. High-resolution submillimetre observations of optically thin molecular lines should enable the detection of such accretion disks. The relation between the inner disk and the outer large-scale inhomogeneity is not fully understood. Pudritz and Norman (1986) proposed that the ionized winds originate from disks with radii of $r \leq 10^{13}$ m, whereas the molecular outflows arise from the cool disk envelope at radii up to 10^{15} m.

Scoville *et al.* (1983) found both a dense ionized stellar wind and an inflow of warm molecular gas with the BN object. This is a first hint that both outflow and inflow of matter may co-exist with young stellar objects.

Kolesnik and Kravchuk (1983) proposed a classification scheme for young objects that is based on a critical mass inflow rate for which the stellar luminosity stops the accretion. For this they investigated the motion of a dust particle under the influence of radiation pressure and gravitation in a one-fluid approximation. Since non-radiative-driven bipolar outflows are common phenomena with BN objects their simplified description is not an adequate representation of the real conditions. For instance, the object GL 490 is misinterpreted as being in the critical state where infalling matter does not reach the star any longer but cannot be blown away. Observations of CO lines give, however, clear evidence that GL 490 does lose matter. Henning (1986) arrived at a mass loss rate for the molecular gas of about $10^{-3} M_\odot$ yr^{-1} and for the ionized wind of $4.5 \times 10^{-7} M_\odot$ yr^{-1}.

In what follows we will give some criteria which should enable us to segregate the BN objects from compact H II regions.

(1) Hydrogen emission lines with relative intensities that cannot be explained by classical recombination theory (Genzel Case B). In Figure 1 we illustrate the analysis of the line ratios for two compact H II regions and two BN objects.

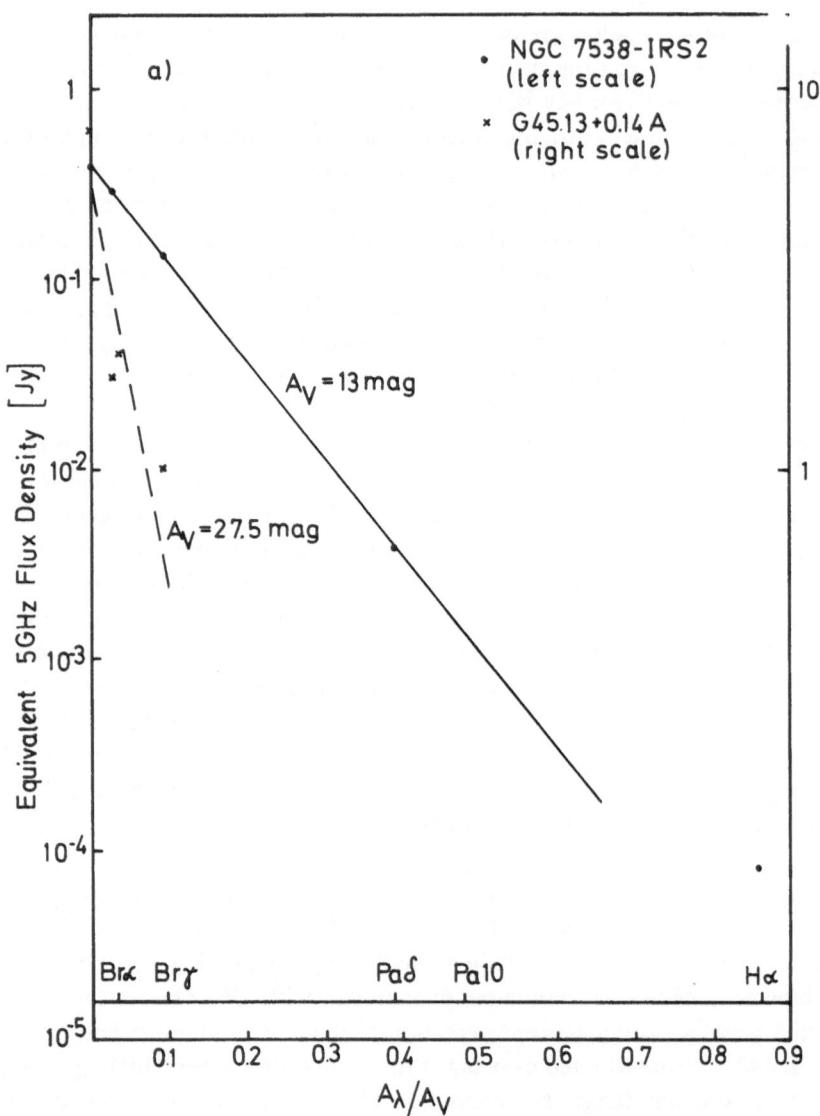

Fig. 1. Equivalent flux density at 5 GHz for a particular hydrogen emission line assuming no dust absorption and Genzel Case B (electron temperature: 10^4 K; electron density: 10^{10} m^{-3}) as a function of the normalized extinction coefficient A_λ/A_V of interstellar dust at the wavelength of the line. The extinction curve is based on the data of Koorneef (1983) and Landini *et al.* (1984). (a) Diagram for the compact H II regions NGC 7538–IRS2 and G 45.13 + 0.14A. The observations are good explained by Genzel Case B

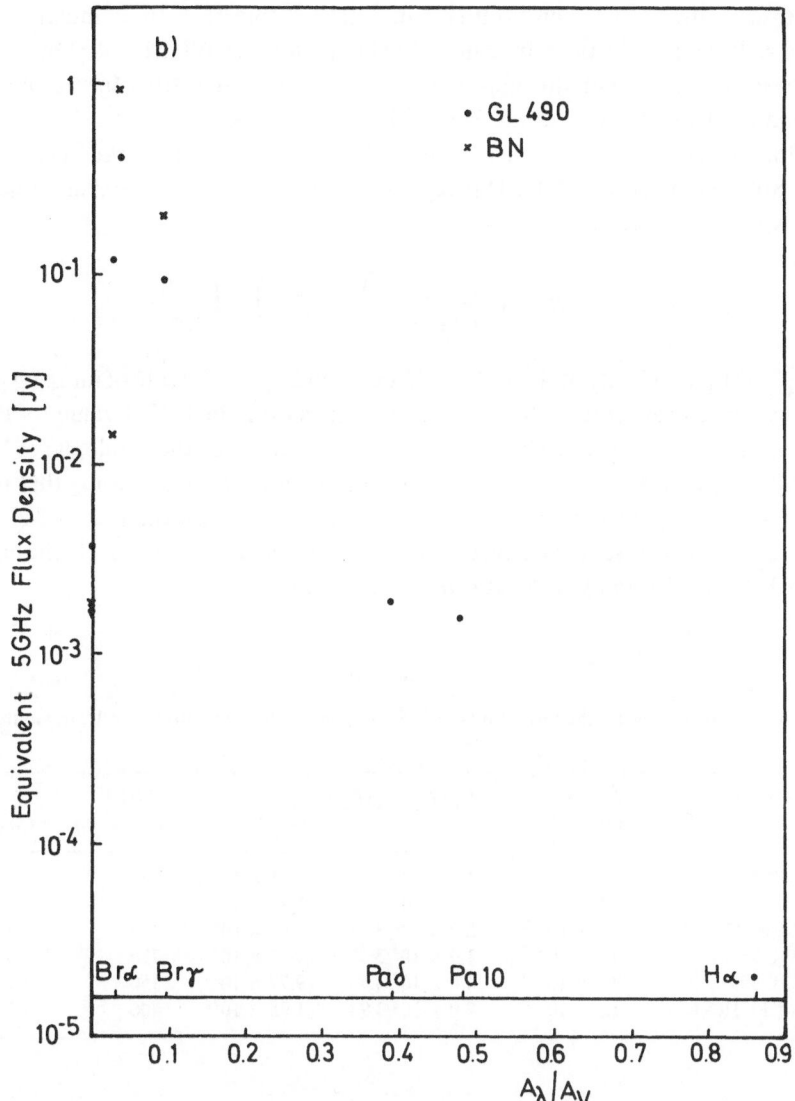

model. The anomalous strength of Hα in the case of NGC 7538–IRS2 is most likely due to foreground emission. The visual extinction A_v can be calculated from the slope of the straight line. Data are taken from Martin (1973), McGregor *et al.* (1984), Werner *et al.* (1979), Fischer *et al.* (1980), Krassner (1982), and Matthews *et al.* (1977). (b) Diagram for the BN objects GL 490 and BN. These objects do not satisfy the Menzel Case B model and are much better described by wind models. Data are from Hall *et al.* (1978), Simon *et al.* (1983), Moran *et al.* (1983), Persson *et al.* (1984), and McGregor *et al.* (1984).

As already mentioned, the observations are best explained by assuming a dense ionized stellar wind. Optically-thick Brackett and Pfund emission lines are typical of the spectra of BN objects.

Moreover, the Brα linewidths in BN objects are much broader (FWHM ≈ 150 km s^{-1}) than in compact H II regions (FWHM ≈ 40 km s^{-1}). The difference is consistent with the emission from a dense wind in BN objects and normal recombination from a rather static H II region, respectively.

(2) Mass loss rate too large for a normal H II region to form. Yorke (1984) showed that the Strömgren radius of the H II region is of the order of the stellar radius if the mass loss rate is greater than

$$\dot{M}_{cr} = 10^{-5} M_{\odot} \text{ yr}^{-1} \left(\frac{v_*}{10^6 \text{ m s}^{-1}} \right) \left(\frac{r_*}{10^9 \text{ m}} \right)^{1/2} \left(\frac{J_{UV}}{10^{49} \text{ s}^{-1}} \right)^{1/2}, \qquad (2)$$

where v_* is the gas velocity at the stellar radius r_* and J_{UV} is the rate of ionizing photons emitted by the central source. The gas velocity comes from the FWZI values of Brα lines of Persson *et al.* (1984) and the luminosity of the sources from the catalogue of Henning *et al.* (1984). The quantity J_{UV} was taken from Panagia (1973), assuming that the stars belong to luminosity class V. Table II summarizes the relevant data for 6 BN objects. In all cases the mass loss rate derived from the recombination lines $\dot{M}_i$ (Henning, 1986) exceeds the critical rate by at least one order of magnitude.

TABLE II

Comparison of critical rates for the formation of an H II region and the mass loss rates by the ionized stellar wind

No.	Source	$\dot{M}_i$ ($M_{\odot}$ yr^{-1})	$L_*(L_{\odot})/r_*(R_{\odot})$	J_{UV} (s^{-1})	FWZI (km s^{-1})	$\dot{M}_{cr}$ ($M_{\odot}$ yr^{-1})
6	GL 490	4.5×10^{-7}	$1.4 \times 10^3/3.63$	8.81×10^{43}	450	1.1×10^{-8}
10	BN	7.5×10^{-7}	$4.6 \times 10^3/4.68$	1.78×10^{45}	300	3.6×10^{-8}
16	Mon R2–IRS2	2.7×10^{-7}	$2.0 \times 10^3/3.94$	1.95×10^{44}	200	7.3×10^{-9}
19	GL 961–E	7.7×10^{-7}	$1.0 \times 10^3/3.36$	4.47×10^{43}	250	4.0×10^{-9}
20	GL 989	4.3×10^{-7}	$4.0 \times 10^3/4.56$	9.77×10^{44}	350	3.1×10^{-8}
29	M 17–IRS1	1.5×10^{-6}	$2.0 \times 10^3/3.94$	1.95×10^{44}	400	1.5×10^{-8}

(3) Very weak or no radio-continuum emission. While a reason for that is simply the small extent of the ionized region, the decisive criterion is whether or not the intensity ratio of Brα and radio continuum is consistent with the classical recombination theory. In each case one has to check whether the radio source is optically thin at the considered frequency. The spectral index for an optically thick source is $\alpha = 2$ (if $S_\nu \propto \nu^\alpha$). An optically thin stellar wind has $\alpha = 0.6$ if we have an electron density distribution which is proportional to r^{-2}, where r is the radial distance from the centre.

(4) Association with H$_2$O masers. As mentioned above, our sample contains much more water masers than hydroxyl masers. This supports the opinion of Habing and Israel (1979) that hydroxyl masers are generally associated with young compact H II regions and only in a few cases related to radio-quiet IR sources. The opposite is true

for H_2O masers. Matthews *et al.* (1985) found in a pilot systematic survey that only about one-third of the sources are associated with main-line OH emission.

(5) High polarization in the near IR. Dyck and Capps (1978) found that the IR sources with the smallest polarization mostly are compact H II regions. As the objects evolve, the conditions for producing polarization apparently worsen.

(6) It seems that the H_2 emission from the environment of BN objects is weaker than from ultracompact radio-loud H II regions (Oliva and Moorwood, 1986) and is lacking for the youngest objects at all. This may be interpreted as an evolutionary effect. But it cannot be excluded that large extinction has prevented the detection of the emission so far. This may be especially true for the probably youngest objects in our sample, W 33A, W 3–IRS5, and G 331.51–0.10, because they possess the largest optical depths in the 10-μm band.

3.3. COMPACT H II REGIONS

BN objects do evolve eventually in compact H II regions if the mass loss dies down. First the objects are optically thick at radio wavelengths, implying a continuum with a spectral index $\alpha = 2$. The diameter is of the order 10^{13} m. Hydroxyl masers occur in the surrounding molecular cloud; water masers may still be present. When the H II region has expanded to a radius of about 10^{15} m water masers fade away, giving a life-time of $5 \times 10^{4 \pm 0.5}$ yr for H_2O masers (Genzel and Downes, 1982). As the H II region continues to expand it becomes optically thin at longer and longer wavelengths. The spectral index is then $\alpha = -0.1$. Finally the OH masers vanish, too. The further evolution runs into the standard evolution of an OB star surrounded by an H II region.

4. Source Models for BN Objects

As a next step it seems important to reveal the linear dimensions of the BN objects. One way is the fitting of model spectra to the observations. Our theoretical spectra were computed with a radiative-transfer model described in some detail by Henning (1983). Early results of the computations are presented by Henning (1985). The basic model is a spherically-symmetric configuration with characteristic dimensions shown in Figure 2. The dust temperature at the inner boundary of the circumstellar dust shell is assumed as 1000 K. An ice-melting temperature of 120 K was used to estimate the extent of the ice-free zone in the shell. The radial density distribution is assumed to follow a power law with the exponent varying between 0 and -2. A constant density may be expected for a sporadic mass loss, whereas the distribution with the exponent -2 corresponds to a stationary outflow of matter. Interestingly the model computations yielded a temperature distribution throughout the shell that can be approximated satisfactorily by a power law too. The exponents depend on the model parameters and vary from -0.5 to -0.7.

The location of the H_2O and OH masers is indicated in accordance with the mean distance derived by Gürtler *et al.* (1985). The model of Figure 2 is very similar to the onion-skin model of a protostar as developed by Yorke (1984) on the base of

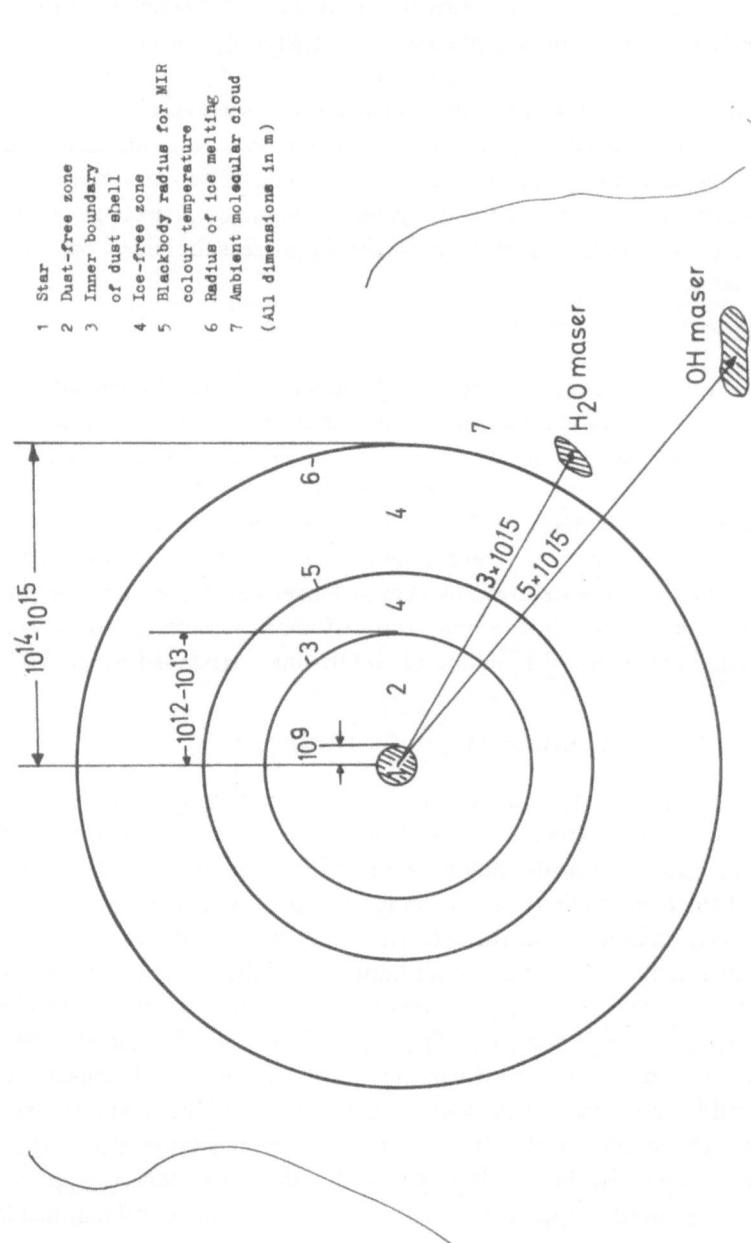

1 Star
2 Dust-free zone
3 Inner boundary
   of dust shell
4 Ice-free zone
5 Blackbody radius for MIR
   colour temperature
6 Radius of ice melting
7 Ambient molecular cloud

(All dimensions in m)

Fig. 2.   Schematical picture of the general structure of a **BN** object. The radii are only approximate; they vary with age and mass of the object.

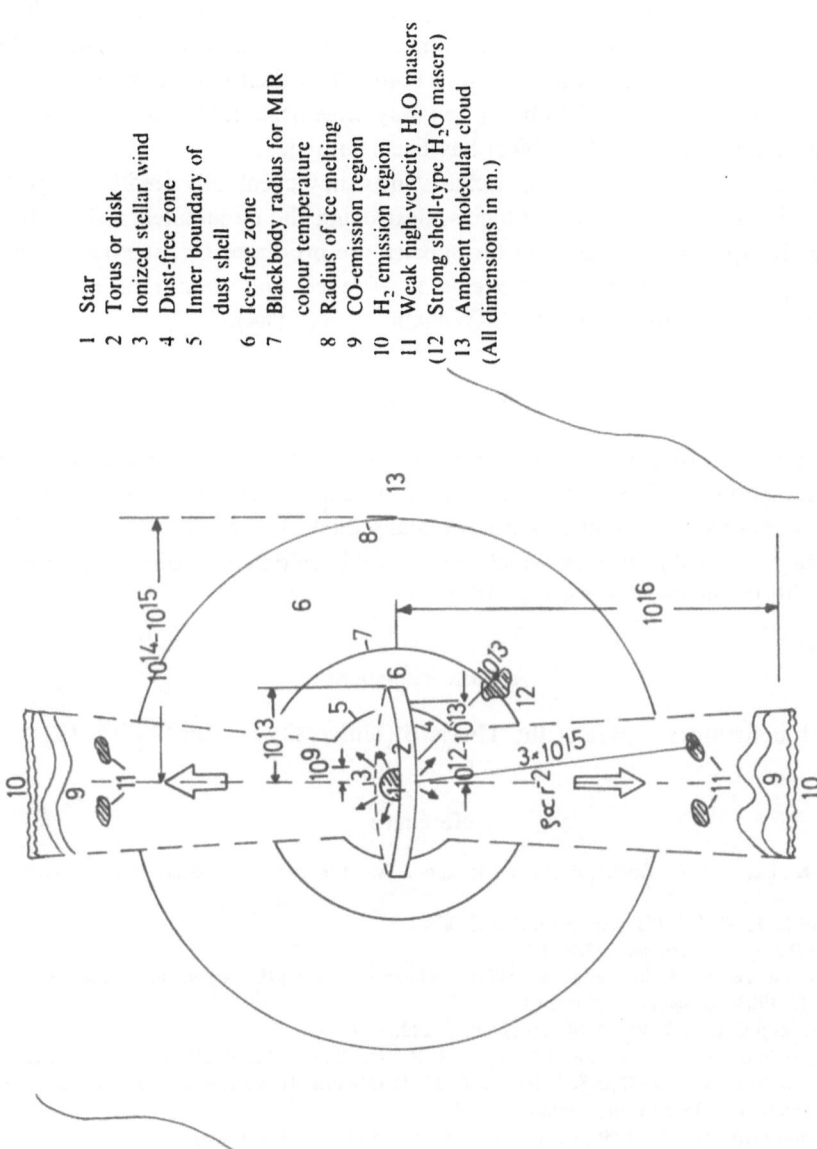

1   Star
2   Torus or disk
3   Ionized stellar wind
4   Dust-free zone
5   Inner boundary of
    dust shell
6   Ice-free zone
7   Blackbody radius for MIR
    colour temperature
8   Radius of ice melting
9   CO-emission region
10 H₂ emission region
11 Weak high-velocity H₂O masers
(12 Strong shell-type H₂O masers)
13 Ambient molecular cloud
(All dimensions in m.)

Fig. 3.   Sketch of possible configuration for a bipolar outflow from a BN object.

hydrodynamical calculations. This similarity supports the view that BN objects and protostars are generically related. The linear dimensions given, of course, vary with the mass and the age of the objects.

The detection of bipolar outflows made clear that a spherically-symmetric model can be a very crude representation of the reality only. A more realistic model is shown in Figure 3. As Rowan-Robinson (1982) pointed out, an object like this is hardly distinguishable from an object as shown in Figure 2 by their mid-IR spectra. We wish to emphasize that an additional absorption may arise from cold matter in the ambient molecular cloud where the BN object has been formed.

It seems possible that the strong stellar winds may eventually create a large cavity (diameter about 10^{15} m) in some objects, destroying the circumstellar shell. In these cases the absorption bands seen in the spectra are produced by intervening optically thick clumps. This conceptional schema may reflect the conditions of the Kleinmann–Low nebula rather well (Wynn-Williams et al., 1984).

5. Conclusions

Summarizing the observational properties and the results of theoretical modelling of these observations we have to conclude that the BN objects form a separate class of very young massive stars embedded in optically thick dust shells. Very probably they link the stage of a real protostar, which has not yet been definitely observed for massive objects, with the stage of a normal OB star.

Acknowledgement

This work constitutes a part of the Thesis B (University of Jena) of Th. H.

References

Aitken, D. K., Bailey, J. A., Roche, P. F., and Hough, J. M.: 1985, *Monthly Notices Roy. Astron. Soc.* **215**, 813.
Bally, J. and Lada, C. J.: 1983, *Astrophys. J.* **265**, 824.
Cantó, J.: 1980, *Astron. Astrophys.* **239**, L17.
Downes, D., Genzel, R., Bedijn, E. E., and Wynn-Williams, C. G.: 1981, *Astrophys. J.* **244**, 869.
Draine, B. T.: 1983, *Astrophys. J.* **270**, 519.
Dyck, H. M. and Capps, R. W.: 1978, *Astrophys. J.* **220**, L49.
Fischer, J., Righini-Cohen, G., Simon, M., Joyce, R. R., and Simon, Th.: 1980, *Astrophys. J.* **240**, L95.
Genzel, R. and Downes, D.: 1982, in R. S. Rodger and P. E. Dewdney (eds.), *Regions of Recent Star Formation*, D. Reidel Publ. Co., Dordrecht, Holland, p. 251.
Gürtler, J. and Henning, Th.: 1986, *Astrophys. Space Sci.* **128**, 163 (this issue).
Gürtler, J., Henning, Th., Dorschner, J., and Friedemann, C.: 1985, *Astron. Nachr.* **306**, 311.
Habing, H. J. and Israel, F. P.: 1979, *Ann. Rev. Astron. Astrophys.* **17**, 345.
Hall, D. N. B., Kleinmann, S. G., Ridgway, S. T., and Gillett, F. C.: 1978, *Astrophys. J.* **223**, L47.
Heckert, P. A. and Zeilik, M.: 1985, *Astron. J.* **90**, 2291.
Henning, Th.: 1983, Thesis, University of Jena.
Henning, Th.: 1985, *Astrophys. Space Sci.* **114**, 401.
Henning, Th: 1986, *Astron. Nachr.* **307**, 119.

Henning, Th. and Gürtler, J.: 1985, *Sterne* **61**, 195.

Henning, Th., Friedemann, C., Gürtler, J., and Dorschner, J.: 1984, *Astron. Nachr.* **305**, 67.

Kawabe, R., Ogawa, H., Fukui, Y., Takano, T., Takaba, H., Fujimoto, Y., Sugitani, K., and Fujimoto, M.: 1984, *Astrophys. J.* **282**, L73.

Kolesnik, I. G. and Kravchuk, S. G.: 1983, *Astron. Zh.* **60**, 889.

Koorneef, J.: 1983, *Astron. Astrophys.* **128**, 84.

Landini, M., Natta, A., Oliva, E., Salinari, P., and Moorwood, A. F. M.: 1984, *Astron. Astrophys.* **134**, 284.

Martin, A. H. M.: 1973, *Monthly Notices Roy. Astron. Soc.* **163**, 141.

Matthews, H. E., Goss, W. M., Winnberg, A., and Habing, H. J.: 1977, *Astron. Astrophys.* **61**, 261.

Matthews, H. E., Olnon, F. M., Winnberg, A., and Baud, B.: 1985, *Astron. Astrophys.* **149**, 227.

McGregor, P. J., Persson, S. E., and Cohen, J. E.: 1984, *Astrophys. J.* **286**, 609.

Moran, J., Garay, G., Reid, M. J., Genzel, R., Wright, M. C. H., and Plambeck, R. L.: 1983, *Astrophys. J.* **271**, L31.

Nadeau, D., Geballe, T. R., and Neugebauer, G.: 1982, *Astrophys. J.* **253**, 154.

Oliva, E. and Moorwood, A. F. M.: 1986, *Astron. Astrophys.* **164**, 104.

Panagia, N.: 1973, *Astron. J.* **78**, 929.

Persson, S. E., Geballe, T. R., McGregor, P. J., Edwards, S., and Lonsdale, C. J.: 1984, *Astrophys. J.* **286**, 289.

Pudritz, R. E. and Norman, C. A.: 1986, *Astrophys. J.* **301**.

Rowan-Robinson, M.: 1982, *Monthly Notices Roy. Astron. Soc.* **201**, 289.

Scoville, N., Kleinmann, S. G., Hall, D. N. B., and Ridgway, S. T.: 1983, *Astrophys. J.* **275**, 201.

Simon, M., Righini-Cohen, G., Fischer, J., and Cassar, L.: 1981, *Astrophys. J.* **251**, 552.

Simon, M., Felli, M., Cassar, L., Fischer, J., and Massi, M.: 1983, *Astrophys. J.* **266**, 623.

Takano, T., Fukui, Y., Ogawa, H., Takaba, H., Kawabe, R., Fujimoto, Y., Sugitani, K., Fujimoto, M.: 1984, *Astrophys. J.* **282**, L69.

Torbett, M. V.: 1984, *Astrophys. J.* **278**, 318.

Werner, M. W., Becklin, E. E., Gatley, I., Matthews, K., Neugebauer, G., and Wynn-Williams, C. G.: 1979, *Monthly Notices Roy. Astron. Soc.* **188**, 463.

Wynn-Williams, C. G.: 1982, *Ann. Rev. Astron. Astrophys.* **20**, 587.

Wynn-Williams, C. G., Genzel, R., Becklin, E. E., and Downes, D.: 1984, *Astrophys. J.* **281**, 172.

Yorke, H. W.: 1979, *Astron. Astrophys.* **80**, 308.

Yorke, H. W.: 1980, *Astron. Astrophys.* **85**, 215.

Yorke, H. W.: 1984, in R. D. Wolstencroft (ed.), *Star Formation Workshop*, Occ. Rep. of the Royal Observatory, Vol. 13, Edinburgh, p. 63.

Yorke, H. W. and Shustov, B. M.: 1981, *Astron. Astrophys.* **98**, 125.

Discussion

J. Staude: My problem is that we always have hints on such huge cavities but we have never seen them.

Th. Henning: I think that the study of Wynn-Williams *et al.* (1984) showed that such a cavity really exists in the environment of OMC I–IRc2.

J. Staude: Is the meaning that a model that explains the bipolar outflows requires such a huge cavity, such a huge empty space, around the star?

Th. Henning: No, I think that such a cavity is merely the result of the outflow. I mentioned two groups of models in my lecture to explain the bipolar outflows. I prefer the models with intrinsic collimation, e.g., by accretion disks, for several reasons.

J. Staude: This is exactly what we need for our objects.

Th. Henning: But, of course, our hypothesis must be proved. Some evidence came from infrared maps of several T Tauri stars, e.g., HL Tau. Another possibility would be observations of optically thin molecular lines with high spatial resolution.

P. G. Mezger: It may be that we are able to observe this with the 30-m telescope in Spain.

G. Rüdiger: It seems important to recall that the protostars have a lot of problems with the magnetic flux. Is there any model in this direction?

Th. Henning: Yes, Draine (1983) proposed a model in which rotation of the star winds up a frozen-in magnetic field above the poles of the star. The increased magnetic energy causes the bipolar outflow.

E. Krügel: May I ask what is your favoured explanation for the final energy which drives the outflow?

Th. Henning: I think no final conclusion can be drawn. But I prefer the model of Torbett (1984). Accretion from a disk produces an energetic boundary layer. The thermalization of orbital kinetic energy together with the disk structure leads to a pressure-driven wind which is perpendicular to the plane of the disk. But an exact hydrodynamical investigation of this mechanism has been lacking yet.

W. Pfau: If I understood you correctly the OB stars have dust shells when they are on the Main Sequence.

Th. Henning: Yes, I think that BN objects are OB stars at or approaching the Main Sequence. But, of course, OB stars will not have such dust shells during their whole Main-Sequence lifetime. The BN objects will appear later as compact H II regions.

P. G. Mezger: So, I think that compact H II regions have stopped the accretion. The ratio of compact H II regions as compared to the rest of O stars implies that about 10% of the O stars are being associated with compact H II regions. Now, one should know what is the ratio of O stars with compact H II regions to such stars which are only IR sources.

Th. Henning: Yes, I agree with you. But our catalogue of BN objects is not complete in a statistical sense.

H. J. Habing: Perhaps, we should look in the IRAS data bank.

Th. Henning: There may be some problems with spatial isolation of the objects. In addition you have to make some ground-based follow-up observations, e.g., the determination of the radio fluxes, for a clear identification of the IR sources.

CENTERS OF ACTIVITY IN DUST CLOUDS*

ROLF CHINI

Max-Planck-Institut für Radioastronomie, Bonn, F.R.G.

(Received 27 June, 1986)

Abstract. Recent submm observations of dust emission from star forming regions are presented and combined with NIR and IRAS data. The spectra cover the range 0.3 to 1300 µm and origin from ZAMS stars, compact H II regions, low-luminosity objects in dark clouds and external galaxies. The dust emission spectra are interpreted in terms of density and temperature distributions around central heating sources. In addition, general properties like IR luminosity, optical depth, wavelength dependence of dust opacity and, in the case of galaxies, star formation efficiency are discussed.

1. Introduction

The amount of dust associated with a young star is closely related to its evolutionary stage in the sense that the younger the object, the higher the optical depth of the surrounding dust cloud. From the observational point of view, this means that the detectable radiation shifts to longer wavelengths when looking towards younger stars. The extreme cases are, on one hand, the Main-Sequence star, which emits most of its energy in the UV and visible region, and, on the other hand, the protostar, which shows up primarily in the submm range. The present paper gives an overview on continuum observations in the wavelengths range from 0.3 to 1300 µm of young objects, which are still embedded in their parental dust clouds. The scenario includes ZAMS B-stars, which have driven away most of their protostellar dust shells (Section 2), compact H II regions, deeply burried within interstellar dust (Sections 3 and 4) and low-luminosity objects in dense regions of molecular clouds (Section 5). These different evolutionary stages of star formation contribute to the total FIR spectrum of a galaxy. Therefore, the investigation of the energy distributions from external galaxies enables us to study the properties of star formation and its efficiency on large scales (Section 6).

2. Cocoon Stars in M17

The young and luminous H II region M17 is bordered by an extended massive molecular cloud to the southwest. During the investigation of the stellar content of this complex, a number of heavily reddened stars have been found near the interface H II region-molecular cloud (Chini, 1982). Their energy distributions from 0.3 to 2.2 µm showed evidence for circumstellar dust shells. Observations out to 20 µm indicated that these stars exhibit the largest IR excesses known for optically visible stars today and that these excesses originate from hot dust close to the stars (Chini and Krügel, 1985). From

---

* Paper presented at a Workshop on 'The Role of Dust in Dense Regions of Interstellar Matter', held at Georgenthal, G.D.R., in March 1986.

Astrophysics and Space Science **128** (1986) 217–228.
© 1986 *by D. Reidel Publishing Company*

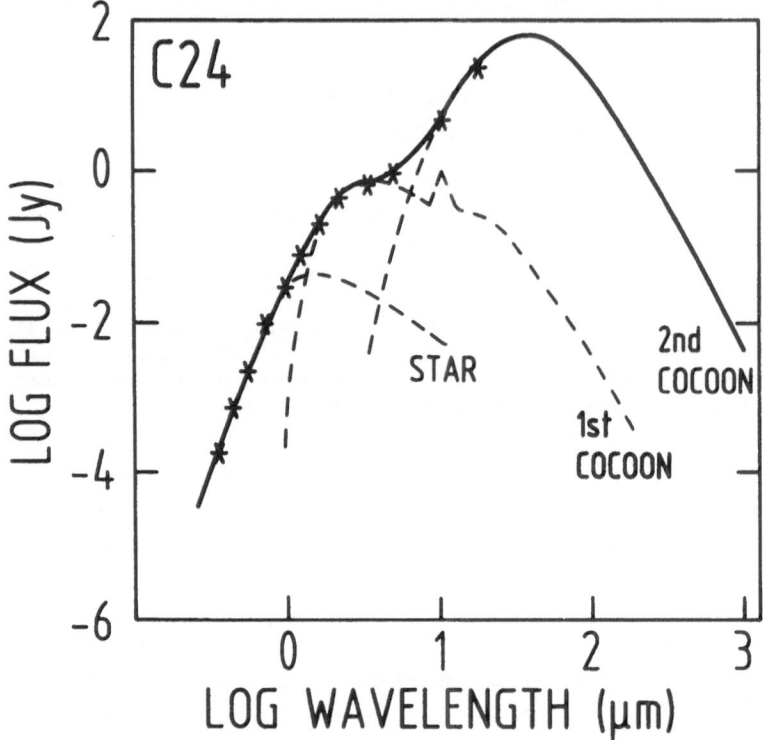

Fig. 1.   Observations (crosses) and model fit (solid line) for a cocoon star in M17. The spectrum consists
of three components (dashed lines), viz., stellar radiation, hot dust, and cool dust.

spectroscopy their spectral types could be estimated to be B1 or B2. Figure 1 shows the
continuum observations for one of the stars, which will be discussed in the following
in more detail. The observed energy distribution and the derived properties for this star
are representative for the entire sample.

Figure 1 shows that the energy distribution keeps rising from 0.3 to 20 μm;
unfortunately no information exists at longer wavelengths. An attempt was made to fit
this spectrum by a spherical dust distribution around the central B2 star (see Krügel,
this issue, for details of dust model and calculation procedure). The solid line in Figure 1
shows the result of a radiative transfer calculation, which fits the observations over the
entire range. In particular there are three components: (a) Below 1 μm the spectrum is
dominated by heavily reddened stellar radiation. (b) Between 1 and 4 μm the radiation
comes from a very hot (470–1360 K) dust cocoon close to the star (3–4.7 × 10^{14} cm).
(c) Above 4 μm the flux is due to a cooler dust shell of 55–255 K at a distance of
3–5.7 × 10^{16} cm. The inner cocoon produces 0.10 mag of visual extinction, the outer
one 0.62 mag.

At first glance it may seem artificial to fit this spectrum with two distinct dust cocoons
and no matter in between them. However, in spherical symmetry the only simple

alternatives to such a density distribution are power laws of the form $\rho(r) \propto r^{-n}$, with $n = 0$ (constant density), 1.5 (protostellar collapse), and 2 (stellar wind). The observations clearly rule out $n = 0$ because in a constant density model the grain temperatures vary smoothly with the distance from the star and, therefore, the model spectrum does not follow the observed wiggle around 4 µm. Models with $n = 1.5$ and 2 are essentially similar to the inner cocoon but fail to explain the observations at 10 and 20 µm. Therefore, it is plausible to explain the inner dust shell as being formed by incident matter from the protostellar cloud at a distance where the grains melt; this occurs at about 2×10^{14} cm. The second density maximum, which is required to interpret the observations at longer wavelengths, is probably an effect of radiation pressure on the ice-coated dust grains: at the melting radius of the ices ($\sim 3 \times 10^{16}$ cm) the infalling grains are halted and form the outer cocoon. Thus in both cases the IR emitting matter can be explained as a remnant of the protostellar envelope.

The low optical depths of the two dust shells indicate that the accretion period must be close to its end. Consequently these stars are in the latest stage of early stellar evolution. This interpretation seems to fit very well into the picture of large-scale star formation within the M17 complex. Star formation proceeds from the H II region towards the neutral cloud: in the ionized gas there are dozens of bright, weakly reddened stars, whereas in M17 SW only two very obscured IR sources are known. If this proposal is correct, one expects at the interface H II region-molecular cloud stars in their final stage of formation.

3. Dust Emission Spectra from Star-Forming Regions

The knowledge about several star-forming regions has considerably increased during recent years since IR astronomy has extended its wavelength regime out to about 200 µm. Most recently ground-based observations at 350 and 1300 µm completed the spectral coverage of 12 well-known regions (Chini *et al.*, 1986d) so that it is worth while readdressing the problem of dust emission spectra. The ultimate information one can get from fitting the observations from about 1 to 1300 µm is a 3-dimensional picture of the dust distribution around the embedded star as well as the temperature of the grains. In the following, two examples of well-known star forming regions are discussed in more detail. The results concerning the density and temperature distribution of dust are qualitatively the same for the remaining regions (Chini *et al.*, 1986d).

NGC 2071

About 4' north of the reflection nebula NGC 2071, at 500 pc from the Sun, there is an active region with a total luminosity of 900 $L_\odot$. At least four IR sources can be identified within an area 20" across, of which IRS 1 seems to be the dominant source of excitation; it is the strongest object at 10 µm and also at radio wavelengths. The radio spectrum is flat and is interpreted to arise from a weak compact H II region. Figure 2(a) shows the observations, indicated by dots; the solid line has been drawn to give a better impression of the energy distribution. The model computations fit the observed

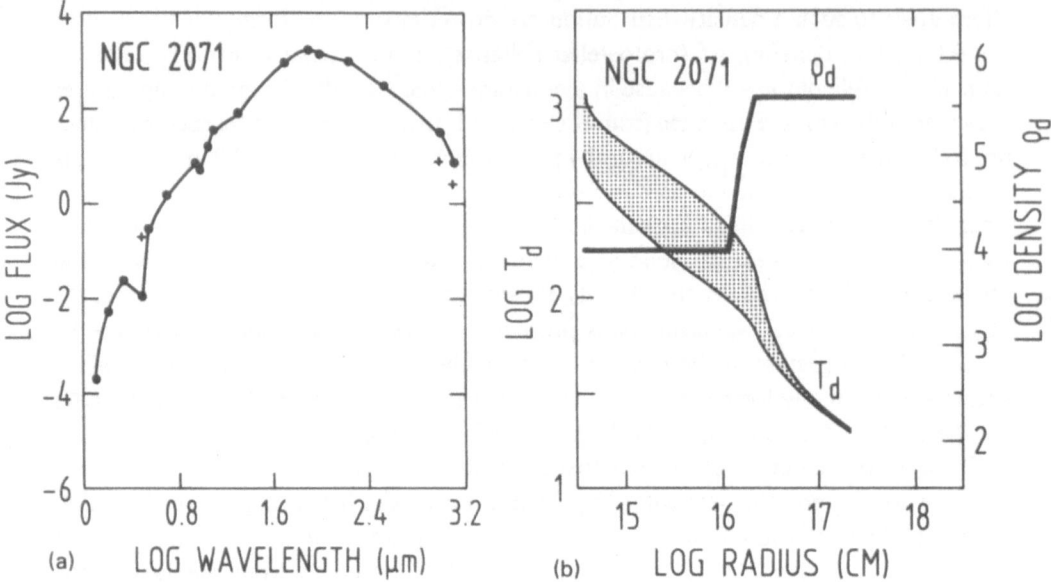

Fig. 2. (a) The spectrum of NGC 2071. The measurement at 3.1 μm includes the ice feature, which is not considered in the calculations. Therefore, the model predicts too large a flux. (b) The density and temperature structure of dust in NGC 2071.

spectrum very well so that it is not shown explicitly in Figure 2(a). Only at 3 wavelengths there are significant deviations, which are marked by crosses. The first deviation occurs at 3.1 μm because the ice-feature was not included in these calculations. The deviations at the long-wavelengths end of the spectrum are due to cold dust contained within the line-of-sight of the submm beam, which gives an excess of the observations over the fit values; this dust is not directly related to the source and must be situated beside or behind the star. Otherwise it would heavily influence the NIR part of the spectrum. Figure 2(b) summarizes the density distribution and the range of temperatures as a function of distance from the star. Surprisingly the dust is mainly located within two shells, of which the inner produces a visual extinction of only 2 mag, the outer of about 40 mag. Near the star the temperatures of the various grains attain about 1000 K whereas the outer boundary of the model they have only 25 K. In spherical symmetry the distribution of Figure 2(b) is a unique solution; quantitatively it might be changed slightly by choosing different dust properties but qualitatively it is not possible to fit the observations by, e.g., a constant density model or even a distribution where the dust density increases towards the star.

K3–50

At a distance of 8.8 kpc, K3–50 is a luminous H II region of $2.2 \times 10^6 L_\odot$. Its dust emission spectrum is shown in Figure 3(a), the results of the radiative transfer calculations are summarized in Figure 3(b). Again the model for K3–50 has a shell structure

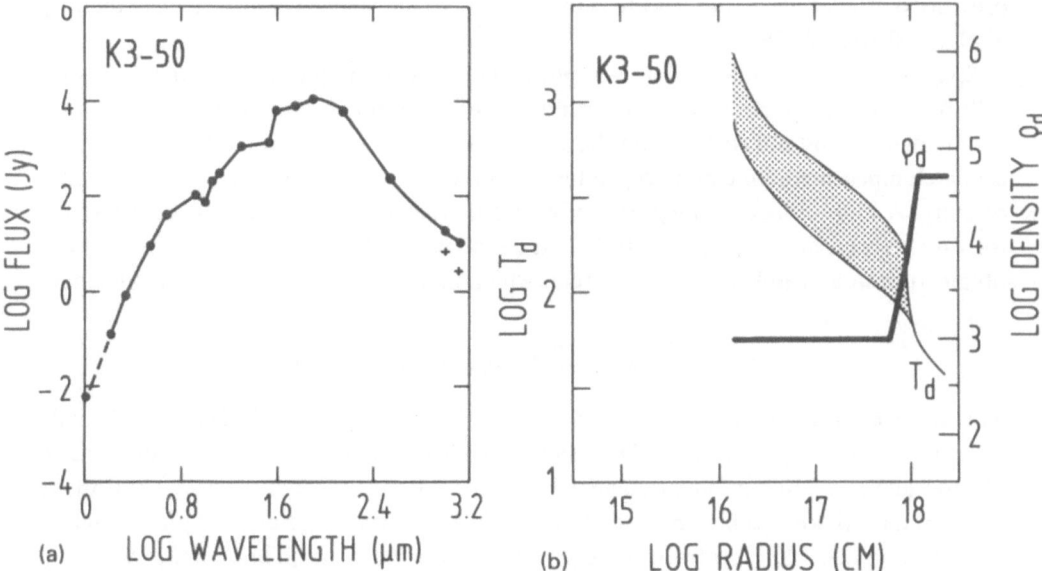

Fig. 3.   (a) The specturm of K3–50. The observed 1 μm point probably does not refer to the IR object and is, therefore, only connected with a dashed line. (b) The density and temperature structure of dust in K3–50.

with a total visual optical depth $\tau_V = 29$ to the embedded star. In addition, strong dust depletion (depletion factor $f \approx 10$) within the H II region is necessary to explain the NIR observatins.

In summary the star forming regions investigated so far over the wavelengths range from 1 to 1300 μm span more than three orders of magnitude in luminosity. Nevertheless, they show common features, which must be characteristic of very young stars. All objects are still deeply burried within the interstellar medium. This medium is formed by the remnants of the protostellar cloud out of which the central star collapsed and by the ambient molecular cloud into which the protostellar envelope is embedded. The collapse of a protostellar cloud proceeds very nonhomologously. The density sharply peaks at the center and in the early phases the total visual extinction may exceed 1000 mag. For the group of objects discussed in this section a very different structure is present and, therefore, a later evolutionary stage is suggested. The mean visual optical depth for all sources is $\tau_V = 55 \pm 33$. All models consist of two shells. The inner shell has a mean τ_V of 2 ± 1 and generates the flux below 20 μm. The outer shell produces the bulk of extinction and the emission longward of 20 μm. The observed flux densities beyond 350 μm, which are larger than those of the models, are obtained with broad beams and thus include the ambient molecular cloud. As mentioned before, this matter is not obscuring the star, but lies beside or behind it. Such a situation is expected if star formation occurs at the edge of a molecular cloud facing the Sun. The masses of the model clouds are typically a few times $10^3 \, M_\odot$. The surprising similar structure of all sources indicates a common mechanism. One could imagine that radiation

pressure on the grains could lead to a low density in the stellar vicinity as described by
Yorke and Krügel (1977).

Spherical symmetry can probably approximate many non-spherical geometries rather
well because the spectrum depends primarily on how much energy is absorbed at a
certain density and not so much on the angular distribution of the dust. We even expect
that a clumped medium can be reproduced correctly if the clumps have a filling factor
of unity. A more difficult configuration would be a ring or disk-like dust distribution,
for which in certain extreme orientations the models would fail. Luckily, the results are
obtained from a sample of objects, a fact, which increases their statistical significance.

4. FIR Spectra of Compact H II Regions

The strongest IR sources contained in the *IRAS Point Source Catalogue* are heavily
obscured compact H II regions. They show increasing energy distributions from 12 to
100 μm and have 100 μm flux densities larger than 1000 Jy. Most recently 87 of these
sources, all with known kinematic distances, have been observed at 1300 μm so that an
almost complete FIR spectrum for each of the objects exists (Chini *et al.*, 1986a, b).
None of these H II regions is visible on the POSS prints and so far no observations
below 12 μm are available. Figure 4(a) shows the averaged spectrum of the 87 H II
regions normalized to a 100 μm flux density of 1000 Jy. The error bars depict the scatter
of the individual regions around the mean spectral points. The observations can be easily

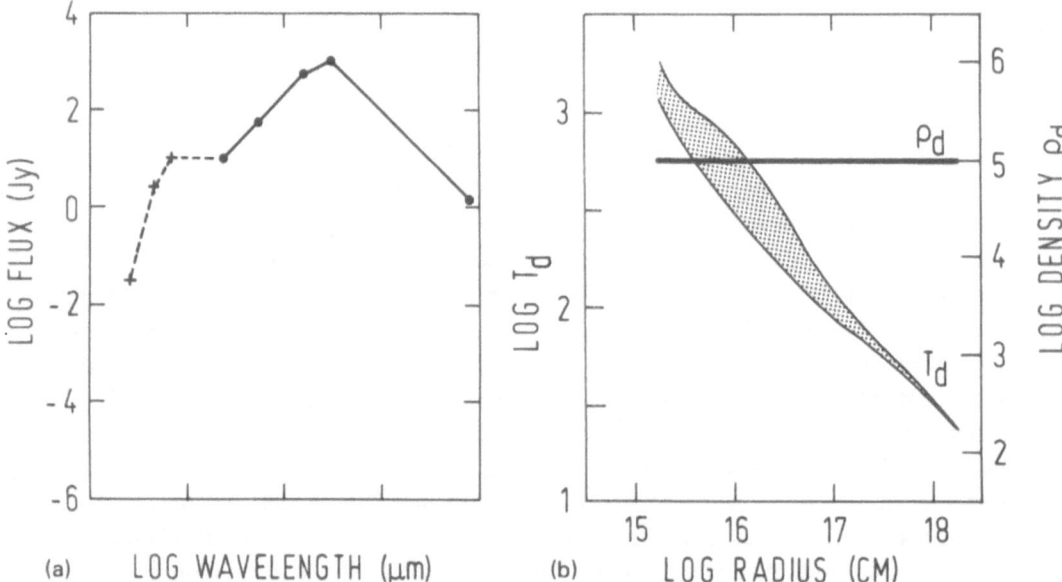

Fig. 4.   (a) The averaged spectrum of 87 H II regions normalized to 1000 Jy at 100 μm; the observations
are marked as dots, the predicted NIR fluxes are indicated as crosses. (b) The mean density and
temperature structure of dust in 87 compact H II regions.

interpreted as emission from a dust cocoon of constant density as shown in Figure 4(b). For illustration the expected NIR flux densities have been marked as crosses in Figure 4(a). It will be very important to get these NIR observations because only they allow a final interpretation whether the constance of density is real or simply due to the paucity of data.

The integration of the individual spectra allow to estimate the luminosity of the embedded stellar sources; because the NIR part is missing the resulting luminosity is a limit, which will be about 20% too low. Nevertheless a comparison with radio data (Wink *et al.*, 1982) indicates that the dominant heating source must be a single early-type star for most of the regions. Therefore, it is possible to convert the observed luminosities into spectral types according to Panagia (1973). The result is shown in Figure 5 which gives the distribution of spectral types throughout the sample. There is clearly a peak around 05.5; the rapid decrease towards early B-type stars might be an effect of incompleteness caused by the sensitivity of the observations. Additional observing material and a thorough investigation of the limits of completeness may help to determine the IMF for massive stars created in dense galactic dust clouds.

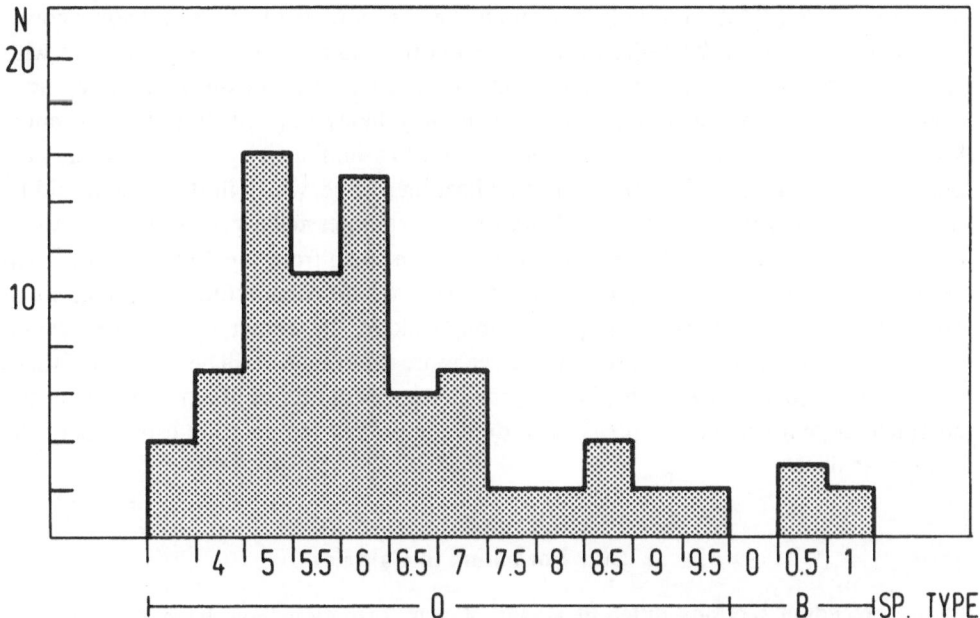

Fig. 5. The distribution of spectral types as derived from the observed IR luminosity between 12 and 1300 μm. It is assumed that each region contains only one dominant heating source.

As shown by Chini *et al.* (1986a, b) the observations between 100 and 1300 μm are well suited to determine the wavelength dependence of dust opacity in the submm range. Approximating the dust emission by a modified Planck curve of the form $\nu^m B_\nu (T_d)$ a lower limit for m of 1.7 is established by the compact H II regions in the present sample.

5. Dense Cores of Dark Clouds and Globules

The earliest stages in stellar evolution are certainly clumps and regions of high density within massive molecular clouds. So far these dense cores could only be detected by radio molecular line observations where they show up by enhanced molecular emission. Recently, 40 of the densest cores within dark clouds and globules have been searched for dust emission at 350 and 1300 μm (Chini et al., 1986e). From the ratio of observed flux densities at these wavelengths one can precisely determine the dust temperature involved. Preliminary computations yield temperatures of 8 ± 2 K for those cores for which no embedded source is known. Dense cores where IRAS found embedded sources have mean temperatures of 15 ± 3 K. From the flux densities obtained at 1300 μm it is straightforward to estimate the optical depth at that wavelength and to convert it to a visual extinction. For the regions under consideration A_v lies between 20 and 100 mag. These values are averaged over a beam size of 90 arc sec, which means that in reality there might occur considerably higher peak column densities. A final answer to this question can be given when high-resolution 1300 μm data from the 30 m MRT on Pico Veleta are available. From the temperatures of the dust derived above and under the assumption that the emission can be described by a modified Planck curve of the form $v^2 B_v(T_d)$ it is possible to estimate the FIR luminosity of the individual cores. The values range from 0.02–0.1 $L_\odot$ for empty cores and from 0.3–5 $L_\odot$ for cores with embedded IRAS sources. At present it is purely speculative where the luminosity comes from. In the case of the IRAS sources it is very likely that inside the dense cores low-mass stars have been already created. The low-luminosity cores, however, may obtain their energy as well from an internal heating source, which in this case would be definitely a collapsing protostar, as from the general interstellar radiation field. Again this answer must await the high-resolution 1300 μm data from the 30 m telescope. An extended emission in the order of the core size would favour the interpretation of external heating, whereas a sharp emission peak in the center of the core would corroborate the existence of a protostar. Even more information will be available if these regions are observed at 800 μm; besides the better resolution of about 7 arc sec at the 30 m telescope it will also be possible to derive improved temperature estimates of the protstellar dust.

6. Star Formation in Galaxies

In the preceding sections different stages of star formation have been described. A summarizing view is obtained if one looks at the FIR emission of an entire galaxy. Then the various contributions are averaged and the large-scale properties of star formation and its efficiency can be observed. For that purpose, submm data at 350 and 1300 μm have been combined with IRAS observations from 12 to 100 μm for 26 galaxies (Chini et al., 1986c). To illustrate the similarity of the energy distributions, the spectra of 18 Sb and Sc galaxies have been normalized to a flux density of 100 Jy at 100 μm and the data points at the remaining wavelengths have been averaged. The result is shown in Figure 6;

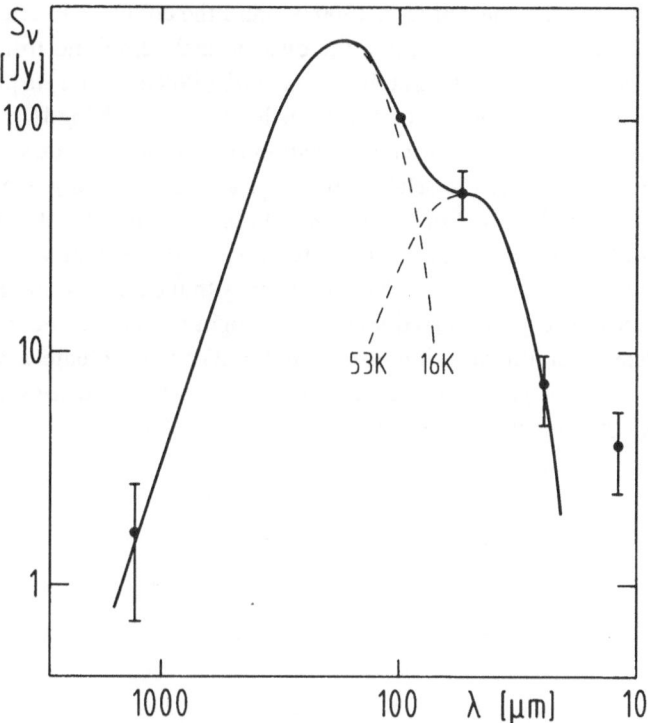

Fig. 6.   The average spectrum of 18 Sb, c galaxies normalized to a flux density of 100 Jy at 100 μm. The error bars depict the range of individual galaxies. The solid curve is the sum of two modified Planck curves of the form $v^2 B_v(T_d)$ with T_d = 16 and 53 K, respectively.

the error bars do not represent observational uncertainties but indicate slight variations in the spectra of the individual objects. The spectra of the remaining 8 galaxies, SB, pec., or unclassified types, have a similar structure.

To fit the FIR part of this spectrum (25 < λ < 1300 μm), which is due to emission from dust grains of various temperatures, modified Planck curves of the form $v^2 B_v(T_d)$ have been superimposed. As shown in Figure 6 it is in fact possible to represent the observations from 25 to 1300 μm very well by just two dust emission spectra with temperatures of 16 K (cold dust) and 53 K (warm dust), respectively. This can also be done for each galaxy separately and the combinations are unique, as long as one explain the entire FIR spectrum by a minimum number of two dust components. A detailed analysis of the dust emission spectrum of our Galaxy by Cox *et al.* (1986) has shown that the interpretation of the FIR spectrum as being composed of contributions of cold and warm dust is not an arbitrary one. Cold dust represents emission from dust associated with diffuse atomic hydrogen and with molecular hydrogen in dense clouds. For heating the warm dust stars of type O and early B are necessary. In this sense the warm luminosity L^w, which is obtained by integrating the corresponding Planck curve from Figure 6, is immediately connected with the star formation rate of massive stars.

From the flux densities measured at 1300 μm and the colour temperature of the cold component in Figure 6, one can obtain the dust optical depth and from that the total gas mass M_H for each galaxy separately. Chini *et al.* (1986c) find a surprisingly narrow relation between these quantities of the form $\log L''' = c' + k \log M_H$. As mentioned above L''' can be replaced by ψ, the star formation rate of massive stars. Schmidt (1959) formulated for the first time the relation between ψ and the total gas mass M_H in the form $\psi = cM_H^k$. From Figure 2 of Chini *et al.* (1986c) one derives that the exponent k is very constant for all galaxies and is close to unity. Only the factor of proportionality seems to vary with the morphological type in the way that c is four times lower in normal Sc galaxies than it is, e.g., in barred spirals. The tight correlation between L''' and the gas mass implies a stationary situation; star bursts would create a wide spread of luminosities for a given gas mass. Remarkably this seems to be true even for the galaxy with the highest luminosity in the present sample, Arp 220, which is considered to be an outstanding active galaxy.

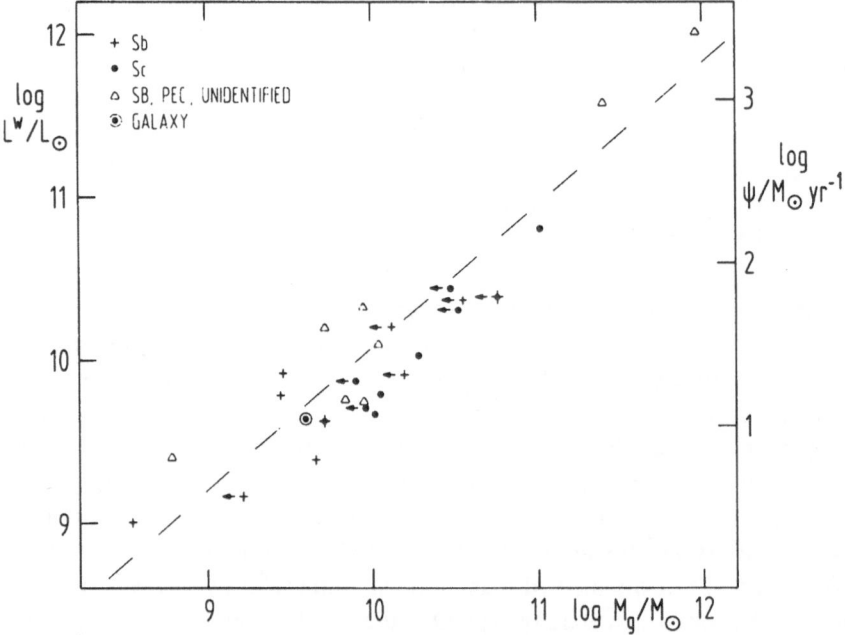

Fig. 7. The luminosity L''' of the warm dust component is plotted as a function of the total gas mass Mg. The dashed line is a least square fit to all data points. The right-hand ordinate is expressed in SFRs.

7. Conclusions

The present paper has described different stages of early stellar evolution from the observational point of view. Starting from ZAMS stars, which are still surrounded by remnants of their protostellar clouds, it was attempted to find observational evidence for even earlier phases till the very beginning of a protostar. It has become clear that the amount of dust and, as a consequence, the role of submm observations increases considerably when approaching the actual formation of a star. The submm data, on which most of the present results are based, have been obtained very recently. Some of them are still incomplete or need additional support by high-resolution observations. This will be done in the near future at the 30 m MRT on Pico Veleta, Spain. Nevertheless, some of the problems concerning the conditions of dust around very young objects could be investigated by modeling their observed FIR spectra. A major result is the finding that most of the well-known regions of star formation are in a comparatively late stage of evolution, as witnessed by the density distribution of dust, which increases with distance from the star. The heavily obscured compact H II regions might be slightly younger because there is evidence for constant density within the surrounding dust cloud; here NIR observations will help to refine the interpretation. The dense and empty cores of dark clouds and globules, finally, represent the earliest stages of stellar evolution observed so far at submm wavelengths.

References

Chini, R.: 1982, *Astron. Astrophys.* **110**, 332.
Chini, R. and Krügel, E.: 1985, *Astron. Astrophys.* **146**, 175.
Chini, R., Kreysa, E., Mezger, P. G., and Gemünd, H.-P.: 1986a, *Astron. Astrophys.* **154**, L8.
Chini, R., Kreysa, E., Mezger, P. G., and Gemünd, H.-P.: 1986b, *Astron. Astrophys.* **157**, L1.
Chini, R., Kreysa, E., Krügel, E., and Mezger, P. G.: 1986c, *Astron. Astrophys.* **166**, L8.
Chini, R., Krügel, E., and Kreysa, E.: 1986d, *Astron. Astrophys.* (in press).
Chini, R., Kreysa, E., and Mezger, P. G.: 1986e, in preparation.
Cox, P., Krügel, E., and Mezger, P. G.: 1986, *Astron. Astrophys.* **155**, 380.
Panagia, N.: 1973, *Astron. J.* **78**, 929.
Schmidt, M.: 1959, *Astrophys. J.* **129**, 243.
Wink, J. E., Altenhoff, W. J., and Mezger, P. G.: 1982, *Astron. Astrophys.* **108**, 227.
Yorke, H. and Krügel, E.: 1977, *Astron. Astrophys.* **98**, 125.

Discussion

H. J. Habing: There will be a second IRAS conference on star formation in galaxies in June in California. I saw all the abstracts as a member of the organizing committee and I was surprised that about half of the papers concern the fact that one should not judge immediately that IR excess means star formation because in a lot of cases there are good indications that there are active nuclei. There are several authors who say that Arp 220 is not at all a star-forming galaxy. It is just a thing of a very active nucleus.

R. Chini: We find that Arp 220 fits well into a relation where total gas mass and star formation rate are proportional; an additional mechanism to explain the high IR luminosity is, therefore, not required.

P. G. Mezger: If you plot our own Galaxy in the diagram it falls on that line for peculiar galaxies. So, apparently the star formation is higher than in other Sc galaxies.

THE SIMULATION OF THE INITIAL MASS FUNCTION AND STAR FORMATION EFFICIENCY*

MAREK WOLF and VLADIMÍR VANÝSEK

Department of Astronomy and Astrophysics, Charles University, Prague, Czechoslovakia

(Received 27 June, 1986)

Abstract. The fragmentation of a molecular cloud is modelled as a random process by the Monte Carlo method. The probability of the fragmentation is a function of the cloud initial mass and decreases rapidly for mass lower than critical mass, which is a defined parameter. The modelled IMF is compared with the mean mass function in open clusters assumed here as the observed IMF. The best fit was found for initial mass $3 \times 10^3 \, M_s$ and for the critical mass range 0.4 to 0.6 M_s. It also implies the star formation efficiency to be about 0.3.

1. Introduction

Present evidence suggests that the initial mass function for stars has nearly a power law form for large masses. It deviates, however, from this form for masses below $1 \, M_s$, reaching its peak at a few tenths of M_s and then declining rapidly toward smaller masses.

One possible way how to determine IMF is the simulation of the fragmentation process by a stochastic method. Monte Carlo simulation of the hierarchical fragmentation was first developed by Elmegreen and Mathieu (1983) as a generalization of the simple probabilistic theory of the fragmentation by Larson (1973).

2. The Fragmentation Model

In our study we used this model to derive the IMF on the whole interval of mass and compare it with observed mass function. According to Elmegreen and Mathieu we assumed that the integer number of fragments N_F, which form in every fragmentation step, is a random variable with normal probability distribution:

$$P_F(N_F) = \frac{1}{s_F \sqrt{2\pi}} \int_{N_F - 1}^{N_F} \exp\left[\frac{-(N - N_{F_0})^2}{2s_F^2}\right] dN .$$

(1)

Relative masses m_i/m_F of all N_F fragments were allowed to be random variable with Gaussian distribution:

$$P_m(m_i/m_F) = P_0 \exp\left[-\frac{[(m_i/m_F) - 1]^2}{2s_m^2}\right], \quad m_i > 0 .$$

(2)

---

* Paper presented at a Workshop on 'The Role of Dust in Dense Regions of Interstellar Matter', held at Georgenthal, G.D.R., in March 1986.

These two probability distribution functions were added by assumption that probability of fragmentation P_F is a function of fragment mass. We introduced a critical fragment mass m_c as follows. Let fragmentation probability for mass $m_F \gg m_c$ be approximately equal 1, respectively, for mass $m_F \ll m_c$ be near 0. Mathematical formulation of these conditions leads to the general Cauchy distribution with free parameter x, of the form

$$P_F(m_F) = \frac{1}{(m_F/m_c)^x + 1} \, , \quad x < 0 \, . \tag{3}$$

Figure 1 shows the dependence of the probability P_F on the fragment mass for some different x.

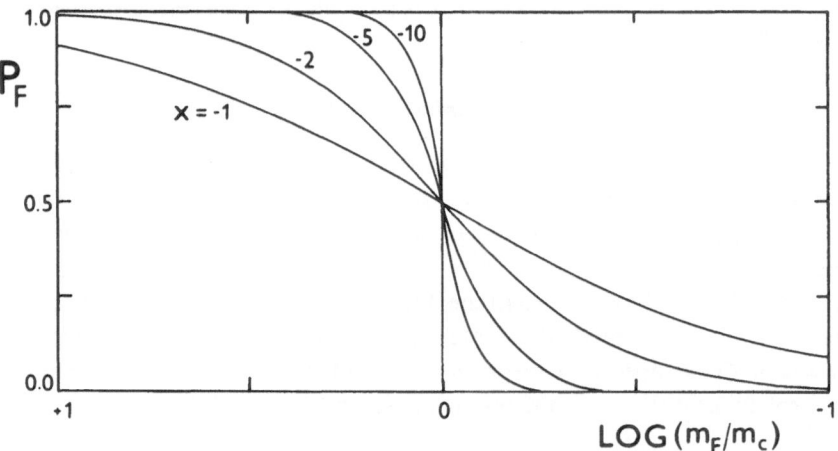

Fig. 1.   Dependence of fragmentation probability P_F on the fragment mass for $x = -1, -2, -5$, and -10.

Each simulation was initiated by choosing the mean number of fragments per event N_{F_0} and its standard deviation s_F, the mass dispersion s_m, the critical mass m_c, and the slope x. These parameters with the total cloud mass were taken to be constant throughout the simulation. Our Monte Carlo procedure iterated similar to the procedure of $E + M$. During each event a value of P_F was choosen for each existing fragment according to its mass. The fragmentation process was halted, when each fragment had smaller probability of fragmentation than was probability according to Equation (3). The final number of originated fragments was always less than 10^4. The variations of the final mass distribution with model parameters N_{F_0}, s_F, and s_m were studied in the above mentioned work. We concerned about the influence of critical mass on resulting mass function.

For each set of fragments being formed via the Monte Carlo method just described, the IMF by standard process was computed. Figure 2 shows the model net of mass function for different critical mass and initial cloud mass. In all pictures the mean mass

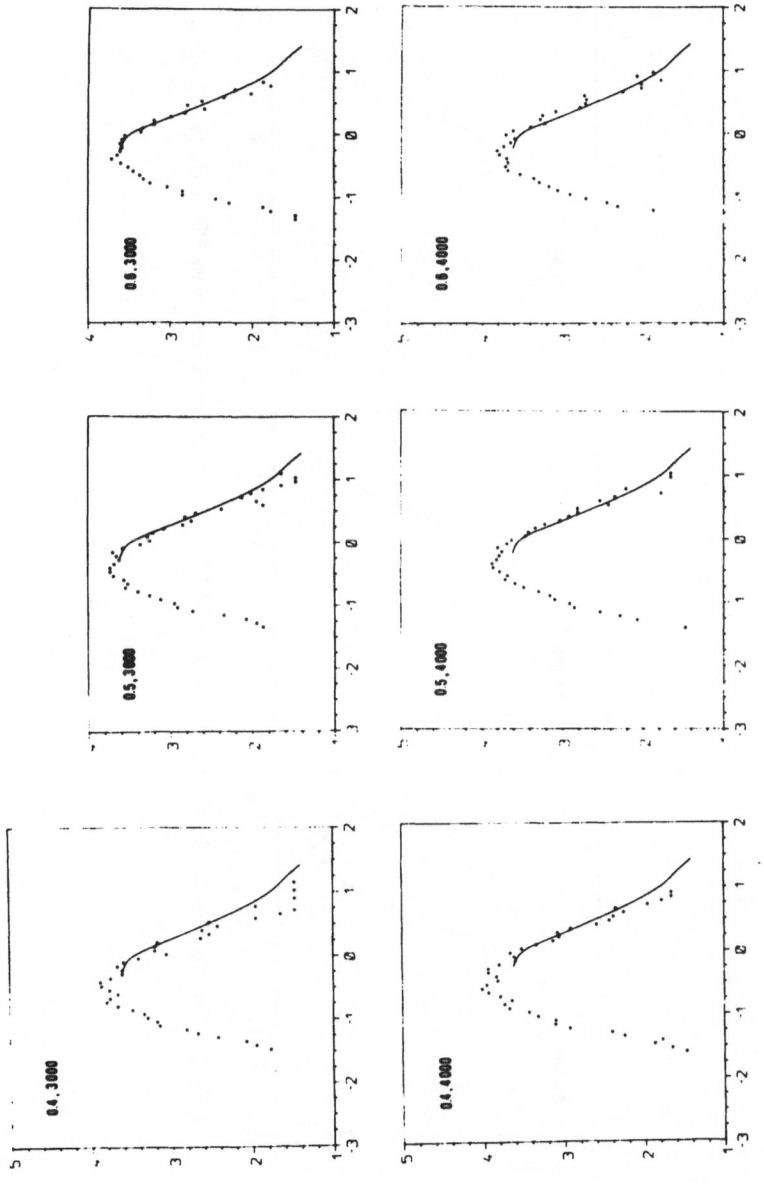

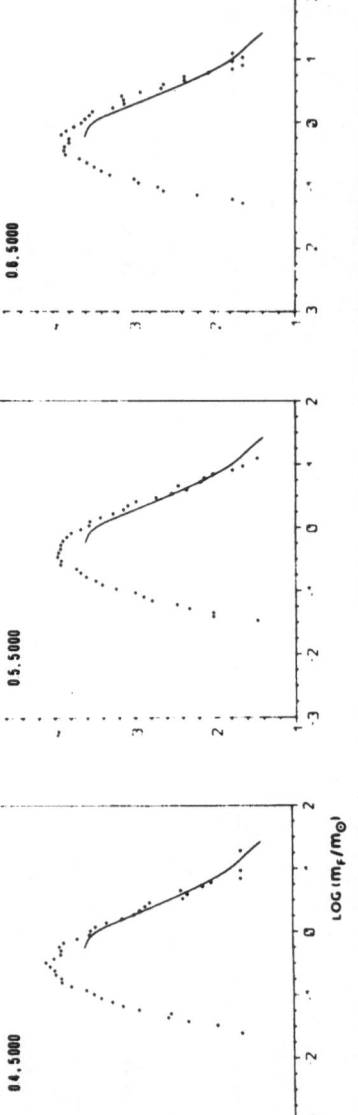

Fig. 2.   The set of mass function for critical mass m_c = 0.4, 0.5, and 0.6 M_s and initial cloud mass 3000, 4000, and 5000 M_s.

function in open clusters according to Taff (1974) is plotted. Figures 3(a–c) show in detail the best fit of modelled and observed IMF for the initial cloud mass 3000 M_s and critical mass 0.4, 0.5, and 0.6 M_s.

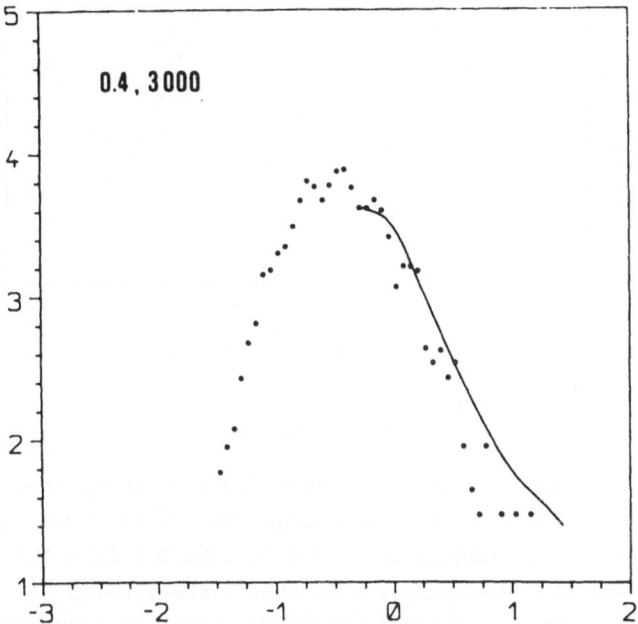

Fig. 3a.   The best fit of modelled and observed IMF for the initial cloud mass 3000 M_s and critical mass 0.4 M_s.

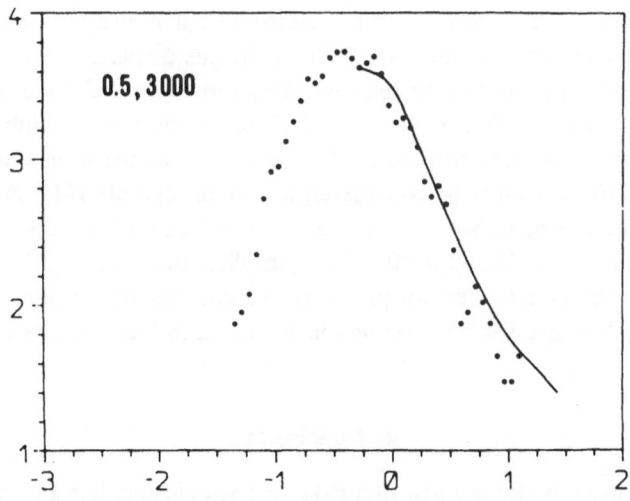

Fig. 3b.   The same as Figure 3(a), but for $m_c = 0.5 M_s$.

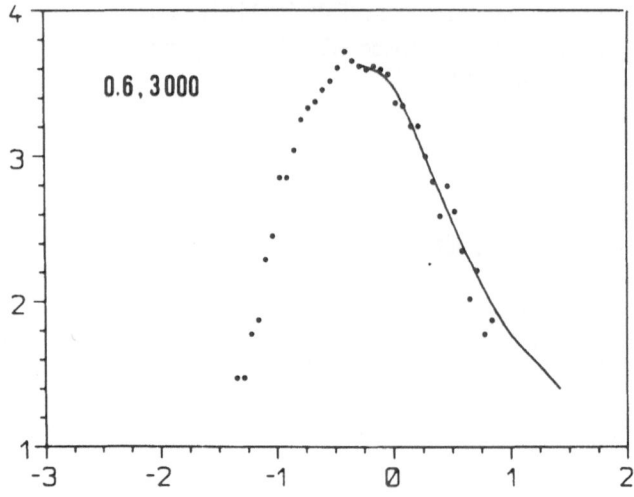

Fig. 3c.   The same as in Figure 3(a), but for $m_c = 0.6\,M_s$.

3. The Influence of SFE

It is necessary to remark that many observational reasons show that not all initial mass of cloud are turned into protostars. For example, from the total mass of Orion stellar association and the surrounding neutral gas has been estimated only 0.3% of the original gaseous mass should be turned into stars (Duerr *et al.*, 1982). The star formation efficiency (SFE) defined as ratio of the initial cloud mass and final mass of the stars formed from this cloud, is SFE ≪ 1. Thus the star formation seems to be an inefficient process. The higher value of SFE follows from the recent studies of the ρ Ophiuchi dark cloud made by Wilking and Lada (1983), from which 0.34–0.47 has been obtained. The value of the SFE is an increasing function of the ratio τ_J/τ_R, where τ_J is the characteristic time-scale of fragmentation steps (i.e., it is proportional to the Jeans' length of fragments) and τ_R is the time-scale of the gas dispersion (i.e., removal) from the cloud by the energy input of the massive stars formed in the cloud. If $\tau_R \gg \tau_J$, then the formation efficiency could be close to 1. The τ_R is decreasing with the increasing number of the massive stars formed in the early stage of the fragmentation process. Therefore, the SFE < 1 must be considered in any acceptable IMF model. Since the average mass of a typical dense cloud is about $10^4\,M_s$ and our results indicate the best fit for the initial mass $M \simeq 3 \times 10^3\,M_s$, it implies the SFE $\simeq 0.3$. Therefore, the formation efficiency factor does not place serious constraints for our approach to the IMF modelling, because the SFE reduction factor is, in fact, included in the 'best fit' initial mass.

4. Conclusions

The Monte Carlo simulations show that IMF in open clusters has a form of log-normal distribution on whole mass range. The maximum of this distribution depends on the value of the critical mass, i.e., about $0.5\ M_s$. The results of this model confirm that hierarchical fragmentation does not continue spontaneously during the star formation, but is broken up. This fact is well comprehended by definition of critical mass. The smallest fragment masses created via the Monte Carlo method are about $0.03\ M_s$. This value agrees with the results of Rees (1976), who for opacity limited fragmentation derived the smallest fragment mass about $0.01\ M_s$.

References

Duerr, R., Imhoff, C. L., and Lada, C. J.: 1982, *Astrophys. J.* **261**, 135.
Elmegreen, B. G. and Mathieu, R. D.: 1983, *Monthly Notices Roy. Astron. Soc.* **203**, 305.
Larson, R. B.: 1973, *Monthly Notices Roy. Astron. Soc.* **161**, 133.
Rees, M. J.: 1976, *Monthly Notices Roy. Astron. Soc.* **176**, 483.
Taff, G. L.: 1974, *Astron. J.* **79**, 1280.
Wilking, B. A. and Lada, C. J.: 1983, *Astrophys. J.* **274**, 698.

Discussion

P. G. Mezger: What do you define as star formation efficiency?

M. Wolf: The star formation efficiency is the ratio of the mass of the stars to the total mass of the cloud.

SELF-REGULATED STAR FORMATION AND THE EVOLUTION
OF STELLAR SYSTEMS*

TH. HENNING and B. STECKLUM

Universitäts-Sternwarte Jena, G.D.R.

(Received 23 June, 1986)

Abstract. In this paper we estimate the star formation efficiency using the assumption that star formation continues until the radiation pressure disrupts the cloud. The results suggest that in the case of low/medium-mass star formation the efficiency could be about five times higher than in the case of high-mass star formation.

For a three-component star-forming system (low/medium-mass stars, high-mass stars, gas) we investigate the temporal behaviour and the final star formation efficiency. We can show that the efficiency in $10^4 \, M_\odot$ clouds is higher than in $10^6 \, M_\odot$ clouds. This supports our view that bound stellar systems form from medium-mass clouds, whereas OB associations form in the cores of giant molecular clouds. Furthermore, the effect of induced high-mass star formation may cause a change of the mass spectrum during the formation of an OB association.

1. Introduction

Recently evidence is increasing that the initial mass spectrum (IMS) in any particular region of star formation depends on the physical properties of the parental molecular cloud, e.g., on the mass, the density, the temperature, and the turbulent state of the cloud. It seems that low-mass star formation takes place in cold ($T \approx 10$ K) and dispersed clouds and that the stars are in scattered small groups. High-mass stars, on the other hand, appear to form in the warm ($T \approx 30$ K) centrally condensed cores of massive clouds. Larson (1982) suggested from this picture that star formation in large molecular clouds is an ongoing process over which time both the structure of a molecular cloud and its stellar content evolve systematically. This suggestion is closely connected with the scenario of sequential star formation proposed by Herbig (1962), Eggen (1976), and Elmegreen (1983). In this scenario the massive stars form later than the low-mass stars and terminate the star formation process due to their disruptive power. Spatial bimodality was advocated by Güsten and Mezger (1982) with only high- and medium-mass stars forming in spiral arms and stars in the whole stellar mass range in the interarm regions.

Star formation is a process which alters its own conditions by the interaction of newly born stars with the molecular cloud. This feedback may be stimulating as well as restraining in dependence on stellar mass. The formation of stars may induce further star formation, an assumption on which the theory of stochastic self-propagating star formation (Gerola *et al.*, 1980) is based. On the other hand, the formation of low-mass

---

* Paper presented at a Workshop on 'The Role of Dust in Dense Regions of Interstellar Matter', held at Georgenthal, G.D.R., in March 1986.

Astrophysics and Space Science **128** (1986) 237–252.

stars may be suppressed as a result of the cloud heating by the radiation field of newly formed massive stars as shown by Fleck (1982) and Larson (1985).

In a galactic plane survey Solomon *et al.* (1985) found two molecular cloud populations: molecular clouds characterized by warm molecular cloud cores located inside of spiral arms and clouds with cold molecular cores located both inside and outside of spiral arms. It is very probable that the warm cloud cores are the result of the heating by recently formed or presently forming OB-stars. This supports the common view that massive stars form mainly in the spiral arms.

Unfortunately, our knowledge of the physical conditions and processes in a star-forming region are far from being complete. Therefore, there is no detailed theory of star formation in a molecular cloud up to now. Nevertheless, several models in context with the estimation of the star formation efficiency (SFE) were developed because this quantity is closely connected with both the conditions in the parental molecular cloud and the formation of bound clusters. The SFE is defined by

$$\text{SFE} = M_* / (M_* + M_g), \tag{1}$$

where M_* is the mass of the stellar population and M_g the gas mass.

The models for the formation and very early evolution of stellar systems employ either a pure statistical approach or evolutionary equations. Using N-body calculations Lada *et al.* (1984) showed that bound clusters form if the SFE exceeds 30% and the gas removal time is about 3 million years.

Elmegreen and Clemens (1985) applied a Monte-Carlo procedure to compute SFEs. They obtained the result that the average SFE for large molecular clouds is much lower than in the case of small molecular clouds and concluded that at least 10% of small molecular clouds should form bound clusters. This value results in a far too high open cluster formation rate if the clouds would be in free fall. To solve this problem Elmegreen and Clemens proposed either a rather long mean formation time of bound clusters of $\approx 10^8$ yr (cf. Wielen, 1971; Miller and Scalo, 1978), or that such clouds avoid forming clusters by mechanisms unrelated to internal stellar disruption. Before drawing final conclusions, several parts of the model, e.g., the adopted mass-luminosity relation for high-mass stars and the description of the disruption process should be improved.

Models using evolutionary equations were at first developed to describe the evolution of galaxies (Shore, 1981; Ferrini and Marchesoni, 1984). The equations reflect the temporal evolution of a system consisting of several phases of matter, e.g, diffuse gas, clouds, and stars. Using a three-component system Bodifée and de Loore (1985) studied a star-forming region by this method. The rate coefficients entering into the equations depend on the details of the star-forming process. So, a crude estimation of these quantities is unavoidable at present. Bodifée and de Loore found solutions with an oscillating star formation rate. The reason for these oscillations was the rather arbitrary assumption of a reservoir from which cool atomic gas enters the system. Nevertheless, their approach to the problem is of some importance for our study.

In this paper we investigate several models of self-regulated star formation. After introducing the definitions of the formation rates and other relevant equations in

Section 2 we present in Section 3 a simple estimations of the SFE. In Section 4 we study a three-component star-forming region (high-mass stars, low/medium-mass stars, and gas) and the results are summarized in Section 5.

2. Definitions and Basic Relations

To describe the temporal behaviour of star formation we have to introduce several birthrates which are similar to those of Franco and Cox (1983). The stellar birthrate is the number of stars created per unit time

$$\dot{S}_* = dN/dt \quad (\text{stars yr}^{-1}) . \tag{2}$$

The rates per unit volume $\dot{S}_V$ and per unit gas mass $\dot{S}_M$ become

$$\dot{S}_V = \dot{S}_*/V_g \quad (\text{stars pc}^{-3}\,\text{yr}^{-1}) \quad \text{and} \quad \dot{S}_M = \dot{S}_*/M_g \quad (\text{stars } M_\odot^{-1}\,\text{yr}^{-1}), \tag{3}$$

respectively, where M_g and V_g are the mass and the volume of the cloud gas. The star formation rate is

$$\dot{S} = \langle m \rangle \, \dot{S}_* \quad (M_\odot\,\text{yr}^{-1}) , \tag{4}$$

where $\langle m \rangle$ is the mean mass of the stars. Now, we can introduce the mass consumption rates per unit volume $\dot{M}_V$ and per unit mass $\dot{M}_M$

$$\dot{M}_V = \langle m \rangle \, \dot{S}_V = \dot{S}/V_g \quad (M_\odot\,\text{pc}^{-3}\,\text{yr}^{-1}) ;$$

$$\dot{M}_M = \langle m \rangle \, \dot{S}_M = \dot{S}/M_g \quad (\text{yr}^{-1}) . \tag{5}$$

Extrapolating the Taurus Auriga star formation rate to the galaxy as a whole, Franco and Cox (1983) arrived at a total star formation rate of $\approx 3\text{--}16\,M_\odot\,\text{yr}^{-1}$. This is similar to the results of Smith et al. (1978), who found $\approx 5\text{--}10\,M_\odot\,\text{yr}^{-1}$.

If we want to proceed to other galactic rates we have to divide the different rates by πR^2, where R is the galactic radius. So, the surface density of the galactic birthrate is

$$\sigma_n = \dot{S}_*/(\pi R^2) . \tag{6}$$

This quantity is usually assumed to go with the average total density of atomic hydrogen if the direct conversion of the gas phase into stars is considered (Schmidt, 1959): i.e.,

$$\sigma_n \propto \rho^n \propto [M_g/(\pi R^2)]^n , \tag{7}$$

with n between 1 and 2. This is only a convenient assumption, since it is now well known that stars form from molecular clouds. We will use an analogous relation between the mass of the molecular cloud and the spontaneous star formation rate

$$\dot{S}_s \propto M_g^n . \tag{8}$$

The rate of induced star formation $\dot{S}_i$ depends on both the mass of the molecular cloud and the mass of the stars S_{act} which cause the birth of new stars

$$\dot{S}_i \propto S_{\text{act}} M_g^n . \tag{9}$$

The rate coefficients for spontaneous and induced star formation are still be left unspecified. Some constraints on these will be discussed later.

It is assumed that the form of the stellar mass distribution – the IMS – is invariant during star formation. We use the IMS proposed by Miller and Scalo (1979) which is a log-normal density distribution of the stellar masses

$$n(m) = \frac{1}{m\sigma(2\pi)^{1/2}} \exp\{-[\ln m - a]^2/(2\sigma^2)\},$$ (10)

with the parameters $a = \ln 0.1$ and $\sigma^2 = \ln 10$. The quantity m is given in units of the solar mass, and normalized so that

$$\int_0^\infty n(m)\, dm = 1.$$ (11)

To account for the stellar mass limits we introduce both a low-mass cut-off at m_l and a high-mass cut-off at m_{max}. The lower mass limit m_l is taken to be 0.05 $M_\odot$. The upper mass limit m_{max} is chosen according to a relation which was suggested by Larson (1982)

$$m_{max} = 0.33 M_g^{0.43}.$$ (12)

According to this relation, the mass of the most massive star in a star-forming region is dependent on the mass of the molecular cloud. We have to mention that the existence of such a relation is not well-established because of difficulties in assigning and estimating the mass of the most massive star in a star-forming region. Furthermore, it is questionable if this mass should be related to the mass of the whole molecular cloud or to the mass of the molecular cloud core. In our computations we consider two cases of molecular clouds, massive clouds having a mass of $10^6\ M_\odot$ and smaller clouds with a mass of $10^4\ M_\odot$.

In Section 3 we subdivide the whole stellar mass range in two parts with the assumption that in both parts the stellar masses are distributed according to Equation (10), but with a different normalization.

The mass-luminosity relation used in the determination of the rate coefficients is from Heintze (1973): i.e.,

$$l(m) = \begin{cases} m^{4.85(1 - 0.179 \log m)}, & m > 0.7\ M_\odot, \\ m^{2.48}, & m \le 0.7\ M_\odot. \end{cases}$$ (13)

In Section 3 we use the mass-luminosity relations of Straižys and Kuriliene (1981) and Schmidt-Kaler (1982), respectively, for the sake of comparison. All these relations are for Main-Sequence stars.

3. Estimation of the Star Formation Efficiency

A first crude estimation of the SFE is possible using the assumption that star formation continues until the radiation pressure disrupts the cloud. We assume that all stars are in the centre of the cloud core, so, the radiation pressure acts radially against gravitation. The radiation pressure force on a dust grain of radius a in a distance r from the centre is in the case of an optically thin core

$$F_r^d = \frac{\pi a^2 L_\odot N \langle l_{pr} \rangle}{4\pi r^2 c} \, , \tag{14}$$

where N is the total number of stars, $L_\odot$ the solar luminosity, and $\langle l_{pr} \rangle$ the radiation pressure weighted mean luminosity given in units of the solar luminosity

$$\langle l_{pr} \rangle = \frac{\displaystyle\int_{m_l}^{m_{max}} n(m) l(m) \langle Q_{pr}[a, T(m)] \rangle \, dm}{1 - \displaystyle\int_0^{m_l} n(m) \, dm - \int_{m_{max}}^{x} n(m) \, dm} \, , \tag{15}$$

where $\langle Q_{pr}[a, T(m)] \rangle$ is the Planck mean of the radiation pressure efficiency factor. T is the effective temperature which depends on the mass m of the star. The denominator accounts for the absence of stars with masses lower than m_l and greater than m_{max}. The molecular hydrogen is accelerated by gas-dust collisions. The radiation pressure force on one hydrogen molecule is given by

$$F_r = F_r^d N^d / N^{H_2} \, , \tag{16}$$

where N^d and N^{H_2} are the number densities of dust and gas, respectively. Using $a = 0.1$ μm and a ratio of the mass densities between dust and gas of 6×10^{-3} it follows that

$$F_r = N m_{H_2} \langle l_{pr} \rangle 1.384 \times 10^{18} / r^2 \, . \tag{17}$$

On the other hand, the gravitational force on a hydrogen molecule at distance r from the centre is

$$F_g = GM m_{H_2} / r^2 \, , \tag{18}$$

where $M = M_* + M_g$ is the total mass inside of the sphere of radius r. Equating the two forces leads to a limiting condition of star formation. Turning from SI to astronomical units it follows that

$$M_* + M_g = 1.04 \times 10^{-2} N \langle l_{pr} \rangle \quad M_\odot \, . \tag{19}$$

The total mass of stars is given by $M_* = N \langle m \rangle$, where $\langle m \rangle$ is the mean mass of stars in the considered mass range

$$\langle m \rangle = \frac{\displaystyle\int_{m_l}^{m_{max}} n(m) \, dm}{1 - \displaystyle\int_0^{m_l} n(m) \, dm - \int_{m_{max}}^{x} n(m) \, dm} \, . \tag{20}$$

Now we can replace N and use the definition of the SFE to arrive at

$$\text{SFE} = 95.9 \langle m \rangle / \langle l_{pr} \rangle . \tag{21}$$

For the computation of the radiation pressure weighted mean luminosity we used the Planck means of silicates by Gilman (1974). The numerical integration of Equation (15) was performed applying mass-luminosity relations according to Straižys and Kuriliene (1981) as well as Schmidt-Kaler (1982). The results are given in Table I. The absolute

TABLE I

Results for the SFE in the simple model case

M_R ($M_\odot$)	m_{max} ($M_\odot$)	SFE (%)	Mass-luminosity relation
10^6	120	24	Straižys and Kuriliene
10^6	120	18	Schmidt-Kaler
7200	15	60	Straižys and Kuriliene
7200	15	78	Schmidt-Kaler

values of the SFE should be not overinterpreted because of the simplicity of the model and the uncertainty of some input parameters. They are somewhat larger than the derived values for different star-forming regions. Nevertheless, they suggest that in the case of low-mass star formation the efficiency could be about five times higher than in the case of high-mass star formation.

From the observational point of view the determination of the SFE seems to be very difficult because the possibility of the estimation of the initial gas content in the region which is now occupied by newly formed stars is questionable. In addition, the SFE is a time-dependent quantity which reaches its ultimate value not until all gas is converted.

4. A Three-Component Star-Forming System

In this chapter we study the behaviour of a three-component star-forming system. The components of the system are low- and medium-mass stars S_1 with masses between $0.05\,M_\odot$ and $15\,M_\odot$, high-mass stars S_2 in the mass range $15\,M_\odot < m \leq 125\,M_\odot$, and the gas of total mass M_g. The mass limit of $15\,M_\odot$ between the components S_1 and S_2 is chosen because only stars with initial masses greater than $15\,M_\odot$ create H II regions and show strong stellar winds during their whole life. Abbott (1982) demonstrated that these stellar winds dominate the energy balance of the interstellar medium in the environment of OB associations.

The system of evolutionary equations can be established using the following assumptions:

– both low/medium- and high-mass stars are formed spontaneously with different rates,

– high-mass stars induce further formation of high-mass stars,

– high-mass stars may suppress the formation of low/medium-mass stars by increasing the cloud temperature.

Both groups of stars cause an outflow of matter by the momentum of their radiation. The gas removal is given by the expression S_i/M_g ($i = 1, 2$) in accordance with a radiation driven outflow.

It should be mentioned that the effects of other outflow mechanisms (see, e.g., Henning, 1986) have to be considered in more elaborated models. So, in a detailed study of low-mass star formation the T Tauri winds must be taken into account and induced formation of low-mass stars has to be investigated. Our scenario points especially at the formation of stellar systems consisting of massive stars.

The equations are:

$$\dot{S}_1 = K_2/(1 + K_1 S_2)M_g^n , \tag{22}$$

$$\dot{S}_2 = K_3 M_g^n + K_5 S_2 M_g^n , \tag{23}$$

$$\dot{M}_g = - K_2/(1 + K_1 S_2)M_g^n - K_3 M_g^n - K_5 S_2 M_g^n - K_6 S_2/M_g - K_7 S_1/M_g . \tag{24}$$

Equation (22) described the spontaneous formation of low/medium-mass stars with a rate K_2 and the suppression of the formation of these stars by high-mass stars. The high-mass stars effectively decrease K_2 with the rate coefficient K_1. The next equation reflects the spontaneous and induced formation of high-mass stars. The conversion of gas into stars and the gas removal is given by Equation (24).

Now, we introduce dimensionless variables by $S_1/M_0 = s_1$, $S_2/M_0 = s_2$, $M_g/M_0 = m_g$, and $\tau = (K_3 M_0^{(n-1)} + K_5 M_0^n)t$, where $M_0 = M_g(t = 0)$ is the initial gas mass. Then we obtain:

$$ds_1/d\tau = \frac{k_2}{(k_3 + k_5)(1 + k_1 s_2)} m_g^n , \tag{25}$$

$$ds_2/d\tau = \frac{k_3}{(k_3 + k_5)} m_g^n + \frac{k_5}{(k_3 + k_5)} s_2 m_g^n , \tag{26}$$

and

$$dm_g/d\tau = - \frac{k_2}{(k_3 + k_5)(1 + k_1 s_2)} m_g^n - \frac{k_3}{(k_3 + k_5)} m_g^n -$$

$$- \frac{k_5}{(k_3 + k_5)} s_2 m_g^n - \frac{k_6}{(k_3 + k_5)} s_2/m_g - \frac{k_7}{(k_3 + k_5)} s_1/m_g , \tag{27}$$

where is $k_1 = K_1 M_0$, $k_2 = K_2 M_0^{(n-1)}$, $k_5 = K_5 M_0^n$, $k_3 = K_3 M_0^{(n-1)}$, $k_6 = K_6/M_0$, and $k_7 = K_7/M_0$.

We will use the following symbols for the ratios of the rate coefficients: $A = k_5/(k_3 + k_5)$, $B = k_3/(k_3 + k_5)$, $C = k_2/(k_3 + k_5)$, $F = k_6/(k_3 + k_5)$, $G = k_7/(k_3 + k_5)$, and $H = k_1$.

Thus the final forms of the equations are

$$ds_1/d\tau = C/(1 + Hs_2)m_g^n , \tag{28}$$

$$ds_2/d\tau = Bm_g^n + As_2m_g^n , \tag{29}$$

$$dm_g/d\tau = -C/(1 + Hs_2)m_g^n - Bm_g^n - As_2m_g^n - Fs_2/m_g - Gs_1/m_g . \tag{30}$$

For the solution of the equations the rate coefficients A (induced high-mass star formation), B (spontaneous high-mass star formation), C (spontaneous low/medium-mass star formation), F (gas removal by high-mass stars), G (gas removal by low/medium-mass stars), and H (suppression coefficient) have to be specified.

The ratio C/B corresponds to the ratio of the total stellar masses within the low/medium and high-mass part of the IMS: i.e.,

$$\frac{C}{B} = \frac{\displaystyle\int_{m_l}^{15\,M_\odot} mn(m)\,dm}{\displaystyle\int_{15\,M_\odot}^{m_{max}} mn(m)\,dm} . \tag{31}$$

The ratio of the outflow rates G/F is determined by the ratio of the mass-averaged luminosities of the low/medium- and high-mass part of the IMS:

$$\frac{G}{F} = \frac{\displaystyle\int_{m_l}^{15\,M_\odot} l(m)n(m)\,dm \displaystyle\int_{15\,M_\odot}^{m_{max}} mn(m)\,dm}{\displaystyle\int_{15\,M_\odot}^{m_{max}} l(m)n(m)\,dm \displaystyle\int_{m_l}^{15\,M_\odot} mn(m)\,dm} , \tag{32}$$

where F is assumed to be 1, which means that the time-scale of outflow caused by high-mass stars is of the same order of magnitude as the formation rate of these stars. The ratio between induced and spontaneous high-mass star formation A/B was chosen as 1, 10, ..., 10^4 ($A + B = 1$ from the normalization) and H is adopted to be 0, 1, 100, and 10^4. For the calculation of C/B and G/F we have to determine m_{max} from M_0 by Equation (12). We used $M_0 = 10^6\,M_\odot$ (giant molecular clouds) and $10^4\,M_\odot$ (medium-mass clouds). This results in $m_{max} = 125\,M_\odot$ and $17.3\,M_\odot$, respectively. For cloud masses smaller than $7200\,M_\odot$ no high-mass stars will be formed and our system degenerates to a two-component system. Current values for the Schmidt exponent n in the literature range from ≈ 0 to 2. We adopted 1 and 2. The case $n = 0$ would decouple the star formation rate from the gas mass and is not considered here.

The initial conditions for the differential equations are $s_1(0) = 0$, $s_2(0) = 0$ (or $s_2(0) = 0.05$ for $B = 0$), and $m_g(0) = 1$. The solutions were obtained using a fourth-order

TABLE II

Results obtained from the solution of the evolutionary equations
(without suppression of the formation of low/medium-mass stars)

Model	M_0 ($M_\odot$)	n	H	A/B	s_1	s_2	SFE (%)	N_1/N_2
1	10^4	1	0	10^4	0.097	0.161	26.1	22
2	10^4	1	0	10^3	0.476	0.033	50.9	520
3	10^4	1	0	10^2	0.777	0.008	78.5	3614
4	10^4	1	0	10	0.907	0.007	91.5	4718
5	10^4	1	0	1	0.953	0.007	96.1	4846
6	10^4	2	0	10^4	0.095	0.114	21.0	30
7	10^4	2	0	10^3	0.414	0.021	43.5	727
8	10^4	2	0	10^2	0.674	0.006	68.1	2314
9	10^4	2	0	10	0.822	0.006	82.9	4734
10	10^4	2	0	1	0.891	0.007	89.8	4845
11	10^6	1	0	10^4	0.019	0.204	22.4	5
12	10^6	1	0	10^3	0.128	0.158	28.8	45
13	10^6	1	0	10^2	0.498	0.061	56.0	452
14	10^6	1	0	10	0.779	0.036	81.5	1199
15	10^6	1	0	1	0.876	0.035	91.1	1387
16	10^6	2	0	10^4	0.019	0.154	17.3	7
17	10^6	2	0	10^3	0.120	0.114	23.5	58
18	10^6	2	0	10^2	0.432	0.045	47.7	532
19	10^6	2	0	10	0.681	0.031	71.3	1218
20	10^6	2	0	1	0.793	0.031	82.5	1418

predictor-corrector scheme. We do not find periodic or chaotic solutions because we have no input of recycled or 'new' gas.

The results are summarized in Table II for clouds with 10^4 and 10^6 $M_\odot$ in the case without suppression of the formation of low/medium-mass stars ($H = 0$).

From this table we can conclude that the SFE is lower in the case of a stronger induced star formation and a higher value of n. Considering the influence of different cloud masses we see that the efficiency is higher in the 10^4 $M_\odot$ cloud. This supports our view that bound stellar systems form from medium-mass clouds. The similarity in the IMS of T Tauri associations (Cohen and Kuhi, 1979) and open clusters (Stecklum, 1985) indicates that clouds with similar masses are the birthplaces of these stellar systems.

Nevertheless, it may be that different physical conditions in the low/medium-mass clouds can cause different gas removal times and initial velocities of the stars resulting in the formation of bound systems (open clusters) and unbound systems (T Tauri associations). Independent of these remarks, the high-mass clouds should be the birthplace of OB associations. However, the results for the 10^6 $M_\odot$ cloud given in Table II show that the number of low/medium-mass stars N_1 exceeds the number of high-mass stars N_2 by far. This fact is not observed in real OB associations. The contradiction could be solved either by the assumption that the low/medium-mass stars leave the association due to dynamical effects which seems rather unlikely or by the suppression of the formation of these stars according to the temperature argument.

TABLE III

Results obtained from the solution of the evolutionary equations
(massive cloud with suppression)

Model	M_0 ($M_\odot$)	n	H	A/B	s_1	s_2	SFE (%)	N_1/N_2
21	10^6	2	1	10^4	0.018	0.154	17.2	6
22	10^6	2	1	10^3	0.118	0.115	23.3	57
23	10^6	2	1	10^2	0.428	0.045	47.3	527
24	10^6	2	1	10	0.678	0.031	71.0	1212
25	10^6	2	1	1	0.790	0.032	82.3	1368
26	10^6	2	10^2	10^4	0.012	0.157	16.9	4
27	10^6	2	10^2	10^3	0.063	0.134	19.8	26
28	10^6	2	10^2	10^2	0.227	0.088	31.5	143
29	10^6	2	10^2	10	0.441	0.073	51.4	335
30	10^6	2	10^2	1	0.560	0.087	64.7	358
31	10^6	2	10^4	10^4	0.003	0.162	16.4	1
32	10^6	2	10^4	10^3	0.006	0.162	16.8	2
33	10^6	2	10^4	10^2	0.012	0.174	18.6	4
34	10^6	2	10^4	10	0.016	0.223	23.9	4
35	10^6	2	10^4	1	0.020	0.307	32.8	4

In Table III we summarize the results for the $10^6\ M_\odot$ cloud and $n = 2$ for models with suppression ($H = 1, 100, 10^4$). From a comparison between Table II and Table III it is obvious that the SFE and N_1/N_2 decrease if the value of the suppression coefficient is increased. Such models seem to be more likely for the formation of OB associations as models without suppression.

The figures show the temporal evolution of several models where the stellar masses are indicated by full lines and the star formation rates by dashed lines. Figure 1 displays the behaviour of model 7.

It can be seen that the formation rate of low/medium-mass stars declines nearly linear whereas the high-mass star formation rate remains nearly flat. This picture is typical for the temporal behaviour of the star formation in the case of the $10^4\ M_\odot$ cloud. The following figures show the results for the high-mass cloud and increasing suppression of the low/medium-mass star formation. In Figure 2 the temporal evolution of model 17 can be seen. According to the higher upper mass limit the effect of induced high-mass star formation comes into play. This results in a maximum of the star formation rate s_2 which, in turn, leads to a rapid gas removal. As a consequence, both star formation rates fall off. The next figure shows the star formation history for model 27 where the formation of low/medium-mass stars is moderately suppressed ($H = 100$). This results in a faster drop of the low/medium-mass star formation rate, a higher peak of the formation rate of high-mass stars, and a final mass fraction of high-mass stars which is about twice that of the low/medium-mass stars. If the suppression is further increased then this tendency strengthens, i.e., the onset of the high-mass star formation terminates the formation of low/medium-mass stars. Therefore, these stars can form only very early in the system's history which can be seen in Figure 4. Furthermore, the effect of induced star formation is very pronounced which results in a sudden gas removal.

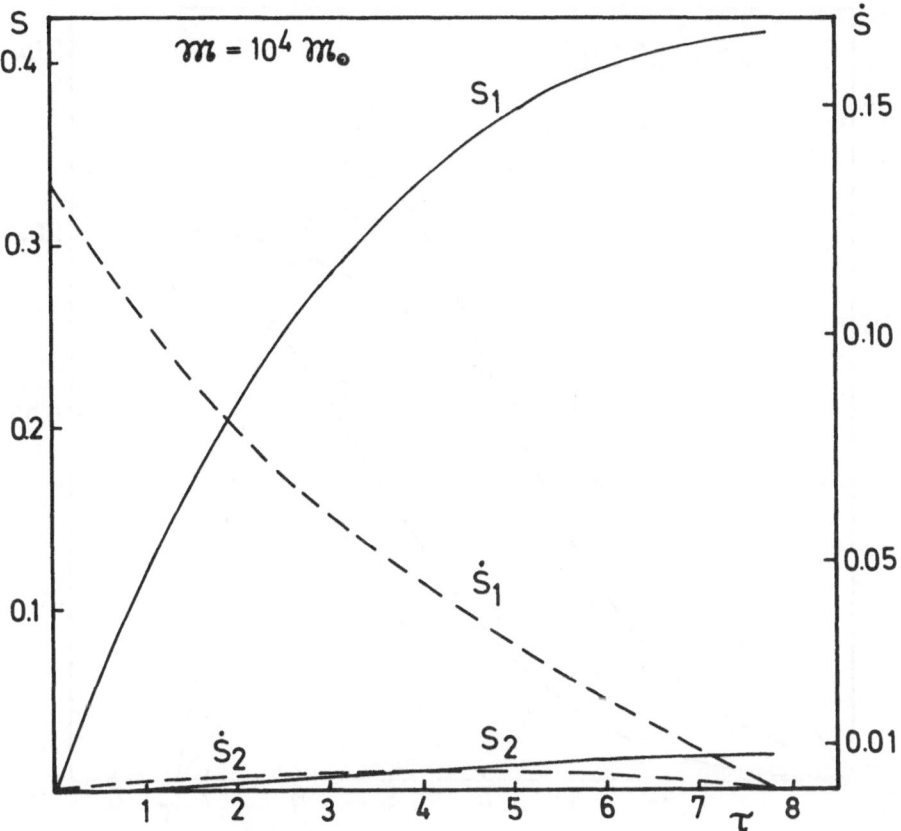

Fig. 1. Time-dependence of the stellar masses s_1, s_2 (full lines) and the star formation rates $\dot{s}_1$, $\dot{s}_2$ (dashed lines) for model 7.

Despite the relatively high ultimate SFEs which can be changed by adjusting the parameters our results allow a comparison between the formation history of low/ medium- and high-mass stars for different initial conditions. Such a comparison needs mass-resolved investigations of the star formation rate of associations and young open clusters. Studies of that kind have been carried out by a number of authors (cf. Adams *et al.*, 1983; and references therein). A quick look at these results provides some support for our model because it was found that the birthrate of low-mass stars is declining whereas the birthrate of medium-mass stars shows evidence for a maximum which reflects in principle the influence of induced star formation. However, recently there was strong criticism of these findings by Stahler (1985). He remarked that the age estimation procedure was incorrect. In his opinion, the appearance of sequential star formation is due to the fact that all stars have been treated as pre-Main-Sequence stars. The corrected ages show no correlation with mass. Stahler's conclusion is that the star formation process may be continuous and consistent with the simple picture that stars of all stellar masses are formed over the same broad time interval. If we compare these

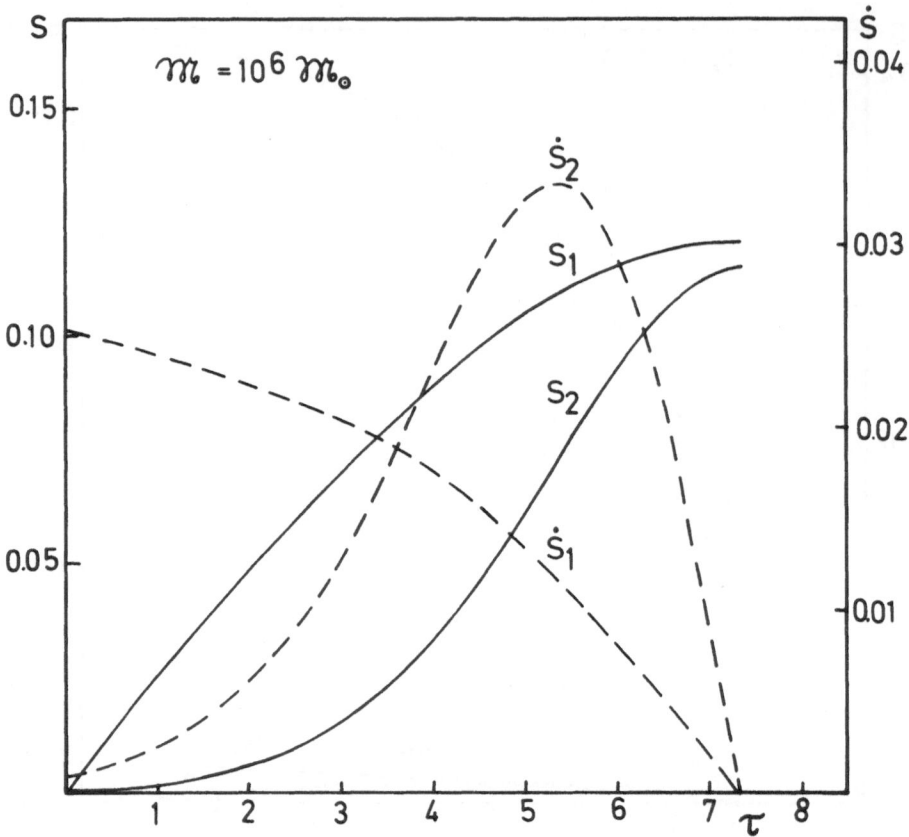

Fig. 2.   Temporal evolution of model 17 (designation as in Figure 1).

suggestions with our results it has to be taken into consideration that the mass range investigated by Stahler is included in our low/medium-mass stellar component. The derivation of the time-dependent star formation in OB associations by Doom *et al.* (1985) allows a better confrontation with our calculations. They found strong evidence for the hypothesis that the most massive stars form later than the medium-mass stars by the investigation of the mass range 8–32 $M_\odot$. It should be noted that for the study of Doom *et al.* the arguments by Stahler do not hold because they assumed that all OB stars within the Main-Sequence band are evolving from the ZAMS. These results together with our theoretical findings support the view that a temporal shift in the mass spectrum exists during the formation of an OB association.

5. Summary

We constructed a simple model to estimate the SFE for two species of molecular clouds, massive clouds which should contain massive stars, and medium-mass clouds where mainly low/medium-mass stars are forming. Taking the radiation pressure efficiency

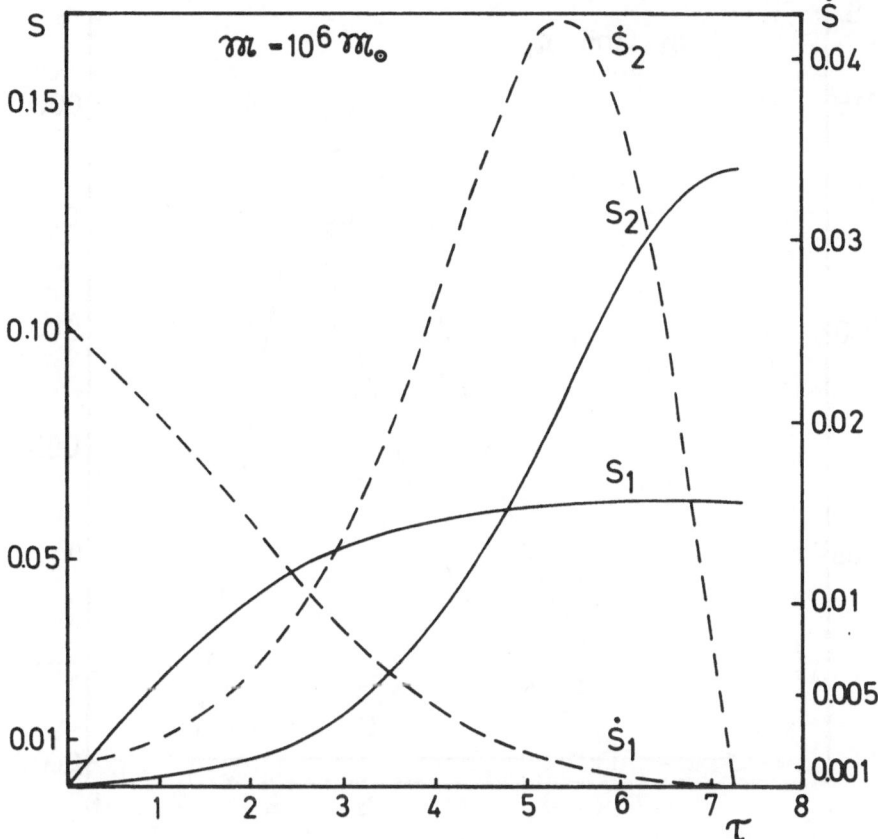

Fig. 3.   Temporal evolution of model 27 (designation as in Figure 1).

factor of the dust grains into account, we derived the force balance to obtain a stability condition for the star formation process. It turned out that the SFE is several times higher in the case of a medium-mass cloud than in the case of a high-mass cloud. The former one, therefore, seems to be the birthplace of a bound stellar system whereas the latter one appears to create unbound stellar systems of massive stars.

To investigate the temporal evolution of the star formation we simulated a star-forming region using evolutionary equations for a three-component model which consists of massive stars, low/medium-mass stars and gas. We established the equations using the assumptions that spontaneous and induced formation occurs for high-mass stars. Low/medium-mass stars form only spontaneously and their formation may be suppressed as a consequence of the cloud heating by high-mass stars. The gas is removed by the action of the newly born stars. The rate coefficients were specified either on the base of the Miller and Scalo IMS or using reasonable values. The results indicate that:

– the SFE is higher in the case of the medium-mass cloud,
– the SFE tends to decrease with increasing induced star formation and a higher value of the Schmidt exponent,

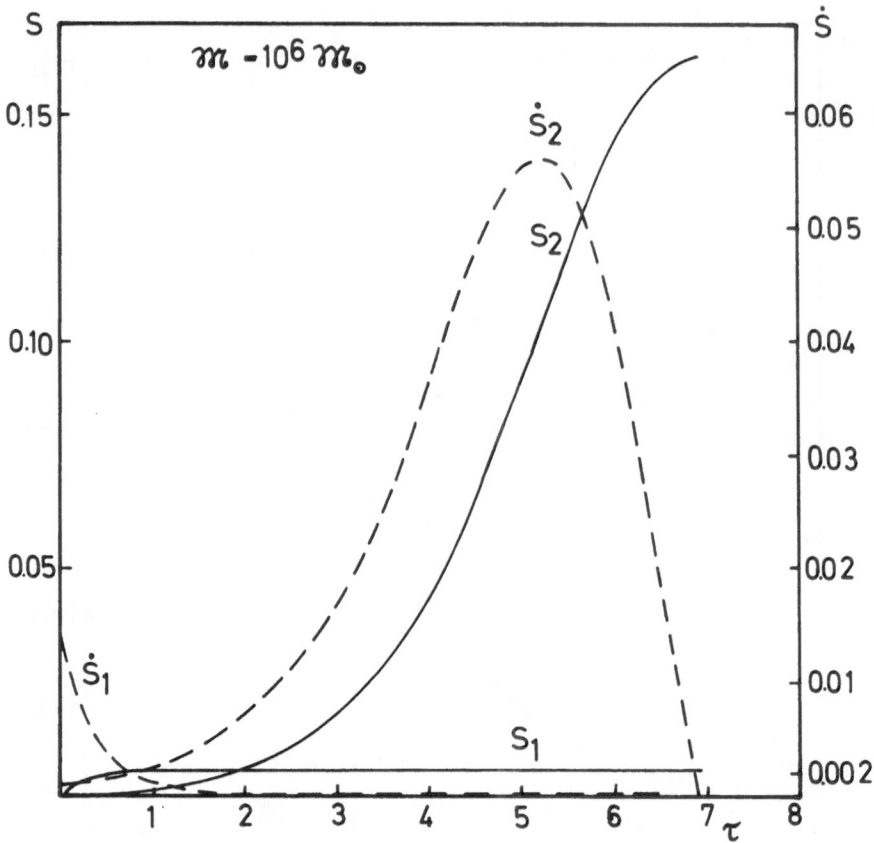

Fig. 4.   Temporal evolution of model 32 (designation as in Figure 1).

– the suppression of the formation of low/medium-mass stars is a possible way to form unbound stellar systems of massive stars – that means OB associations.

Considering the temporal evolution we found that in the case of a medium-mass cloud the star formation rate of the low/medium-mass stars declines nearly linear until all gas is converted/removed. On the other hand, for the high-mass cloud the effect of induced star formation in combination with the suppression of the formation of low/medium-mass stars leads to an entire different picture of the star formation history. In this case the massive stars form later with a pronounced maximum in their formation rate, whereas the low/medium-mass stars are the oldest ones. The results are in agreement with the recent findings concerning sequential star formatin in OB associations.

References

Abbott, D. C.: 1982, *Astrophys. J.* **263**, 723.
Adams, M. T., Strom, K. M., and Strom, S. E.: 1983, *Astrophys. J. Suppl. Ser.* **53**, 893.
Bodifée, G. and de Loore, C.: 1985, *Astron. Astrophys.* **142**, 297.
Cohen, M. and Kuhi, L. V.: 1979, *Astrophys. J. Suppl. Ser.* **41**, 743.

Doom, C., de Greve, J. P., and de Loore, C.: 1985, *Astrophys. J.* **290**, 185.
Eggen, O. J.: 1976, *Quart. J. Roy. Astron. Soc.* **17**, 472.
Elmegreen, B. G.: 1983, *Monthly Notices Roy. Astron. Soc.* **203**, 1011.
Elmegreen, B. G. and Clemens, C.: 1985, *Astrophys. J.* **294**, 523.
Ferrini, F. and Marchesoni, F.: 1984, *Astrophys. J.* **287**, 17.
Fleck, C. F., Jr.: 1982, *Monthly Notices Roy. Astron. Soc.* **201**, 551.
Franco, J. and Cox, D. P.: 1983, *Astrophys. J.* **273**, 243.
Gerola, H., Seiden, P. E., and Schulman, L. S.: 1980, *Astrophys. J.* **242**, 513.
Gilman, R. C.: 1974, *Astrophys. J. Suppl. Ser.* **28**, 397.
Güsten, R. and Mezger, P. G.: 1982, *Vistas Astron.* **26**, 159.
Heintze, J. R. W.: 1973, in B. Hauk and B. E. Westerlund (eds.), *Problems of Calibration of Absolute Magnitudes and Temperatures of Stars*, D. Reidel Publ. Co., Dordrecht, Holland, p. 265.
Henning, Th.: 1986, *Astron. Nachr.* **307**, 119.
Herbig, G. H.: 1962, *Astrophys. J.* **135**, 736.
Lada, C. J., Margulis, M., and Dearborn, D.: 1984, *Astrophys. J.* **285**, 141.
Larson, R. B.: 1982, *Monthly Notices Roy. Astron. Soc.* **200**, 159.
Larson, R. B.: 1985, *Monthly Notices Roy. Astron. Soc.* **214**, 379.
Miller, G. E. and Scalo, J. M.: 1978, *Publ. Astron. Soc. Pacific* **90**, 506.
Miller, G. E. and Scalo, J. M.: 1979, *Astrophys. J. Suppl. Ser.* **41**, 513.
Schmidt, M.: 1959, *Astrophys. J.* **129**, 243.
Schmidt-Kaler, Th.: 1982, in K. Schaifers and H. H. Voigt (eds.), *Landolt–Börnstein Tables*, New Series, Group VI, Vol. 2b, Springer-Verlag, Berlin.
Shore, S. N.: 1981, *Astrophys. J.* **249**, 93.
Smith, L. F., Biermann, P., and Mezger, P. G.: 1978, *Astron. Astrophys.* **66**, 65.
Solomon, P. M., Sanders, B. B., and Rivolo, A. R.: 1985, *Astrophys. J.* **292**, L19.
Stahler, S. W.: 1985, *Astrophys. J.* **293**, 207.
Stecklum, B.: 1985, *Astron. Nachr.* **306**, 45.
Straižys, V. and Kuriliene, G.: 1981, *Astrophys. Space Sci.* **80**, 353.
Wielen, R.: 1971, *Astron. Astrophys.* **13**, 309.

Discussion

P. G. Mezger: There is a semantic problem. This type of sequential star formation was once invented by Elmegreen and Lada. They used it for another purpose because they claimed that H II regions and supernova explosions induce star formation. The application of this term to a time sequence of star formation in one cloud may lead to confusion.

The second thing is that this quantity of SFE, I think, it is not of very much use because it is a quantity which you cannot get out from observation in my opinion. Normally star formation takes place in a certain area of an extended cloud. If you could somehow measure the first condensation which decouples from the cloud this would be a useful thing. But you cannot measure how much gas has been there originally.

Did you compute the IMS because this is something which can be observationally verified? You must get a distribution where the ratio of the number of high-mass stars to the number of low-mass stars varies.

B. Stecklum: To your first remark. Herbig, among others, was the first who realized that the average stellar mass being produced at any time increases in a smooth fashion. The picture which emerged from that was characterized by Stahler as a picture of 'sequential star formation'. This is what we refer to.

I agree with you that the SFE is very difficult to estimate from the observational point of view. There is not only the problem how much gas has been there but also how much stars will be formed in the future, e.g., for the ρ Oph cloud we have current estimates of SFE but not the whole gas has been consumed yet. This means that the SFE is a time-dependent quantity.

Indeed we obtain in our models an IMS where the number of high-mass stars to the number of low/medium-mass stars varies. This is a consequence of the subdivision of the whole stellar mass range in low/medium- and high-mass stars and different star formation rates for both groups. Nevertheless, I have to mention that the stars in each interval are distributed according to the Miller–Scalo IMS.

P. G. Mezger: That was what I meant. We made a similar prediction unvoluntarily in the case of bimodal star formation. We predicted that there is a hump in the IMS between 1 and 3 $M_\odot$ and in fact the latest things show this hump very clearly. Probably, you would have to change the parameter of the critical mass to the point where you match the observations.

B. Stecklum: Our parameter of the critical mass has nothing to do with the observed hump in the IMS of field stars as it merely reflects the different action of low/medium-mass stars and high-mass stars. We choose the value of 15 $M_\odot$ because our aim was to simulate the formation of OB associations in GMCs. We believe that the hump in the IMS of field stars is caused by the superposition of the individual IMS of stars forming in clouds of different masses weighted by the frequency of the clouds.

J. Palouš: I could not find differences in age for members of the Hyades and Sirius super-cluster. But in the case of Pleiades super-cluster (cf. Eggen, 1985) there is some significant spread in age. It would be interesting to derive the mass spectrum for these age groups.

THE PAST STAR-FORMATION RATE AND THE INITIAL
MASS-FUNCTION IN THE SOLAR NEIGHBOURHOOD*

HELMUT MEUSINGER

Zentralinstitut für Astrophysik der AdW der DDR, Karl-Schwarzschild-Observatorium, Tautenburg, G.D.R.

(Received 27 June, 1986)

Abstract. The present-day mass function and the kinematical data of the nearby stars are discussed as constraints on the time-behaviour of the star-formation rate and of the initial mass function during the evolution of the galactic disk. A recently suggested dependence of the initial mass-function slope on the heavy element abundance can not be excluded from this study. The average past star-formation rate is in the order of the present-day rate.

The star formation rate (SFR) and the initial mass function (IMF) of field stars are of crucial importance in the context of galaxy evolution and are interrelated with molecular clouds by the star formation process and by secular chemical and kinematical evolution of the galactic disk.

The SFR is the number of all stars forming at the time t per year in a reference volume. It is reasonable to define the reference volume by the integration perpendicular to the galactic plane. The IMF is the distribution of the masses of the newly-born field stars and is usually described by a power law

$$\xi(\log M) = c(M_l, M_u, \gamma)M^{-\gamma}, \tag{1}$$

where M_l and M_u are the lower and the upper mass limit, respectively.

A 'classical' approach to the history of the SFR and to the IMF bases on the construction of the present-day mass-function (PDMF) of the Main-Sequence field stars. The PDMF is determined by the run of the SFR and of the IMF during that time stars can spend on the Main Sequence till now according to their mass. Therefore, the PDMF is a constraint on the past history of the SFR and of the IMF. To obtain the SFR there are assumptions necessary on the IMF and vice versa. Unfortunately, this is a property of all approaches to the past SFR and a major source of uncertainty. It is such an assumption that the IMF has not been variable during the evolution of the galactic disk. The IMF seems to be variable in the time-scale of the formation of a stellar aggregate (e.g., Doom *et al.*, 1985). In our context temporal variations are secular variations in a galactic time-scale.

Investigations to find out systematic variations in the IMF provided a rather contradictory picture (e.g., Scalo, 1978; Miller and Scalo, 1979). For instance, frequently it has been assumed that the metal abundance Z is of decisive importance if the IMF has been variable in the past. In consideration of the galactic abundance gradient the distribution of O-stars within a few kpc around the Sun (Germany *et al.*,

---

* Paper presented at a Workshop on 'The Role of Dust in Dense Regions of Interstellar Matter', held at Georgenthal, G.D.R., in March 1986.

1982) suggests a higher percentage of high-mass stars in more metal rich regions. However, the opposite relationship is suggested by the studies of the evolutionary history of the elements C, O, N, and Fe. There is growing evidence that the production of these elements has varied during the galactic lifetime. In very metal-poor stars a significant overabundance of O (Sneden *et al.*, 1979) and of N (Bessel and Norris, 1982; Barbuy, 1983) with respect to iron was found. Furthermore, the O-abundance seems to vary much more slowly than iron (Clegg *et al.*, 1981). This can be taken as an indication that the IMF has changed during the galactic lifetime (Chiosi and Matteucci, 1984) favouring the formation of the more massive stars from metal-poor material.

A lot of further arguments for an increasing and for a decreasing fraction of high-mass stars with age came from studies of open and globular clusters (Scalo, 1978; Da Costa, 1982; Stecklum, 1985), from infrared and radio surveys of the galactic disk (Serra *et al.*, 1980; Boissé *et al.*, 1981), from studies of the Magellanic Clouds (Dennefeld and Tammann, 1979), from chemical evolution models of the solar neighbourhood (Truran and Cameron, 1971). Of course, it is impossible that the fraction of high-mass stars in the IMF increases and decreases at the same time.

Recently, Terlevich and Melnick (1983) pointed out that the observed variation in the ionizing flux per unit mass with chemical abundances in giant H II regions and in H II galaxies must be interpreted by a systematic change of the slope γ of the ionizing-star cluster IMF as a function of the heavy element abundance Z of the form

$$\gamma = a \log Z + 3.05 , \tag{2}$$

with $a = 1$ and constant mass limits of the IMF. Terlevich and Melnick (1983) supposed that such an IMF is also the reason for the CNO abundances in old stars. An alternative explanation for the [O/Fe] ratios was investigated by Matteucci and Tornambe (1985); they assumed a modification of the classical nucleosynthesis scenario. Good agreement with the observations was found if both effects were considered, but then $0 < a < 1$ is what is expected. An IMF–Z relation much shallower than $a = 1$ was confirmed by Matteucci and Tosi (1985) from the investigation of N and O evolution in dwarf irregular galaxies. After all, the problem of the CNO abundances is not completely clear and a Z-dependent IMF is not necessarily required (Matteucci and Greggio, 1985; Matteucci, 1986).

We will now discuss three points: (i) the history of the SFR assuming a time-independent IMF; (ii) an argument against secular variability in the IMF; (iii) the history of the SFR assuming a Terlevich–Melnick-like IMF. The investigation is based on the data material of the immediate solar neighbourhood in the *Catalogue of Nearby Stars II* (Gliese, 1969).

(i) *The Case of the Time-Independent IMF*

The first observational foundation of the present investigation is the PDMF of the nearby Main-Sequence field stars. This has been constructed from the luminosity function taking into account for Main Sequence and post-Main Sequence evolution and for kinematical evolution of the stellar disk (Miller and Scalo, 1979; Meusinger, 1983).

If the field star IMF is known it is possible to obtain the past SFR as a function of time from the PDMF in the intermediate mass range. From the low-mass range and the high-mass range together the relative SFR

$$S_r(T) = \int_0^T S(t)\, dt / TS(T)\,, \tag{3}$$

can be estimated.

Can we know the IMF of the field stars? The slope of the low-mass IMF is immediately given by the PDMF. The slope of the IMF in the high-mass range can be estimated from the high-mass PDMF assuming a nearly constant SFR during the last few 10^8 yr. The results (Figure 1) show that the slope of the IMF seems to be smaller in the low-mass range than in the high-mass range. Assuming that the IMF is a smooth function over the intermediate mass range, the IMF-exponent can be approximated by a linear function of $\log M$ (the linear regressions in Figure 1). This is identical with a log-normal IMF instead of a simple power law.

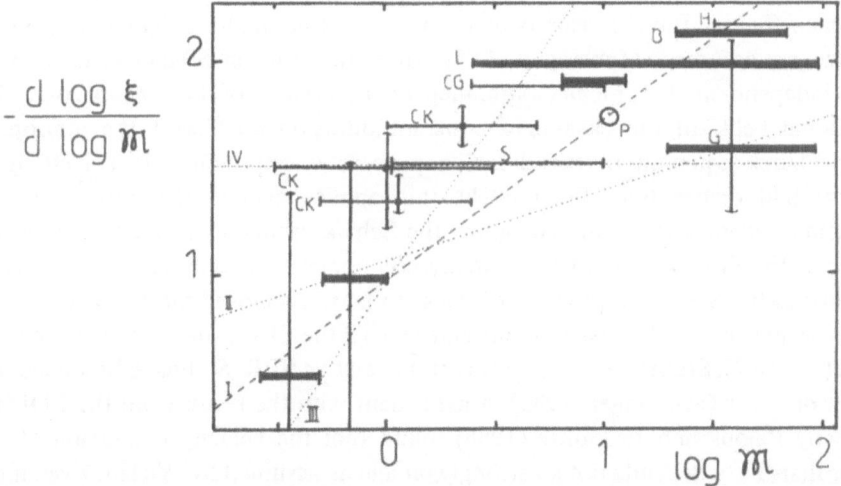

Fig. 1. The IMF slope *versus* $\log M$. The width of the bar indicates the special mass range under consideration. (Thick lines: derived from the field star PDMF; thin lines: open clusters or associations; dashed and dotted lines: linear regressions I, II, III, and IV; symbols: CK: Cohen and Kuhi, 1979; CG: Claudius and Grosbøl, 1980; S: Stecklum, 1985; L: Lequeux, 1979; P: Pyatunina, 1985; G: Garmany *et al.*, 1982; B: Bisiacchi *et al.*, 1983; H: Ho and Haschick, 1981.)

Using such a log-normal IMF a relative SFR was derived between 0.1 and 4 from the PDMF within their uncertainties ($6 \leq T/\text{Gy} \leq 15$). The detailed study of the intermediate mass range yielded results in good agreement with that (cf. Meusinger, 1983). However, the continuity constraint on the IMF is justified essentially only by the lack of any proof against it. Especially, if star formation is a bimodal process (e.g., Mezger and Smith, 1977) it might break down.

A completely different approach to the past SFR is possible because of the kinematical evolution of the galactic disk indicated by the age-dependence of the velocity dispersion of star groups. The observed velocity distribution of a kinematical unbiased sample of nearby late-type Main-Sequence stars, the McCormick K- and M-dwarfs of the Gliese *Catalogue*, is determined by the evolution of the SFR and by the velocity dispersion-age relation.

We do not know the dominant-physical process responsible for the increase of the velocity dispersion with increasing ages. But it is not unreasonable to describe the whole process as a diffusion in the velocity space of the stars caused by the irregular part of the galactic gravitational field. The sources of this irregular field may be giant molecular clouds (Spitzer and Schwarzschild, 1953; Lacey, 1984a), recurrent transient spiral waves (Barbanis and Woltjer, 1967; Carlberg and Sellwood, 1985) or massive black holes from the dark galactic corona (Lacey, 1984b).

The basic idea of the diffusion concept developed by Wielen (1977) and Wielen and Fuchs (1983a, b) is to describe the diffusion process by a diffusion coefficient and to find out this coefficient from the observed velocity dispersion-age relation. The latter has been derived by Wielen (1974) from the data of the Gliese *Catalogue* assuming a constant SFR. In this case the observations are well-fitted by an isotropic and constant diffusion coefficient. But the velocity dispersion-age relation alters significantly if other SFR laws are assumed (Meusinger, 1985). Then the diffusion coefficient turns out to be time-independent. This means a time-dependent strength of the irregular part of the gravitational field. In the case of a constant diffusion coefficient the appropriate Fokker–Planck equation of the diffusion process is approximately solved by the Schwarzschild distribution. Fortunately, this applies also in the case of a time-dependent diffusion coefficient. Therefore, the Schwarzschild distribution can be used to describe the distributions of the velocity components for different SFR laws. We parametrized the SFR by a power n of the gas mass in a closed model. The observed and the calculated velocity distributions agree well if the SFR is not strongly dependent on time (Figure 2). Statistical tests show that the relative SFR, S_r, has to be smaller than about 5, or $n \lesssim 1$ (Meusinger, 1985) in agreement with the result from the PDMF.

Recently Palouš and Piskunov (1985) found that the velocity dispersion of stars younger than 1 Gy depends not so strongly on age as assumed by Wielen. Even if their result is not influenced by systematic effects or kinematical bias in their data base it does not contradict the diffusion concept allowing for time-dependence of the diffusion coefficient. A modification of the procedure described above, where the diffusion coefficient is zero in the last one Gy, results in a negligible modification in the calculated velocity distributions.

Possibly the agreement of the resulting S_r with the result from the PDMF can be regarded as a confirmation of the diffusion concept and of the continuity constraint on the IMF. Note that the metallicity-age relation of the nearby stars is also consistent with a rather time-independent SFR (Twarog, 1980; Tinsley, 1981). However, in each case it is provided that the IMF does not depend on the time.

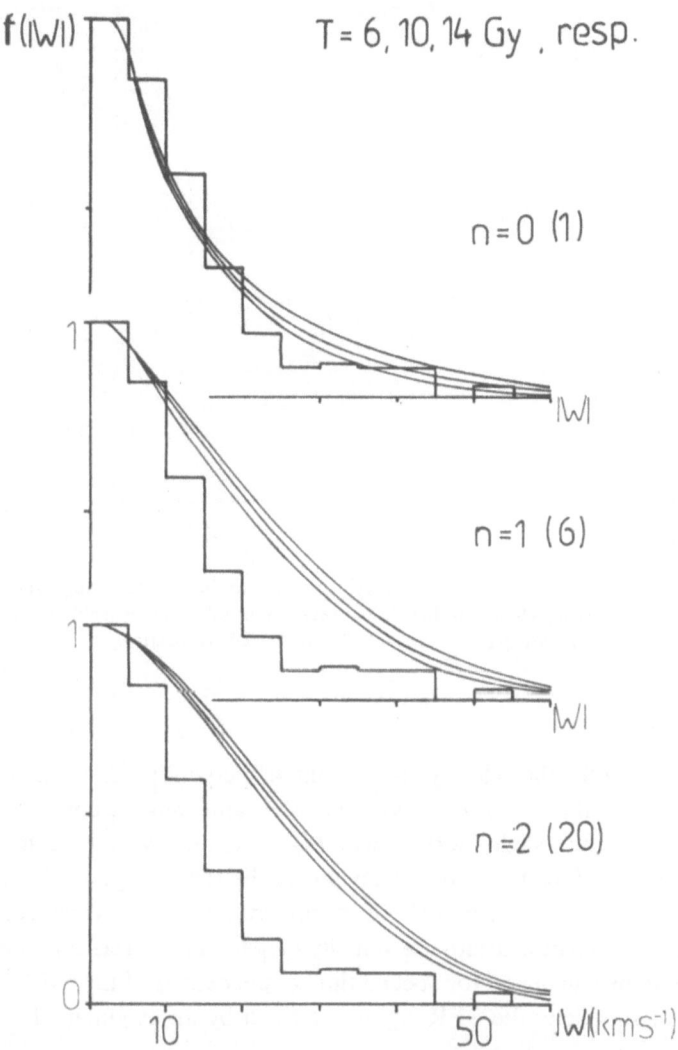

Fig. 2.   Comparison of the observed distribution of W velocities (histogram) with the distribution calculated (curves) for different histories of the SFR parametrized by the exponent n of the gas mass in a closed volume. (In the parenthesis the corresponding relative SFR, S_r.)

(ii) *An Argument Against Secular Variability in the Field Star IMF*

Wielen (1974) concluded that the IMF has been nearly constant in the past since the luminosity functions of stars in different velocity intervals do not differ significantly. This argument is examined in detail in the following way (for a more detailed description, cf. Meusinger, 1986): the McCormick K- and M-dwarfs of the Gliese *Catalogue* are classified according to M_r with about the same number of stars within the groups. For each group the ratio of the number of low-velocity stars to the number of higher-velocity stars is calculated and divided by the average ratio of all groups. Halo-type motions are

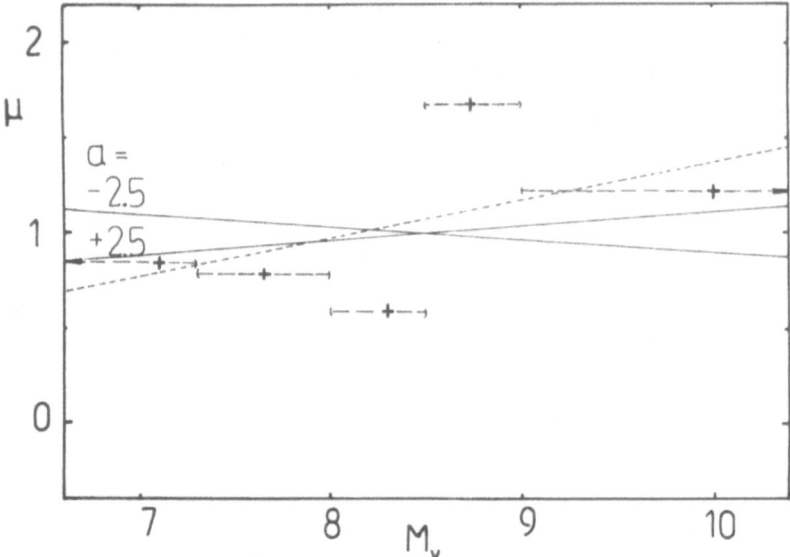

Fig. 3.   The relative ratio μ of the number of low-velocity stars to the number of high-velocity stars in five
M_v intervals for the W velocity. (Short dashes: linear regression; full lines: expected linear regressions for
the IMF parameter $a = -2.5$ and $+2.5$, respectively.)

excluded. The result for the velocity component W perpendicular to the galactic plane
is shown in Figure 3. If the IMF is not variable, the value must be constant and equal 1.
In reality the data show a slight increase with M_v for all three velocity components. This
is in a qualitative agreement with the behaviour of the IMF suggested by Terlevich and
Melnick (1983). However, there are to large uncertainties for a clear conclusion.

If the diffusion concept is useful the velocity dispersion-age relation and the velocity
distributions can be calculated for special time-dependences of the SFR and the IMF.
For the sake of simplicity the SFR is parametrized by an exponential law

$$S(t) = S(T) e^{b(T-t)} . \tag{4}$$

A Terlevich–Melnick-like IMF (Equation (2)) is adopted. The ratio of the number of
low-velocity stars to the number of higher-velocity stars is calculated for each
McCormick star group in dependence on the parameters a and b. The comparison with
the ratios from the observational data is used to confine the range of the variability in
the IMF.

However, within the small mass range under consideration a wide range of the IMF
parameter a is possible (Figure 3). This large uncertainty is partly because of the
behaviour of such an IMF (Figure 4): the major variation in the slope of the IMF is in
the early time. But in the early time the IMF contributes only little to the observed
velocity distribution in the considered mass range.

Miller and Scalo (1979) quoted a paper by Dixon (1970) as a support for a constant

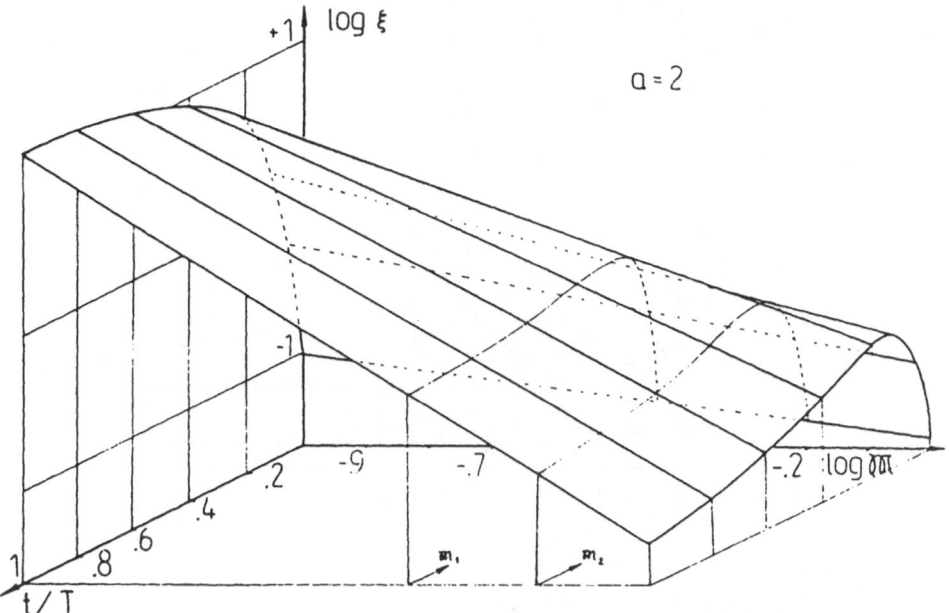

Fig. 4.   Typical secular behaviour of a Terlevich–Melnick-like IMF (Equation (2)) in the mass range under consideration.

IMF. Dixon investigated the velocity distribution of the stars in two distinct age groups in a completely other way. He found that the gradient in the initial luminosity function of the stars younger than 5 Gy may be at most twice as large or as small as the gradient of the general luminosity function for $9 \leq M_v \leq 13$. These limits can be converted into limits for the variability in the IMF in the corresponding mass range. Using a Terlevich–Melnick-like IMF Dixon's result means that the IMF parameter should be smaller than about two.

Therefore, our conclusion is that the kinematical data of the nearby stars are not a support for a constant IMF as suggested by Miller and Scalo. Especially, the Terlevich–Melnick IMF cannot be excluded.

(iii) *The History of the SFR Assuming a Time-Dependent IMF*

The question arises: What behaviour of the past SFR would be required by such a time-dependent IMF in order to reproduce the PDMF and the observed velocity distribution?

Provided the IMF is known, the PDMF is well-determined by two parameters: the ratio of the PDMF at the 'turn-off-point' to the present SFR and the slope of the PDMF behind the 'turn-off-point'. Both are calculated for different combinations of the IMF parameter a and the SFR parameter b and compared with the observational data (Figure 5(a), and (b)).

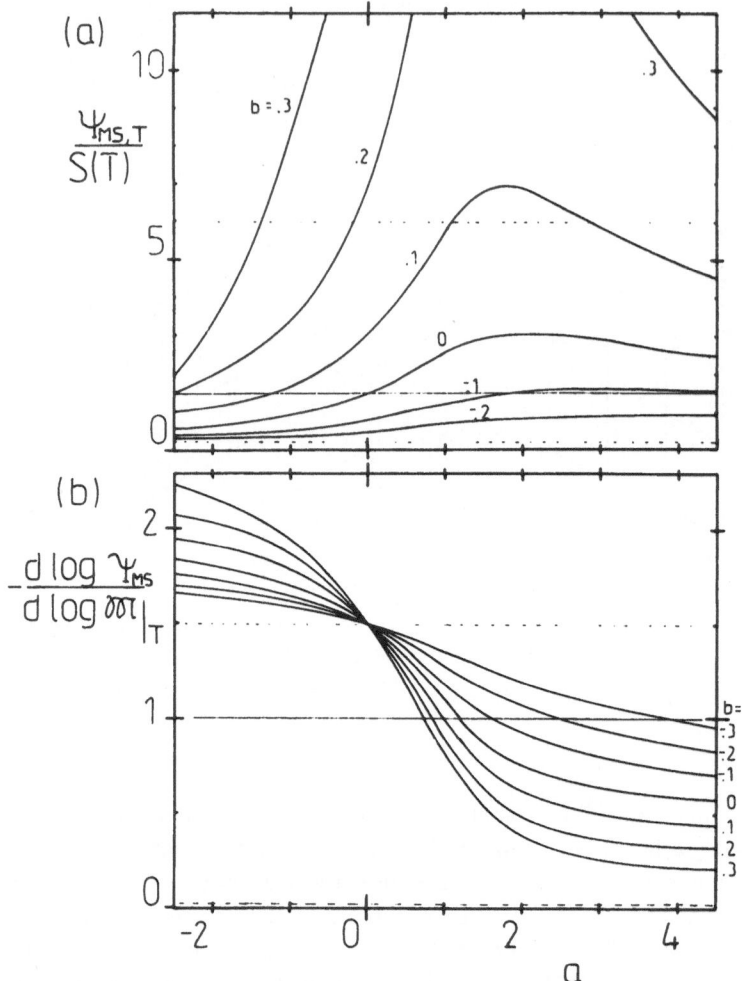

Fig. 5.   Comparison of the PDMF parameters from observations and from calculations for different SFRs and IMFs: (a) the ratio of the PDMF at the turn-off-point to the present SFR; (b) the PDMF slope behind the turn-off-point. The curves are calculated in dependence on the SFR parameter b. The thin horizontal lines are the 'best' values from observations and the dashed lines indicate the range of the corresponding uncertainties.

The following conclusions can be drawn: namely, from Figure 5(b): $a \lesssim 0$.

This implies that a power-law IMF cannot have been significantly steeper in the past than at present in order to be consistent with the PDMF and with the IMF of open clusters and associations.

From Figure 5(a) we deduce that $b \lesssim 0.2$ ($S_r \lesssim 5$); from Figure 5(a) and (b). A constant SFR is in good agreement with the PDMF. In this case, $0 < a < 1$.

Remember that $a = 1$ is just what was predicted by Terlevich and Melnick and that Matteucci and Tornambe (1985) argued for a value of a smaller than 1.

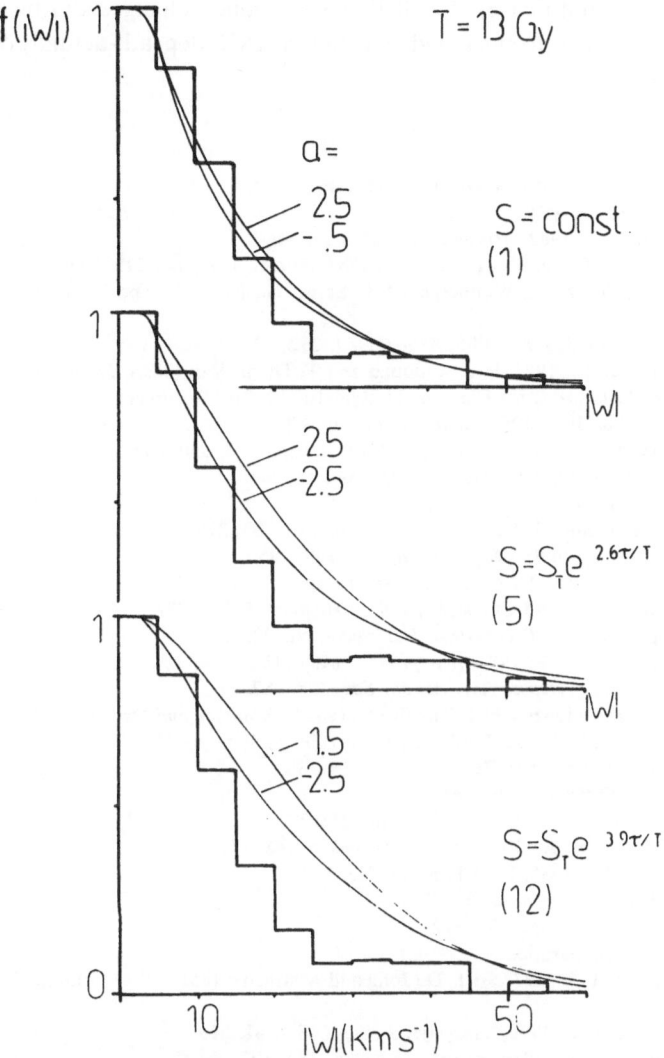

Fig. 6.   The same as in Figure 2, but, for a time-dependent IMF (parameter a) and for an exponential SFR. ($\tau = T - t$ is the age.)

The velocity distribution depends strongly on the SFR $S(t)$ and only weakly on the IMF parameter a. Once more there is no contradiction with the Terlevich–Melnick IMF and a weakly time-dependent SFR is required (Figure 6).

Conclusions

In principle two possibilities exist for the IMF in order to reproduce the observed PDMF of the Main-Sequence field stars and their velocity distribution: a time-independent log-normal distribution or a power law with an exponent proportional to $\log Z$. It is remarkable that in both cases the SFR has not much changed during the galactic lifetime. However, there is no clear evidence that the IMF depends actually on the heavy element abundance Z.

References

Barbanis, B. and Woltjer, L.: 1967, *Astrophys. J.* **150**, 461.

Barbuy, B.: 1983, *Astron. Astrophys.* **123**, 1.

Bessel, M. S. and Norris, J.: 1982, *Astrophys. J.* **263**, L29.

Bissiacchi, G. F., Firmani, C., and Sarmiento, G.: 1983, *Astron. Astrophys.* **119**, 167.

Boissé, P., Gispert, R., Coron, N., Wijnbergen, J. J., Serra, G., Ryter, C., and Puget, J. L.: 1981, *Astron. Astrophys.* **94**, 265.

Carlberg, R. G. and Sellwood, J. A.: 1985, *Astrophys. J.* **292**, 79.

Chiosi, C. and Matteucci, F.: 1984, in J. Audouze and T. Thanh Van (eds.), *Formation and Evolution of Galaxies and Large Structure in the Universe*, D. Reidel Publ. Co., Dordrecht, Holland, p. 417.

Claudius, M. and Grosbøl, P. J.: 1980, *Astron. Astrophys.* **87**, 339.

Clegg, R. E. S., Lambert, D. L., and Tomkin, J.: 1981, *Astrophys. J.* **250**, 262.

Cohen, M. and Kuhi, L.: 1979, *Astrophys. J. Suppl. Ser.* **41**, 743.

Da Costa, G. S.: 1982, *Astrophys. J.* **87**, 990.

Dennefeld, M. and Tammann, G. A.: 1979, *Astron. Astrophys.* **83**, 275.

Dixon, M. E.: 1970, *Monthly Notices Roy. Astron. Soc.* **150**, 195.

Doom, C., DeGrève, J. P., and DeLoore, C.: 1985, *Astrophys. J.* **290**, 185.

Garmany, G. D., Conti, P. S., and Chiosi, C., 1982, *Astrophys. J.* **263**, 777.

Gliese, W.: 1969, *Veröff. Astron. Rechen-Inst. Heidelberg*, No. 22.

Ho, P. T. P. and Haschick, A. D.: 1981, *Astrophys. J.* **248**, 622.

Lacey, C. G.: 1984a, *Monthly Notices Roy. Astron. Soc.* **208**, 687.

Lacey, C. G.: 1984b, in J. Audouze and T. Thanh Van (eds.), *Formation and Evolution of Galaxies and Large Structure in the Universe*, D. Reidel Publ. Co., Dordrecht, Holland, p. 351.

Lequeux, J.: 1979, *Astron. Astrophys.* **71**, 1.

Matteucci, F.: 1986, ESO Preprint, No. 412.

Matteucci, F. and Greggio, L.: 1985, ESO Preprint, No. 391.

Matteucci, F. and Tornambe, A.: 1985, *Astron. Astrophys.* **142**, 13.

Matteucci, F. and Tosi, M.: 1985, ESO Preprint, No. 372.

Meusinger, H.: 1983, *Astron. Nachr.* **304**, 285.

Meusinger, H.: 1985, *Astron. Nachr.* **306**, 123.

Meusinger, H.: 1986, in preparation.

Mezger, P. G. and Smith, L. F.: 1977, in T. De Jong and A. Maeder (eds.), 'Star formation', *IAU Symp.* **75**, 133.

Miller, G. E. and Scalo, J. M.: 1979, *Astrophys. J. Suppl. Ser.* **41**, 513.

Palouš, J. and Piskunov, A. E.: 1985, *Astron. Astrophys.* **143**, 102.

Pyatunina, T. B.: 1985, *Pis'ma Astron. Zh.* **11**, 27.

Scalo, J. M.: 1978, in T. Gehrels (ed.), *Protostars and Planets*, Univ. of Arizona Press, Tucson, p. 265.

Serra, G., Puget, J. L., and Ryter, C.: 1980, *Astron. Astrophys.* **84**, 220.

Sneden, C., Lambert, D. L. and Whitaker, R. W.: 1979, *Astrophys. J.* **234**, 964.

Spitzer, L. and Schwarzschild, M.: 1953, *Astrophys. J.* **118**, 106.

Stecklum, B.: 1985, *Astron. Nachr.* **306**, 45.

Terlevich, R. and Melnick, J.: 1983, ESO Preprint, No. 264.

Tinsley, B. M.: 1981, *Astrophys. J.* **250**, 758.

Truran, J. W. and Cameron, A. G. W.: 1971, *Astrophys. Space Sci.* **14**, 179.

Twarog, B. A.: 1980, *Astrophys. J.* **242**, 242.

Wielen, R.: 1974, *Astron. Rechen-Inst. Heidelberg*, Mitt. Ser. A, No. 86.

Wielen, R.: 1977, *Astron. Astrophys.* **60**, 263.

Wielen, R. and Fuchs, B.: 1983a, in W. L. Shuter (ed.), *Kinematics, Dynamics, and Structure of the Milky Way*, D. Reidel Publ. Co., Dordrecht, Holland, p. 81.

Wielen, R. and Fuchs, B.: 1983b, in H. van Woerden, R. J. Allen, and W. B. Burton (eds.), 'The Milky Way', *IAU Symp.* **106**, 481.

Discussion

J. M. Greenberg: I get the impression you were saying that there were more large stars formed when we have higher heavy element abundances.

H. Meusinger: Terlevich and Melnick found that the formation of high-mass stars seems to be preferred if the heavy element abundance is lower. But in my paper the opposite case is considered too.

H. J. Habing: What is the physical reason for that?

H. Meusinger: I cannot say how the physical mechanism is in detail. I think Terlevich and Melnick also do not know the physical reason for the IMF variability.

E. Krügel: Could it have to do with opacity-limited fragmentation? I mean this only refers to the lower limit, but the lower-mass limit, of course, determines also the mean mass.

J. M. Greenberg: If you look at the Magellanic Clouds you will never find a higher mean mass. You find enormous stars. Is that correct?

H. J. Habing: I think the largest stars are in the Magellanic Clouds.

J. M. Greenberg: Could it be that if you have dust you have a very efficient way to get of the energy, so that you can get contraction occuring for small masses, whereas if you have nothing there and you have no other way of getting off the energy you need a larger mass in order to get the contraction to go. I mean could it be something of this sort?

P. G. Mezger: One favorite idea about star formation and the IMF is that first low- and medium-mass stars form which gradually heat the gas. As a consequence the Jeans mass is increased and more massive stars form. However, that implies of course that the stars through their radiation, not only interact with dust but also heat the gas. This, however, occurs only at gas densities of 10^6 cm^{-3} and higher. It appears to me that gas densities, where stellar radiation may influence the kinetic gas temperature and, hence, the Jeans mass, occur too late in the evolution of protostars in order to influence their final mass.

H. Meusinger: The physics of the IMF was not immediately in the focus of my paper. My topic was to investigate if this type of a time-dependent IMF is in contradiction or not with the kinematical data and with the present-day mass function of the nearby stars.

J. Palouš: May I ask you what would happen if you use the velocity dispersion-age relation derived by R. G. Carlberg *et al.*?

H. Meusinger: I have no idea. Till now only Wielen's relation from the data of Gliese's *Catalogue* and the modification in accordance with your results were considered. I believe that the Gliese *Catalogue* is a very suitable data base for deriving the velocity dispersion-age relation for ages larger than about 10^9 yr. An other point is that these relations are not directly comparable. Wielen's result was derived for groups of stars in the HR diagram whereas Carlberg *et al.* used individual ages of the stars of the Twarog sample and received, therefore, a velocity dispersion-age relation for star generations.

CONCLUDING REMARKS

H. J. STAUDE

This was a very fruitful meeting. I found it especially fortunate that, during these four days, a relatively small number of persons came together, who are approaching our common subject from so many different sides. This is an exception to the present trend, which in numerous meetings brings together many researchers working on the same subject with similar tools.

Impressive progress has been made in defining the properties of interstellar grains and molecules, observationally as well as in the laboratory. We now have models for the dust, which are suitable for a quantitative description of the observations, and which also provide a probably quite realistic picture of the evolution of the grains. However, although detailed studies have been made, the fine structure of the interstellar extinction, i.e., the broad features in the UV and IR and the diffuse interstellar absorption lines, point to the existence of additional particle – or molecule – populations, which have still be identified uniquely.

The study of dense starforming interstellar clouds, circumstellar matter and very young stellar objects is presently in a stage of dramatic development. This in my opinion is due to three factors, which came into play during the very last years. First, the recently achieved closure of the gap between near-infrared and radio range now allows the observational study of the very cold material in which the active star forming regions are embedded. Second, the availability of the IRAS data opens up the way to introduce new criteria into the search for key objects to be studied in detail in the whole wavelength range. Finally, new detectors with high sensitivity and spatial resolution and new ground-based telescopes with very good optical properties are now providing detailed observations in the optical and near-infrared range. This will improve our knowledge about the structure of the immediate surroundings of the newborn stars and of their violent interaction with circumstellar matter within few astronomical units. We stand just at the beginning of these new developments, which will drastically enrich and diversify our field of research in the next future.

But beyond these considerations, let me make a more personal remark. This is the first time I have the opportunity to come to this part of the world and to meet those friend and colleagues, with whom we are together. I have collected a number of very concrete and sizeable ideas in relation with a variety of forms of collaboration with you here. This is the most important thing I will take home with me. I did not expect it, probably because of a lack of imagination: I want to thank you for this.

Astrophysics and Space Science **128** (1986) 265.
© 1986 *by D. Reidel Publishing Company*

ANNOUNCEMENT

The Role of Dust in Dense Regions of Interstellar Matter.
Proceedings of the Jena Workshop, held in Georgenthal, G.D.R.,
March 10–14, 1986

Editors: Thomas Henning and Bringfried Stecklum

Please note that a hardbound edition of this special issue of *Astrophysics and Space Science,* Vol. 128, No. 1 (December (I), 1986), is available from the publishers.

ISBN 90–277–2421–0   Price:Dfl. 178,–/$ 79.00/£ 64.00